Probability and its Applications

Published in association with the Applied Probability Trust

Probability and its Applications

Azencott et al.: Series of Irregular Observations. Forecasting and Model Building. 1986
Bass: Diffusions and Elliptic Operators. 1997
Bass: Probabilistic Techniques in Analysis. 1995
Berglund/Gentz: Noise-Induced Phenomena in Slow-Fast Dynamical Systems: A Sample-Paths Approach. 2006
Biagini/Hu/Øksendal/Zhang: Stochastic Calculus for Fractional Brownian Motion and Applications. 2008
Chen: Eigenvalues, Inequalities and Ergodic Theory. 2005
Costa/Fragoso/Marques: Discrete-Time Markov Jump Linear Systems. 2005
Daley/Vere-Jones: An Introduction to the Theory of Point Processes I: Elementary Theory and Methods. 2nd ed. 2003, corr. 2nd printing 2005
Daley/Vere-Jones: An Introduction to the Theory of Point Processes II: General Theory and Structure. 2nd ed. 2008
de la Peña/Gine: Decoupling: From Dependence to Independence, Randomly Stopped Processes, U-Statistics and Processes, Martingales and Beyond. 1999
de la Peña/Lai/Shao: Self-Normalized Processes. 2009
Del Moral: Feynman-Kac Formulae. Genealogical and Interacting Particle Systems with Applications. 2004
Durrett: Probability Models for DNA Sequence Evolution. 2002, 2nd ed. 2008
Ethier: The Doctrine of Chances. 2010
Feng: The Poisson–Dirichlet Distribution and Related Topics. 2010
Galambos/Simonelli: Bonferroni-Type Inequalities with Equations. 1996
Gani (ed.): The Craft of Probabilistic Modelling. A Collection of Personal Accounts. 1986
Gut: Stopped RandomWalks. Limit Theorems and Applications. 1987
Guyon: Random Fields on a Network. Modeling, Statistics and Applications. 1995
Kallenberg: Foundations of Modern Probability. 1997, 2nd ed. 2002
Kallenberg: Probabilistic Symmetries and Invariance Principles. 2005
Last/Brandt: Marked Point Processes on the Real Line. 1995
Molchanov: Theory of Random Sets. 2005
Nualart: The Malliavin Calculus and Related Topics, 1995, 2nd ed. 2006
Rachev/Rueschendorf: Mass Transportation Problems. Volume I: Theory and Volume II: Applications. 1998
Resnick: Extreme Values, Regular Variation and Point Processes. 1987
Schmidli: Stochastic Control in Insurance. 2008
Schneider/Weil: Stochastic and Integral Geometry. 2008
Serfozo: Basics of Applied Stochastic Processes. 2009
Shedler: Regeneration and Networks of Queues. 1986
Silvestrov: Limit Theorems for Randomly Stopped Stochastic Processes. 2004
Thorisson: Coupling, Stationarity and Regeneration. 2000

Shui Feng

The Poisson–Dirichlet Distribution and Related Topics

Models and Asymptotic Behaviors

Springer

Shui Feng
Department of Mathematics and Statistics
McMaster University
Hamilton, Ontario
L8S 4K1
Canada
shuifeng@mcmaster.ca

ISSN 1431-7028
ISBN 978-3-642-11193-8 e-ISBN 978-3-642-11194-5
DOI 10.1007/978-3-642-11194-5
Springer Heidelberg Dordrecht London New York

Library of Congress Control Number: 2010928906

Mathematics Subject Classification (2010): 60J60, 60J70, 92D15, 60F05, 60F10, 60C05

Cover design: WMXDesign

Printed on acid-free paper

Springer is part of Springer Science+Business Media (www.springer.com)

To Brian, Ronnie, and Min

Preface

The Poisson–Dirichlet distribution, a probability on the infinite-dimensional simplex, was introduced by Kingman in 1975. Since then it has found applications in Bayesian statistics, combinatorics, number theory, finance, macroeconomics, physics and, especially, in population genetics. Several books have appeared that contain sections or chapters on the Poisson–Dirichlet distribution. These include, but are not limited to, Aldous [2], Arratia, Barbour and Tavaré [9], Ewens [67], Kingman [127, 130], and Pitman [155]. This book is the first that focuses solely on the Poisson–Dirichlet distribution and some closely related topics.

The purposes of this book are to introduce the Poisson–Dirichlet distribution, to study its connections to stochastic dynamics, and to give an up-to-date account of results concerning its various asymptotic behaviors. The book is divided into two parts. Part I, consisting of Chapters 1–6, includes a variety of models involving the Poisson–Dirichlet distribution, and the central scheme is the unification of the Poisson–Dirichlet distribution, the urn structure, the coalescent, and the evolutionary dynamics through the grand particle systems of Donnelly and Kurtz. Part II discusses recent progress in the study of asymptotic behaviors of the Poisson–Dirichlet distribution, including fluctuation theorems and large deviations. The original Poisson–Dirichlet distribution contains one parameter denoted by θ. We will also discuss an extension of this to a two-parameter distribution, where an additional parameter α is needed. Most developments center around the one-parameter Poisson–Dirichlet distribution, with extensions to the two-parameter setting along the way when there is no significant increase in complexity. Complete derivations and proofs are provided for most formulae and theorems. The techniques and methods used in the book are useful in solving other problems and thus will appeal to researchers in a wide variety of subjects.

The selection of topics is based mainly on mathematical completeness and connections to population genetics, and is by no means exhaustive. Other topics, although related, are not included because they would take us too far afield to develop at the same level of detail. One could consult Arratia, Barbour and Tavaré [9] for a discussion of general logarithmic combinatorial structures; Barbour, Holst and Janson [11] for Poisson approximation; Durrett [48] and Ewens [67] for comprehensive coverage of mathematical population genetics; Bertoin [12] and Pitman [155] for fragmentation and coagulation processes; and Pitman [155] for connections to combinatorial properties of partitions, excursions, random graphs and forests. References for additional topics, including works on Bayesian statistics, functional inequalities, and multiplicative properties, will be given in the Notes section at the end of every chapter.

The intended audience of this book includes researchers and graduate students in population genetics, probability theory, statistics, and stochastic processes. The contents of Chapters 1–6 are suitable for a one-term graduate course on stochastic models in population genetics. The material in the book is largely self-contained and should be accessible to anyone with a knowledge of probability theory and stochastic processes at the level of Durrett [47].

The first chapter reviews several basic models in population genetics including the Wright–Fisher model and the Moran model. The Dirichlet distribution emerges as the reversible measure for the approximating diffusions. The classical relation between gamma random variables and the Dirichlet distribution is discussed. This lays the foundation for the introduction of the Poisson–Dirichlet distribution and for an understanding of the Perkins disintegration theorem, to be discussed in Chapter 5.

The second chapter includes various definitions and derivations of the Poisson–Dirichlet distribution. Perman's formula is used, in combination with the subordinator representation, to derive the finite-dimensional distributions of the Poisson–Dirichlet distribution. An alternative construction of the Poisson–Dirichlet distribution is included using the scale-invariant Poisson process. The GEM distribution appears in the setting of size-biased sampling, and the distribution of a random sample of given size is shown to follow the Ewens sampling formula. Several urn-type models are included to illustrate the relation between the Poisson–Dirichlet distribution, the GEM distribution, and the Ewens sampling formula. The last section is concerned with the properties of the Dirichlet process.

In Chapter 3 the focus is on the two-parameter Poisson–Dirichlet distribution, a natural generalization of the Poisson–Dirichlet distribution. The main goal is to generalize several results in Chapter 2 to the two-parameter setting, including the finite-dimensional distributions, the Pitman sampling formula, and an urn model. Here, a fundamental difference in the subordinator representation is that the process with independent increments is replaced by a process with exchangeable increments.

The coalescent is a mathematical model that traces the ancestry of a sample from a population. It is an effective tool in describing the genealogy of a population. In Chapter 4, the coalescent is defined as a continuous-time Markov chain with values in the set of equivalence relations on the set of positive integers. It is represented through its embedded chain and an independent pure-death Markov chain. The marginal distributions are derived for both the embedded chain and the pure-death Markov chain.

Two symmetric diffusion processes, the infinitely-many-neutral-alleles model and the Fleming–Viot process with parent-independent mutation are studied in Chapter 5. The reversible measure of the infinitely-many-neutral-alleles model is shown to be the Poisson–Dirichlet distribution and the reversible measure of the

Fleming–Viot process is the Dirichlet process. The representations of the transition probability functions are obtained for both processes and they involve the pure-death process studied in Chapter 4. It is shown that the Fleming–Viot process with parent-independent mutation can be obtained from a continuous branching process with immigration through normalization and conditioning. These can be viewed as the dynamical analog of the relation between the gamma distribution and the Dirichlet distribution derived in Chapter 1. This chapter also includes a brief discussion of the two-parameter generalization of the infinitely-many-neutral-alleles model.

As previously mentioned, the urn structure, the coalescent and the infinite-dimensional diffusions discussed so far, are unified in Chapter 6 under one umbrella called the Donnelly–Kurtz particle representation. This is an infinite exchangeable particle system with labels incorporating the genealogy of the population. The Fleming–Viot process in Chapter 5 is the large-sample limit of the empirical processes of the particle system and the Poisson–Dirichlet distribution emerges as a natural link between all of the models in the first six chapters.

The material covered in the first six chapters concerns, for the most part, well-known topics. In the remaining three chapters, our focus shifts to recent work on the asymptotic behaviors of the Poisson–Dirichlet distributions and the Dirichlet processes. In the general two-parameter setting, α corresponds to the stable component, while θ is related to the gamma component. When θ is large, the role of α diminishes and the behavior of the corresponding distributions becomes nonsingular or Gaussian. For small α and θ, the distributions are far away from Gaussian. These cases are more useful in physics and biology.

Fluctuation theorems are obtained in Chapter 7 for the Poisson–Dirichlet distribution, the Dirichlet process, and the conditional sampling formulas when θ is large. As expected, the limiting distributions involve the Gumbel distribution, the Brownian bridge and the Gaussian distribution.

Chapter 8 discusses large deviations for the Poisson–Dirichlet distributions for both large θ and small θ and α. The large deviation results provide convenient tools for evaluating the roles of natural selection. The large deviations for the Dirichlet processes are the focus of Chapter 9. The explicit forms of the rate functions provide a comparison between standard and Bayesian statistics. They also reveal the role of α as a measurement on the closeness to the large θ limit.

Notes included at the end of each chapter give the direct sources of the material in those chapters as well as some remarks. These are not meant to be an historical account of the subjects. The appendices include a brief account of Poisson processes and Poisson random measures, and several basic results of the theory of large deviations.

Some material in this book is based on courses given by the author at the summer school of Beijing Normal University between 2006 and 2008. I wish to thank Fengyu Wang for the opportunity to visit the Stochastic Research Center of Beijing Normal University. I also wish to thank Mufa Chen and Zenghu Li for their hospitality during my stay at the Center. Chapters 1–6 have been used in a graduate course given in the Department of Mathematics and Statistics at McMaster University during the academic year 2008–2009. I thank the students in those courses, who have helped me with suggestions and corrections.

I wish to express special thanks to Donald A. Dawson for his inspiration and advice, and for introducing me to the areas of measure-valued processes, large deviations, and mathematical population genetics. I would also like to thank Fred M. Hoppe and Paul Joyce for sharing their insight on urn models. Several anonymous reviewers have generously offered their deep insight and penetrating comments on all aspects of the book, from which I benefited immensely. Richard Arratia informed me about the scale-invariant spacing lemma and the references associated with the scale-invariant Poisson process. Sion's minimax theorem and the approach taken to Theorem 9.10 resulted from correspondence with Fuqing Gao. Ian Iscoe and Fang Xu provided numerous comments and suggestions for improvements. I would like to record my gratitude to Marina Reizakis, my editor at Springer, for her advice and professional help. The financial support from the Natural Sciences and Engineering Research Council of Canada is gratefully acknowledged.

Last, but not least, I thank my family for their encouragement and steadfast support.

Hamilton, Canada

Shui Feng
November, 2009

Contents

Part I
Models

Chapter 1
Introduction

In this chapter, we introduce several basic models in population genetics including the Wright–Fisher model, the Moran model, and the corresponding diffusion approximations. The Dirichlet distribution is introduced as the reversible measure of the corresponding diffusion processes. Its connection to the gamma distribution is explored. These will provide the necessary intuition and motivation for the Poisson–Dirichlet distribution and other sophisticated models considered in subsequent chapters.

1.1 Discrete Models

We begin this section with a brief introduction of the genetic terminology used throughout the book.

1.1.1 Genetic Background

All living organisms inherit biological characteristics from their parents. The characteristics that can be inherited are called the *genetic information*. Genetics is concerned with the study of heredity and variation. Inside each cell of an organism, there is a fixed number of *chromosomes*, which are threadlike objects, each containing a single, long molecule called DNA (deoxyribonucleic acid). Each DNA molecule is composed of two strands of nucleotides in which the sugar is deoxyribose and the bases are adenine (A), cytosine (C), guanine (G), and thymine (T). Linked by hydrogen bonds with A paired with T, and G paired with C, the two strands are coiled around to form the famous double helix structure discovered by Watson and Crick in 1953. These DNA molecules are responsible for the storage and inheritance of genetic information. A *gene* is a hereditary unit composed of a portion of DNA. The place that a gene resides on a chromosome is called a *locus* (*loci* in plural form).

S. Feng, *The Poisson–Dirichlet Distribution and Related Topics*,
Probability and its Applications, DOI 10.1007/978-3-642-11194-5_1,

Different forms of a gene are called *alleles*. An example is the ABO blood group locus, admitting three alleles, A, B, and O. The complete set of genetic information of an organism is called a *genome*.

An organism is called *haploid* if there is only one set of chromosomes; if chromosomes appear in pairs, the organism is *diploid*. The set of chromosomes in a *polyploid* organism is at least tripled. Bacteria and fungi are generally haploid organisms, whereas most higher organisms such as mammals are diploid. Human diploid cells have 46 (or 23 pairs) chromosomes and human haploid gametes (egg and sperm) each have 23 chromosomes. Bread wheat and canola have six and four sets of chromosomes, respectively. The *genotype* of an individual at a particular locus is the set of allele(s) presented. For a diploid individual, this is an unordered pair of alleles (one on each chromosome). A diploid individual is *homozygous* (*heterozygous*) at a particular locus if the corresponding genotype consists of two identical alleles (different alleles). A *phenotype* of an organism is any observable characteristic. Phenotypes are determined by an organism's genotypes and the interaction of the genotypes with the environment.

As a branch of biology, population genetics is concerned with the genetic structure and evolution of biological populations, and the genetic forces behind the evolution. The focus is on the population as a whole instead of individuals. The genetic makeup of a population can be represented by the frequency distribution of different alleles in the population. A population is *monoecious* if every individual has both male and female organs. If each individual can only be a male or female, then the population is called *dioecious*. We shall assume in this book that all populations are monoecious.

Many forces influence the evolution of a population. A change in the DNA sequence is referred to as a *mutation*. The cause of a mutation could be a mistake in DNA replication, a breakage of the chromosome, or certain environmental impact. *Fitness* is a measure of the relative breeding success of an individual or genotype in a population at a given time. Those that contribute the most offspring to future generations are the fittest. The *selection* is a deterministic force that favors certain alleles over others and leads to the survival of the fittest. Mutations create variation of allele frequencies. A deleterious mutation results in less fit alleles which can be reduced in frequency by natural selection, while advantageous mutation produces alleles with higher fitness and this combined with the selection force will result in an increase in frequency. A mutation that brings no change to the fitness of an individual is called *neutral*.

In a finite population, changes in allele frequencies may occur because of the random sampling of genes from one generation to the next. This pure random change in allele frequencies is referred to as a *random genetic drift*. It will result in a decay in genetic variability and the eventual loss or fixation of alleles without regard to the survival or reproductive value of the alleles involved. Other forces that play major roles in the evolution process include the mechanism of recombination and migration, which will not be discussed as these topics lie outside the focus of this book.

R.A. Fisher, J.B.S. Haldane and S. Wright provided the theoretical underpinnings of population genetics in the 1920s and 1930s. Over the years population geneticists have developed more sophisticated mathematical models of allele frequency dynamics. Even though these models are highly idealized, many theoretical predictions based on them, on the patterns of genetic variation in actual populations, turn out to be consistent with empirical data. The basic mathematical framework can be described loosely as follows. Consider a large biological population of individuals. We are interested in the allele frequencies of the population at a particular locus. An allele will be called a type and different alleles correspond to different types. The complete set of alleles corresponds a type space which is modeled through a topological space. Both mutation and selection are described by a deterministic process. The random genetic drift corresponds to a random sampling. The population evolution corresponds to a probability-valued stochastic process describing the changes of the frequency distributions over time under the influence of mutation, selection, and random genetic drift. Different structures of mutation, selection, and genetic drift give rise to different mathematical models.

1.1.2 The Wright–Fisher Model

The Wright–Fisher model is a basic model for the reproduction process of a monoecious, randomly mating population. It assumes that the population consists of N diploid or $2N$ haploid individuals, the population size remains the same in each generation, and there is no overlap between generations. We focus attention to a particular locus with alleles represented by the type space E. Time index is $\{0,1,2,3,\ldots\}$. Time 0 refers to the starting generation and positive time n corresponds to generation n. Consider the current generation as an urn consisting of $2N$ balls of different colors. Then, in the original Wright–Fisher model, the next generation is formed by $2N$ independent drawings of balls from the urn with replacement. This mechanism is called *random sampling*. More general models should take into account mutation and selection. Here, for simplicity, we will only consider models with mutation and random sampling.

Two-allele Model with Mutation: $E = \{A_1, A_2\}$

Let $X(0)$ denote the total number of individuals of type A_1 at time zero. The population evolves under the influence of mutation and random sampling.

Assume that a type A_1 individual can mutate to a type A_2 individual with probability u_2, and a type A_2 individual can mutate to a type A_1 individual with probability u_1. Fix the value of $X(0)$, the number of type A_1 individuals after mutation becomes

$$X(0)(1-u_2)+(2N-X(0))u_1.$$

Let $p = \frac{X(0)}{2N}$ denote the initial proportion of type A_1 individuals in the population. After mutation, the proportion of type A_1 individuals becomes

$$p' = (1-u_2)p + u_1(1-p).$$

Let $X(1)$ denote the total number of individuals of type A_1 at time one. Since individuals in the next generation are selected with replacement from a population of size $2N$ with $2Np'$ number of type A_1 individuals, $X(1)$ is a binomial random variable with parameters $2N, p'$. Repeat the same procedure, and let $X(n)$ denote the number of type A_1 individuals in generation n. Then $X(n)$ is a discrete time, homogeneous, finite-state Markov chain. For any $0 \le i, j \le 2N$, the one-step transition probability is

$$P_{ij} = P\{X(1) = j | X(0) = i\} = \binom{2N}{j} (p')^j (1-p')^{2N-j},$$

where $\binom{2N}{j}$ is the binomial coefficient.

If u_1 and u_2 are strictly positive, then the chain is positive recurrent and a unique stationary distribution exists. The original Wright–Fisher model corresponds to the case $u_1 = u_2 = 0$.

***K*-allele Model with Mutation**: $E = \{A_1, \ldots, A_K\}$, $K > 2$

Let $\mathbf{X}(0) = (X_1(0), \ldots, X_K(0))$ with $X_i(0)$ denoting the number of type A_i individuals at time zero. Mutation is described by a matrix $(u_{ij})_{1 \le i,j \le K}$ with non-negative elements. Here, u_{ij} is the mutation probability from type A_i to type A_j for distinct i, j and

$$\sum_{j=1}^{K} u_{ij} = 1, \; i = 1, \ldots, K. \tag{1.1}$$

Let

$$\mathbf{p} = (p_1, \ldots, p_K), \; p_i = \frac{X_i(0)}{2N}, \; i = 1, \ldots, K.$$

The allele frequency after the mutation becomes

$$\mathbf{p}' = (p'_1, \ldots, p'_K),$$

where

$$p'_i = \sum_{j=1}^{K} p_j u_{ji}, \; i = 1, \ldots, K. \tag{1.2}$$

Due to random sampling, the allele count in the next generation

$$\mathbf{X}(1) = (X_1(1), \ldots, X_K(1))$$

follows a multinomial distribution with parameters $2N$ and $\mathbf{p}'$. The nth generation can be obtained similarly from its preceding generation. The process $\mathbf{X}(n)$ is again a discrete time, homogeneous, finite-state Markov chain.

1.1.3 The Moran Model

Similarly to the Wright–Fisher model, the Moran model also describes the change in allele frequency distribution in a randomly mating population. But the population is haploid, and the mechanism of multinomial sampling in the Wright–Fisher model is replaced with a birth–death procedure in the Moran model. At each time step, one individual is chosen at random to die and another individual (may be the replaced individual) is chosen at random from the population to reproduce. Since birth and death occur at the same time, the population size remains constant. If each time step is considered as one generation, then only one reproduction occurs at each generation and, therefore, the generations are allowed to overlap. The Moran model is mathematically more tractable than the Wright–Fisher model because many interesting quantities can be expressed explicitly. When mutation is introduced, the birth probability will be weighted by the mutation probabilities.

Two-allele Model with Mutation: $E = \{A_1, A_2\}$

Assume that the population size is $2N$. Let $X(0) = i$ denote initial population size of A_1 individuals, and $p = i/2N$. The mutation probability from $A_1(A_2)$ to $A_2(A_1)$ is $u_2(u_1)$. Then $X(1)$ can only take three possible values: $i-1, i, i+1$. The case of $X(1) = i-1$ corresponds to the event that an A_1 individual is chosen to die and an A_2 individual is chosen to reproduce with weighted probability $(1-p)(1-u_1)+pu_2$. The corresponding transition probability is

$$\mathbb{P}\{X(1) = i-1 \mid X(0) = i\} = p[(1-p)(1-u_1)+pu_2].$$

Similarly we have

$$\mathbb{P}\{X(1) = i+1 \mid X(0) = i\} = (1-p)[(1-p)u_1 + p(1-u_2)],$$

and

$$\mathbb{P}\{X(1) = i \mid X(0) = i\} = 1 - [p(1-p)(2-u_1-u_2) + p^2u_2 + (1-p)^2u_1].$$

Clearly, $X(n)$, the total number of individuals of type A_1 at time n, is a discrete time, homogeneous, finite-state Markov chain.

***K*-allele Model with Mutation**: $E = \{A_1, \ldots, A_K\},\ K \geq 2$

As before, let $\mathbf{X}(0) = (X_1(0), \ldots, X_K(0))$ with $X_i(0)$ denoting the number of type A_i individuals at time zero and $p_i = \frac{X_i(0)}{2N}$ for $i = 1, \ldots, K$. Let the mutation be described by the matrix $(u_{ij})_{1 \leq i,j \leq K}$ satisfying (1.1). Then the allele frequency distribution after the mutation is given by (1.2).

For any $1 \leq l \leq K$, let e_l denote the unit vector with the lth coordinate being one. Starting with $\mathbf{X}(0) = \mathbf{i} = (i_1, \ldots, i_K)$, the next generation $\mathbf{X}(1) = (X_1(1), \ldots, X_K(1))$ will only take values of the form $\mathbf{j} = \mathbf{i} + e_l - e_k$ for some $1 \leq k, l \leq K$. For any distinct pair k, l, we have

$$\mathbb{P}\{\mathbf{X}(1) = \mathbf{i} + e_l - e_k \mid \mathbf{X}(0) = \mathbf{i}\} = p_k \sum_{m=1}^{K} p_m u_{ml}.$$

For $\mathbf{j} = \mathbf{i}$, the transition probability is

$$\mathbb{P}\{\mathbf{X}(1) = \mathbf{i} \mid \mathbf{X}(0) = \mathbf{i}\} = 1 - \sum_{k \neq l} p_k \left(\sum_{m=1}^{K} p_m u_{ml} \right).$$

1.2 Diffusion Approximation

A continuous time Markov process with continuous sample paths is called a *diffusion process*. By changing the scales of time and space, many Markov processes can be approximated by diffusion processes. Both the Wright–Fisher model and the Moran model are Markov processes with discontinuous sample paths. When the number of alleles is large, direct calculations associated with these models are difficult to carry out and explicit results are rare. In these cases, useful information can be gained through the study of approximating diffusion processes. This section includes a very brief illustration of the diffusion approximation. More detailed discussions on the use of diffusion process in genetics can be found in [48] and [67].

For any $M \geq m \geq 1$, let $(X_1, \ldots, X_m)$ be a multinomial random variable with parameters $M, \mathbf{q}$, where

$$\mathbf{q} = (q_1, \ldots, q_m), \sum_{k=1}^{m} q_k = 1, \; q_i \geq 0, 1 \leq i \leq m.$$

The means and covariances of $(X_1, \ldots, X_m)$ are

$$\mathbb{E}[X_i] = Mq_i, \; Cov(X_i, X_j) = Mq_i(\delta_{ij} - q_j), \; 1 \leq i, j \leq m. \tag{1.3}$$

K-allele Wright–Fisher Model with Mutation

For distinct i, j, change (u_{ij}) to $(\frac{1}{2N} u_{ij})$. Set

$$\mathbf{Y}^N(t) = (Y_1^N(t), \ldots, Y_K^N(t)), Y_i^N(t) = X_i([2Nt]), \tag{1.4}$$

$$\mathbf{p}^N(t) = (p_1^N(t), \ldots, p_K^N(t)) = \frac{1}{2N} \mathbf{Y}^N(t), \tag{1.5}$$

where $[2Nt]$ is the integer part of $2Nt$.

Note that one unit of time in the $\mathbf{Y}^N(\cdot)$ process corresponds to $2N$ generations of the process $\mathbf{X}(\cdot)$. If we choose $\Delta t = \frac{1}{2N}$, then the time period $[t, t + \Delta t]$ corresponds to one generation in the process $\mathbf{X}(\cdot)$. During this period $\mathbf{p}$ becomes $\mathbf{p}'$ due to mutation followed by a multinomial sampling. Let

$$\Delta \mathbf{p}^N(t) = (\Delta p_1^N(t), \ldots, \Delta p_K^N(t)) \\ = \mathbf{p}^N(t+\Delta t) - \mathbf{p}^N(t)$$

denote the change of $\mathbf{p}^N(t)$ over $[t, t+\Delta t]$. Our aim is to approximate $E[\Delta p_i^N(t)]$ and $E[\Delta p_i^N(t) \Delta p_j^N(t)]$ for $i, j = 1, \ldots, K$. Assume that $\mathbf{p}^N(t) = \mathbf{p} = (p_1, \ldots, p_K)$. Then by (1.3) we have

$$\begin{aligned} &\mathbb{E}[\Delta p_i^N(t)] \\ &= \mathbb{E}[p_i^N(t+\Delta t) - p_i' + p_i' - p_i] \qquad (1.6)\\ &= p_i' - p_i = \sum_{j \neq i}^{K} \frac{1}{2N}(p_j u_{ji} - p_i u_{ij}) \\ &= \frac{1}{2N} b_i(\mathbf{p}) \\ &\mathbb{E}[\Delta p_i^N(t) \Delta p_j^N(t)] \\ &= (p_i' - p_i)(p_j' - p_j) + Cov(p_i^N(t+\Delta t), p_j^N(t+\Delta t)) \qquad (1.7)\\ &= \frac{1}{(2N)^2} b_i(\mathbf{p}) b_j(\mathbf{p}) + \frac{1}{2N} a_{ij}(\mathbf{p}) + o\left(\frac{1}{(2N)^2}\right), \end{aligned}$$

where

$$b_i(\mathbf{p}) = \sum_{j \neq i} p_j u_{ji} - \sum_{j \neq i} p_i u_{ij}, \qquad (1.8)$$

$$a_{ij}(\mathbf{p}) = p_i(\delta_{ij} - p_j). \qquad (1.9)$$

Letting N go to infinity, it follows that

$$\lim_{N \to \infty} \frac{1}{\Delta t} \mathbb{E}[\Delta p_i^N(t)] = b_i(\mathbf{p}),$$

$$\lim_{N \to \infty} \frac{1}{\Delta t} \mathbb{E}[\Delta p_i^N(t) \Delta p_j^N(t)] = a_{ij}(\mathbf{p}).$$

Ignoring higher order terms that approach zero faster than Δt, we have shown that when the mutation rate is scaled by a factor of $1/2N$ and time is measured in units of $2N$ generations, the Wright–Fisher model is approximated by diffusion process,

$$dp_i(t) = b_i(\mathbf{p}(t))dt + \sum_{j=1}^{K-1} \sigma_{ij}(\mathbf{p}(t)) dB_j(t)$$

with

$$\sum_{l=1}^{K-1} \sigma_{il}(x(t)) \sigma_{jl}(\mathbf{p}(t)) = a_{ij}(\mathbf{p}).$$

The generator of the diffusion is

$$Lf(\mathbf{p}) = \frac{1}{2}\sum_{i,j=1}^{K} a_{ij}(\mathbf{p})\frac{\partial^2 f}{\partial p_i \partial p_j} + \sum_{i=1}^{K} b_i(\mathbf{p})\frac{\partial f}{\partial p_i}.$$

In the two-allele case, the approximating diffusion solves the following stochastic differential equation:

$$dp(t) = (u_1(1-p(t)) - u_2 p(t))dt + \sqrt{p(t)(1-p(t))}dB_t, \tag{1.10}$$

where u_1, u_2 are the mutation rates of the process. For $a \geq 0$, let

$$\Gamma(a) = \int_0^\infty p^{a-1} e^{-p} dp.$$

For $u_1 > 0, u_2 > 0$, let

$$h(p) = \frac{\Gamma(2(u_1+u_2))}{\Gamma(2u_1)\Gamma(2u_2)} p^{2u_1-1}(1-p)^{2u_2-1},\ 0 < p < 1,$$

which is the density function of the $Beta(2u_1, 2u_2)$ distribution.

By direct calculation we have that for any twice continuously differentiable functions f, g on $[0,1]$,

$$\int_0^1 [g(p)Lf(p)]h(p)dp = 0.$$

Thus the $Beta(2u_1, 2u_2)$ distribution is a reversible measure for the process $p(t)$. When the mutation is symmetric; i.e., $u_1 = u_2 = u$, we write $\theta = 2u$. Recall that $u = 2N\mu$ with μ being the original individual mutation rate, it follows that $\theta = 4N\mu$ which is called the *scaled population mutation rate*.

A particular selection model with a selection factor s is the stochastic differential equation

$$\begin{aligned} dp(t) &= (u_1(1-p(t)) - u_2 p(t) + s p(t)(1-p(t)))dt \\ &\quad + \sqrt{p(t)(1-p(t))}dB_t. \end{aligned} \tag{1.11}$$

If $u_1 > 0, u_2 > 0, s > 0$, this diffusion process will have a reversible measure given by

$$C p^{2u_1-1}(1-p)^{2u_2-1} e^{2sp} dp,\ 0 < p < 1,$$

where C is the normalizing constant. It is clear that this distribution is simply a change-of-measure with respect to the selection-free case.

For the K-allele model with mutation, if

$$u_{ii} = 0,\ u_{ij} = \frac{\theta}{2(K-1)},\ i,j \in \{1,\ldots,K\},\ i \neq j,$$

then the reversible measure is the $Dirichlet(\frac{\theta}{K-1}, \ldots, \frac{\theta}{K-1})$ distribution.

Here, for any $m \geq 2$, the $Dirichlet(\alpha_1, \ldots, \alpha_m)$ distribution is a multivariate generalization of the Beta distribution, defined on the simplex

$$\left\{(p_1, \ldots, p_m) : p_i \geq 0, \sum_{k=1}^{m} p_k = 1\right\},$$

and has a density with respect to the $(m-1)$-dimensional Lebesgue measure given by

$$f(p_1, \ldots, p_{m-1}, p_m) = \frac{\Gamma(\sum_{k=1}^{m} \alpha_k)}{\prod_{k=1}^{m} \Gamma(\alpha_k)} p_1^{\alpha_1 - 1} \cdots p_{m-1}^{\alpha_{m-1}-1} p_m^{\alpha_m - 1},$$

where

$$p_m = 1 - p_1 - \cdots - p_{m-1}.$$

Next we turn to the Moran model with mutation. For distinct i, j, we make the same change from (u_{ij}) to $(\frac{1}{2N} u_{ij})$. Set

$$\mathbf{Z}^N(t) = (Z_1^N(t), \ldots, Z_K^N(t)), \ Z_i^N(t) = X_i([2N^2 t]),$$
$$\mathbf{p}^N(t) = (p_1^N(t), \ldots, p_K^N(t)) = \frac{1}{2N} \mathbf{Z}^N(t).$$

Here the time unit is $2N^2$ generations. Assume that $\mathbf{p}^N(t) = (p_1, \ldots, p_K)$. Then by choosing $\Delta t = \frac{1}{2N^2}$, it follows that for any $1 \leq i \neq j \leq K$,

$$\mathbb{E}[p_i^N(t + \Delta t) - p_i^N(t)] = \frac{1}{2} b_i(\mathbf{p}) \Delta t,$$
$$\mathbb{E}[(p_i^N(t + \Delta t) - p_i^N(t))^2] = [p_i(1 - p_i) + O(\Delta t)] \Delta t,$$
$$\mathbb{E}[(p_i^N(t + \Delta t) - p_i^N(t))(p_j^N(t + \Delta t) - p_j^N(t))] = [-p_i p_j + O(\Delta t)] \Delta t,$$

which leads to

$$\lim_{N \to \infty} \frac{1}{\Delta t} \mathbb{E}[p_i^N(t + \Delta t) - p_i^N(t)] = \frac{1}{2} b_i(\mathbf{p}), \tag{1.12}$$

$$\lim_{N \to \infty} \frac{1}{\Delta t} \mathbb{E}[(p_i^N(t + \Delta t) - p_i^N(t))(p_j^N(t + \Delta t) - p_j^N(t))] = a_{ij}(\mathbf{p}). \tag{1.13}$$

Note that the time is speeded up by $N(2N)$ instead of $(2N)^2$, and the drift is half of that of the Wright–Fisher model. Thus the scaled population mutation rate is given by $\theta = 2N\mu$ in the Moran model.

1.3 An Important Relation

In this section, we will discuss an important relation between the gamma distribution and the Dirichlet distribution. The dynamical analog will be exploited intuitively.

A non-negative random variable Y is said to have a $Gamma(\alpha,\beta)$ distribution if its density is given by

$$f(y) = \frac{1}{\beta^\alpha \Gamma(\alpha)} y^{\alpha-1} e^{-y/\beta},\ y > 0.$$

Theorem 1.1. *For each $m \geq 2$, let $Y_1,\ldots,Y_m$ be independent gamma random variables with respective parameters $(\alpha_i,1), i = 1,\ldots,m$. Define*

$$X_i = \frac{Y_i}{\sum_{k=1}^m Y_k},\ i = 1,\ldots,m.$$

Then $(X_1,\ldots,X_m)$ has a $Dirichlet(\alpha_1,\ldots,\alpha_m)$ distribution and is independent of $\sum_{k=1}^m Y_k$.

Proof. Define a transformation of $(y_1,\ldots,y_m) \in \mathbb{R}^m$, by

$$x_i = \frac{y_i}{\sum_{k=1}^m y_k},\quad i = 1,\ldots,m-1,$$
$$z = \sum_{k=1}^m y_k.$$

Then the determinant $J(y_1,\ldots,y_n)$ of the Jacobian matrix of this transformation is given by $\frac{1}{z^{m-1}}$ and the joint density $g(x_1,\ldots,x_{m-1},z)$ of $(X_1,\ldots,X_{m-1},Z)$, where

$$Z = \sum_{k=1}^m Y_k,$$

is given by

$$\begin{aligned} g(x_1,\ldots,x_{m-1},z) &= \frac{1}{\Gamma(\alpha_1)\cdots\Gamma(\alpha_m)} y_1^{\alpha_1-1}\cdots y_m^{\alpha_m-1} e^{-z} z^{m-1} \\ &= \frac{\Gamma(\sum_{k=1}^m \alpha_k)}{\Gamma(\alpha_1)\cdots\Gamma(\alpha_m)} x_1^{\alpha_1-1}\cdots x_m^{\alpha_m-1} \frac{1}{\Gamma(\sum_{k=1}^m \alpha_k)} z^{\sum_{k=1}^m \alpha_k - 1} e^{-z}, \end{aligned}$$

which implies that Z has a $Gamma(\sum_{k=1}^m \alpha_k, 1)$ distribution, $(X_1,\ldots,X_m)$ is independent of Z, and has a $Dirichlet(\alpha_1,\ldots,\alpha_m)$ distribution.

□

Remark: For any $\beta > 0$ and a $Gamma(\alpha,1)$ random variable Y, the new variable βY is a $Gamma(\alpha,\beta)$ random variable. Thus Theorem 1.1 still holds if $Y_1,\ldots,Y_m$ are independent $Gamma(\alpha_i,\beta)$ random variables, $i = 1,\ldots,m$.

Consider m independent stochastic differential equations:

$$dY_i(t) = \frac{1}{2}(\alpha_i - \beta Y_i(t))dt + \sqrt{Y_i(t)}dB_i(t),\ i = 1,\ldots,m.$$

The diffusion $Y_i(t)$ has a unique stationary distribution $Gamma(\alpha_i,\beta)$. Define

$$X_i(t) = \frac{Y_i(t)}{\sum_{k=1}^m Y_k(t)}.$$

Applying Itô's lemma formally, one obtains

$$\begin{aligned}
dX_i(t) &= \frac{1}{(\sum_{k=1}^m Y_k(t))^2}\left(\frac{1}{2}\left[\sum_{k\neq i} Y_k(t)(\alpha_i - \beta Y_i(t)) - Y_i(t)\sum_{k\neq i}(\alpha_k - \beta Y_k(t))\right] dt \right.\\
&\quad \left. + \left(\sum_{k\neq i} Y_k(t)\right)\sqrt{Y_i(t)}dB_i(t) - \sum_{k\neq i} Y_i(t)\sqrt{Y_k(t)}dB_k(t)\right)\\
&= \frac{1}{(\sum_{k=1}^m Y_k(t))^2}\left(\left[\frac{\alpha_i \sum_{k\neq i} Y_k(t) - Y_i(t)\sum_{k\neq i}\alpha_k}{2}\right] dt \right.\\
&\quad \left. + \left(\sum_{k\neq i} Y_k(t)\right)\sqrt{Y_i(t)}dB_i(t) - Y_i(t)(\sum_{k\neq i}\sqrt{Y_k(t)}dB_k(t))\right).
\end{aligned}$$

Now assume that $\sum_{k=1}^m Y_k(t) = 1$ for all t. Then this SDE becomes

$$\begin{aligned}
dX_i(t) &= \frac{1}{2}\left(\alpha_i - (\sum_{k=1}^m \alpha_k)X_i(t)\right) dt\\
&\quad + \left[(1 - X_i(t))\sqrt{X_i(t)}dB_i(t) - \sum_{k\neq i} X_i(t)\sqrt{X_k(t)}dB_k(t)\right].
\end{aligned}$$

Due to independence of the Brownian motions, the second term can be formally written as $\sqrt{X_i(t)(1-X_i(t))}d\tilde{B}(t)$ with $\tilde{B}(t)$ being a standard Brownian motion, and $(X_1(t),\ldots,X_m(t))$ becomes the m-allele Wright–Fisher diffusion. Thus the structure in Theorem 1.1 has a dynamical analog. Later on, we will see that the rigorous derivation of this result is related to the Perkins disintegration theorem in measure-valued processes (cf. [20]).

1.4 Notes

One can consult [86] for a concise introduction to population genetics. A brief introduction to the mathematical framework can be found in [127] and Chapter 10 in [62]. The most comprehensive references on mathematical population genetics are [67] and [48].

The origin of the Wright–Fisher model is [81] and [185]. The Moran model was proposed in [139]. The existence and uniqueness of the Wright–Fisher diffusion process in Section 1.2 was obtained in [55]. In [62], one can find the rigorous derivation of the diffusion approximation in Section 1.2 and a proof of the existence and uniqueness of the stationary distribution for the two-allele diffusion process with two way mutation and selection. The relation between the gamma distribution and the Dirichlet distribution is better understood in the polar coordinate system with the Dirichlet distribution being the angular part and the total summation as the radial part. The infinite dimensional analog of this structure is the Perkins disintegration theorem, obtained in [143].

Chapter 2
The Poisson–Dirichlet Distribution

The focus of this chapter is the Poisson–Dirichlet distribution, the central topic of this book. We introduce this distribution and discuss various models that give rise to it. Following Kingman [125], the distribution is constructed through the gamma process. An alternative construction in [8] is also included, where a scale-invariant Poisson process is used. The density functions of the marginal distributions are derived through Perman's formula. Closely related topics such as the GEM distribution, the Ewens sampling formula, and the Dirichlet process are investigated in detail through the study of urn models. The required terminology and properties of Poisson processes and Poisson random measures can be found in Appendix A.

2.1 Definition and Poisson Process Representation

For each $K \geq 2$, set

$$\begin{aligned} \nabla_K &= \left\{(p_1,\ldots,p_K) : p_1 \geq p_2 \geq \cdots \geq p_K \geq 0, \sum_{j=1}^{K} p_j = 1\right\}, \\ \nabla_\infty &= \left\{(p_1,p_2,\ldots,) : p_1 \geq p_2 \geq \cdots \geq 0, \sum_{j=1}^{\infty} p_j = 1\right\}, \\ \nabla &= \left\{(p_1,p_2,\ldots,) : p_1 \geq p_2 \geq \cdots \geq 0, \sum_{j=1}^{\infty} p_j \leq 1\right\}. \end{aligned} \tag{2.1}$$

The space ∇_K can be embedded naturally into ∇_∞ and thus viewed as a subset of ∇_∞. The space ∇ is the closure of ∇_∞ in $[0,1]^\infty$, and the topology on each of ∇_∞ and ∇ is the subspace topology inherited from $[0,1]^\infty$. For $\theta > 0$, let $(X_1,\ldots,X_K)$ have a *Dirichlet*$(\frac{\theta}{K-1},\ldots,\frac{\theta}{K-1})$ distribution. Let $\mathbf{Y}^K = (Y_1^K,\ldots,Y_K^K)$ be the decreasing order statistics of $(X_1,\ldots,X_K)$.

S. Feng, *The Poisson–Dirichlet Distribution and Related Topics*,
Probability and its Applications, DOI 10.1007/978-3-642-11194-5_2,

Theorem 2.1. *Let $M_1(\nabla)$ denote the space of probability measures on ∇. Then the sequence $\{\mu_K : K \geq 2\}$ of laws of $\mathbf{Y}^K$ converges weakly in $M_1(\nabla)$.*

Proof. Let $\gamma(t)$ be a gamma process; i.e., a process with stationary independent increments such that each increment, $\gamma(t) - \gamma(s)$ for $0 \leq s < t$, follows a $Gamma(\theta(t-s), 1)$ distribution. We sometimes write γ_t instead of $\gamma(t)$ for notational convenience. Set

$$I_l = \left(\frac{l}{K-1}, \frac{l+1}{K-1}\right), \ l = 0, \ldots, K-1,$$

$$\tilde{X}_l = \frac{\gamma(\frac{l+1}{K-1}) - \gamma(\frac{l}{K-1})}{\gamma(\frac{K}{K-1})},$$

and $\tilde{\mathbf{Y}}^K = (\tilde{Y}_1^K, \ldots, \tilde{Y}_K^K)$ denote the descending order statistics of $(\tilde{X}_1, \ldots, \tilde{X}_K)$. It follows from Theorem 1.1 that $(\tilde{X}_1, \ldots, \tilde{X}_K)$ has the same distribution as $(X_1, \ldots, X_K)$. For $l = 0, \ldots, K-1$, denote the lth highest jump of $\gamma(t)$ over the interval $(0,1)$ by J_l. Since the process $\gamma(t)$ does not have any fixed jump point, for almost all sample paths the jumps occur in $I_1, \ldots, I_{K-1}$. Thus with probability one, we have

$$\tilde{Y}_l^K \gamma\left(\frac{K}{K-1}\right) \geq J_l$$

which implies that

$$\liminf_{K \to \infty} \tilde{Y}_l^K \geq \frac{J_l}{\gamma(\theta)}. \tag{2.2}$$

Fatou's lemma guarantees that the summation over l of the left-hand side of equation (2.2) is no more than one. This, combined with the fact that the right-hand side adds up to one, implies that the inequality in (2.2) is actually an equality for all l and the lower limit is actually the limit. Let $P_l(\theta) = \frac{J_l}{\gamma(\theta)}$ for any $l \geq 0$, $\mathbf{P}(\theta) = (P_1(\theta), P_2(\theta), \ldots)$, and let Π_θ denote the law of $\mathbf{P}_\theta$. Then Π_θ belongs to $M_1(\nabla)$, and we have shown that for any fixed $r \geq 1$ and any bounded continuous function g on ∇_r,

$$\lim_{K \to \infty} \int_\nabla g(p_1, \ldots, p_r) d\mu_K = \int_\nabla g(p_1, \ldots, p_r) d\Pi_\theta. \tag{2.3}$$

It now follows from the Stone–Weierstrass theorem that (2.3) holds for every real-valued continuous function g on ∇.

□

Definition 2.1. The law Π_θ of $\mathbf{P}(\theta)$ in Theorem 2.1 is called the Poisson–Dirichlet distribution with parameter θ.

Let $\varsigma_1 \geq \varsigma_2 \geq \ldots$ be the random points of a Poisson process with intensity measure

$$\mu(dx) = \theta x^{-1} e^{-x} dx, x > 0.$$

Set

$$\sigma = \sum_{i=1}^{\infty} \varsigma_i.$$

Then we have:

Theorem 2.2. *For each positive θ, the distribution of $(\frac{\varsigma_1}{\sigma}, \frac{\varsigma_2}{\sigma}, \ldots)$ is Π_θ. Furthermore, σ has the distribution $Gamma(\theta, 1)$ and is independent of $(\frac{\varsigma_1}{\sigma}, \frac{\varsigma_2}{\sigma}, \ldots)$.*

Proof. Noting that the jump sizes of the process $\gamma(t)$ over the interval $[0,1]$ form a nonhomogeneous Poisson process with intensity measure $\mu(dx)$, it follows from the construction in Theorem 2.1 that the law of $(\frac{\varsigma_1}{\sigma}, \frac{\varsigma_2}{\sigma}, \ldots)$ is Π_θ. It is clear from the definition that σ has the distribution $Gamma(\theta, 1)$. The fact that σ and $(\frac{\varsigma_1}{\sigma}, \frac{\varsigma_2}{\sigma}, \ldots)$ are independent, follows from Theorem 1.1.

□

The next example illustrates the application of the Poisson–Dirichlet distribution in random number theory.

Example 2.1 (Prime factorization of integers). *For each integer $n \geq 1$, let N_n be chosen at random from $1, 2, \ldots, n$. Consider the prime factorization of N_n*

$$N_n = \Pi_p p^{C_p(n)},$$

where $C_p(n)$ is the multiplicity of p. Define

$$\begin{aligned}
K_n &= \sum_p C_p(n), \\
A_i(n) &= ith\ biggest\ prime\ factor\ of\ N_n,\ i = 1, \ldots, K_n, \\
A_i(n) &= 1, i > K_n, \\
L_i(n) &= \log A_i(n), i \geq 1.
\end{aligned}$$

Then, as n goes to infinity, $(\frac{L_1(n)}{\log n}, \frac{L_2(n)}{\log n} \ldots)$ converges in distribution to $(Y_1, Y_2, \ldots)$ whose law is the Poisson–Dirichlet distribution with parameter one.

2.2 Perman's Formula

We start this section with the definition of a stochastic process, the subordinator, which is closely related to the Poisson process and the Poisson random measure.

Definition 2.2. A process $\{\tau_s : s \geq 0\}$ is called a *subordinator* if it has stationary, independent, and non-negative increments with $\tau_0 = 0$.

Definition 2.3. A subordinator $\{\tau_s :\ s \geq 0\}$ is said to have no drift if for any $\lambda \geq 0$, $s \geq 0$,

$$\mathbb{E}[e^{-\lambda \tau_s}] = \exp\left\{ -s \int_0^\infty (1-e^{-\lambda x})\Lambda(dx) \right\},$$

where the measure Λ concentrates on $[0,+\infty)$ and is called the Lévy measure of the subordinator.

All subordinators considered in this book are without drift and each has a Lévy measure Λ satisfying

$$\Lambda((0,+\infty)) = +\infty, \tag{2.4}$$

$$\int_0^\infty x \wedge 1 \, \Lambda(dx) < +\infty, \tag{2.5}$$

$$\Lambda(dx) = h(x)dx, \text{ for some } h(x) \geq 0. \tag{2.6}$$

Let $J_1(\tau_t), J_2(\tau_t), \ldots$ denote the jump sizes of τ_t up to time t, ranked by size in descending order. Then the sequence is infinite due to (2.4), and condition (2.5) ensures that for every $t > 0$, $\tau_t = \sum_{i=1}^\infty J_i(\tau_t) < \infty$ with probability one. The condition (2.6) implies that for every $t > 0$, the distribution of τ_t has a density with respect to the Lebesgue measure (cf. [106]). Set

$$P_i^{\tau_t} = \frac{J_i(\tau_t)}{\tau_t}.$$

Then $(P_1^{\tau_t}, P_2^{\tau_t}, \ldots)$ forms a probability-valued random vector. Change h to $t \cdot h$, and all calculations involving jumps of the subordinator $\{\tau_s : 0 \leq s \leq t\}$ change to those of the subordinator $\{\tilde{\tau}_s : 0 \leq s \leq 1\}$ with Lévy measure $th(x)dx$. Without loss of generality, we will choose $t = 1$. Let g denote the density function of τ_1.

For any Borel set $A \subset [0,+\infty)$, and any $t \geq 0$, define

$$N(t,A) = \sum_{s \leq t} \chi_A(\tau_s - \tau_{s-}). \tag{2.7}$$

Then for each fixed t, $N(t,\cdot)$ is a Poisson random measure with mean measure $t\Lambda(\cdot)$. For simplicity, we write $N(\cdot)$ for $N(1,\cdot)$.

Lemma 2.1. *For distinct*

$$0 < p_1, \ldots, p_n < 1, \ p_1 + \cdots + p_n < 1,$$

we have

$$\begin{aligned} &\mathbb{P}\{\mathit{some}\ P_i^{\tau_1} \in dp_1, \ldots, \mathit{some}\ P_i^{\tau_1} \in dp_n, \tau_1 \in ds\} \\ &= s^n\, h(p_1 s) \cdots h(p_n s)\, g(s(1-p_1-\cdots-p_n)) dp_1 \cdots dp_n ds. \end{aligned}$$

Proof. Let $J_i = J_i(\tau_1)$. Then by properties of the Poisson process,

$$
\begin{aligned}
&\mathbb{P}\{some\ J_i \in dx_1,\ldots,some\ J_i \in dx_n,\ \tau_1 \in ds\}\\
&= \mathbb{E}[N(dx_1)\cdots N(dx_n)\ \chi_{\{\tau_1 \in ds\}}]\\
&= \mathbb{E}\Big[N(dx_1)\cdots N(dx_n)\ \mathbb{E}\big[\chi_{\{\tau_1 \in ds\}} \mid N(dx_1)\cdots N(dx_n)\big]\Big].
\end{aligned}
$$

Since the mean measure of a Poisson process does not have fixed atoms, it follows that, conditioning on fixing the location of finite points, the remaining points form the same Poisson process. Thus

$$
\begin{aligned}
&\mathbb{E}\big[\chi_{\{\tau_1 \in ds\}} \mid N(dx_1)\cdots N(\mathrm{d}x_n)\big]\\
&= g(s - x_1 - \cdots - x_n)ds
\end{aligned}
$$

and

$$
\begin{aligned}
&\mathbb{P}\{some\ J_i \in dx_1,\ldots,\ some\ J_i \in dx_n,\ \tau_1 \in ds\}\\
&= h(x_1)h(x_2)\cdots h(x_n)\ g(s - x_1 - \cdots - x_n)dx_1\cdots dx_n ds
\end{aligned}
$$

Making the change of variable $p_i = \frac{x_i}{s}$ leads to

$$
\begin{aligned}
&\mathbb{P}\{some\ P_i^{\tau_1} \in dp_1,\ldots,some\ P_i^{\tau_1} \in dp_n, \tau_1 \in ds\}\\
&= s^n\ h(p_1 s)\cdots h(p_n s)\ g(s(1 - p_1 - \cdots - p_n))dp_1\cdots dp_n ds.
\end{aligned}
$$

□

Theorem 2.3. *For each* $0 \le p < 1$,

$$\mathbb{P}\{P_1^{\tau_1} > p,\ \tau_1 \in ds\} = \sum_{n=1}^{\infty} \frac{(-1)^{n+1}}{n!} I_n(p,ds),$$

where

$$I_n(p,ds) = \int_{B_p^n} s^n\ h(u_1 s)\cdots h(u_n s)\ g(s\ \hat{u}_n)du_1\cdots du_n ds,$$

$$B_p^n = \left\{(u_1,\ldots,u_n) : p < u_i < 1, \sum_{i=1}^{n} u_i < 1\right\},$$

$$\hat{u}_n = 1 - u_1 - \cdots - u_n.$$

Integrating out the s component, one gets the distribution function of $P_1^{\tau_1}$. *Note that there are only finite non-zero terms in the summation for each fixed* $0 < p < 1$, *the change in the order of integration is justified.*

Proof. It follows from Lemma 2.1 that

$$
\begin{aligned}
&\mathbb{E}[\ \#\{(k_1,\ldots,k_n) : k_i \text{ distinct}, (P_{k_1}^{\tau_1},\ldots,P_{k_n}^{\tau_1}) \in B\}, \tau_1 \in ds]\\
&= \int_B s^n\ h(u_1 s)\cdots h(u_n s)\ g(s(1 - u_1 - \cdots - u_n))du_1\cdots du_n ds
\end{aligned}
$$

for each measurable set $B \subset \mathbb{R}^n$, where the symbol # denotes the cardinality of the set following it. Set

$$K_p^n(ds) = \#\{(k_1,\ldots,k_n) : k_i \text{ distinct}, (P_{k_1}^{\tau_1},\ldots,P_{k_n}^{\tau_1}) \in B_p^n,\ \tau_1 \in ds\}.$$

Then $\mathbb{E}[K_p^n(ds)] = I_n(p,ds)$. For any $i \geq 1$, let $F_i = \{P_i^{\tau_1} > p\}$. Then clearly

$$F_1 = \bigcup_{i=1}^{\infty} F_i.$$

By the inclusion–exclusion formula,

$$\chi_{F_1} = \sum_{i=1}^{\infty} \chi_{F_i} - \sum_{i<j} \chi_{F_i \cap F_j} + \cdots.$$

Discounting the permutations, it follows that

$$\chi_{\{P_1 > p,\ \tau_1 \in ds\}} = \sum_{n=1}^{\infty} \frac{(-1)^{n+1} K_p^n(ds)}{n!}.$$

Taking expectations leads to the result.

□

Next, we present Perman's formulae for the finite dimensional density functions of random vector $(\tau_1, P_1^{\tau_1},\ldots,P_m^{\tau_1})$ for each $m \geq 1$.

Theorem 2.4. (Perman's formula) *For each $m \geq 1$, the vector*

$$(\tau_1, P_1^{\tau_1},\ldots,P_m^{\tau_1}) \in \mathbb{R}_+^{m+1}$$

has a density function $g_m(t,p_1,\ldots,p_m)$ with respect to the Lebesgue measure and

(1) *For $t > 0$ and $p \in (0,1)$,*

$$g_1(t,p) = t h(tp) \int_0^{\frac{p}{1-p} \wedge 1} g_1(t(1-p),z)\,dz. \tag{2.8}$$

(2) *For*

$$t > 0,\ m \geq 2,\ 0 < p_m < \cdots < p_1 < 1,\ \sum_{i=1}^{m} p_i < 1,$$

and

$$\hat{p}_{m-1} = 1 - p_1 - \cdots - p_{m-1},$$

we have

$$g_m(t,p_1,\ldots,p_m) = \frac{t^{m-1} h(tp_1)\cdots h(tp_{m-1})}{\hat{p}_{m-1}} g_1\left(t\hat{p}_{m-1}, \frac{p_m}{\hat{p}_{m-1}}\right). \tag{2.9}$$

Proof. Noting that $(J_1,\ldots,J_m) \equiv (J_1(\tau_1),\ldots,J_m(\tau_1))$ are the m largest jumps of $\{\tau_s : 0 \le s \le 1\}$ with $J_1 > \cdots > J_m$, it follows that

$$\begin{aligned}\mathbb{P}(J_1 < v_1) &= \mathbb{P}(N([v_1,+\infty)) = 0)\\ &= e^{-\Lambda([v_1,+\infty))} = e^{-\int_{v_1}^{+\infty} h(z)\mathrm{d}z},\end{aligned}$$

and

$$\mathbb{P}(J_1 \in dv_1) = e^{-\Lambda([v_1,+\infty))} h(v_1) dv_1.$$

Conditioning on $\{J_1 \in dv_1\}$, we have

$$\mathbb{P}(J_2 \in dv_2 \mid J_1 \in dv_1) = e^{-\Lambda([v_2,v_1))} h(v_2) dv_2,$$

$$\mathbb{P}(J_1 \in dv_1, J_2 \in dv_2) = e^{-\Lambda([v_2,+\infty))} h(v_1)h(v_2) dv_1 dv_2.$$

It follows by induction that for $v_1 > v_2 > \cdots > v_m > 0$,

$$\mathbb{P}(J_1 \in dv_1,\ldots,J_m \in dv_m) = h(v_1)\cdots h(v_m) e^{-\Lambda([v_m,+\infty))} dv_1 \cdots dv_m.$$

Let $f_v(x)$ be the density function of τ_1 given $J_m = v$. Then $f_v(x)$ is the density function of $\tilde{\tau}_1$, where $\tilde{\tau}_s$ is a subordinator with Lévy measure $\tilde{\Lambda}(dx) = h(x)1_{\{x \le v\}}dx$. Hence for $t > v_1 + \cdots + v_m$, $(\tau_1, J_1,\ldots,J_m)$ has joint density function

$$h(v_1)\cdots h(v_m) e^{-\Lambda([v_m,+\infty))} f_{v_m}(t - v_1 - \cdots - v_m).$$

Making the change of variable $v_i = tp_i$ gives the density function

$$g_m(t,p_1,\ldots,p_m) = t^m h(tp_1)\cdots h(tp_m) e^{-\Lambda([tp_m,+\infty))} f_{tp_m}(t(1 - p_1 - \cdots - p_m)).$$

For $m = 1$, we have $g_1(t,p_1) = th(tp_1)e^{-\Lambda([tp_1,+\infty))} f_{tp_1}(t\hat{p}_1)$, which implies that for $m \ge 2$,

$$g_1\left(t, \frac{p_m}{\hat{p}_{m-1}}\right) = th\left(t\cdot\frac{p_m}{\hat{p}_{m-1}}\right) e^{-\Lambda([t\cdot\frac{p_m}{\hat{p}_{m-1}},+\infty))} f_{t\cdot\frac{p_m}{\hat{p}_{m-1}}}\left(t\left(1 - \frac{p_m}{\hat{p}_{m-1}}\right)\right),$$

$$g_1\left(t\hat{p}_{m-1}, \frac{p_m}{\hat{p}_{m-1}}\right) = \hat{p}_{m-1} th(tp_m) e^{-\Lambda([p_m,+\infty))} f_{tp_m}(t\hat{p}_m),$$

and

$$\begin{aligned}g_m(t,p_1,\ldots,p_m) &= \frac{t^{m-1}h(tp_1)\cdots h(tp_{m-1})}{\hat{p}_{m-1}} \hat{p}_{m-1} th(tp_m) e^{-\Lambda([p_m,+\infty))} f_{tp_m}(t\hat{p}_m)\\ &= \frac{t^{m-1}h(tp_1)\cdots h(tp_{m-1})}{\hat{p}_{m-1}} g_1\left(t\hat{p}_{m-1}, \frac{p_m}{\hat{p}_{m-1}}\right).\end{aligned}$$

For $m = 2$,

$$g_2(t,p_1,p_2) = \frac{th(tp_1)}{1 - p_1} g_1\left(t(1-p_1), \frac{p_2}{1-p_1}\right).$$

Integrating out p_2 yields

$$\begin{aligned} g_1(t,p_1) &= \int_0^{p_1} g_2(t,p_1,p_2)dp_2 \\ &= th(tp_1)\int_0^{p_1} \frac{1}{1-p_1} g_1\left(t(1-p_1),\frac{p_2}{1-p_1}\right)dp_2 \\ &= th(tp_1)\int_0^{\frac{p_1}{1-p_1}} g_1(t(1-p_1),z)dz \\ &= th(tp_1)\int_0^{\frac{p_1}{1-p_1}\wedge 1} g_1(t(1-p_1),z)dz. \end{aligned}$$

Finally, we need to show that this equation determines g_1 uniquely. First for $\frac{1}{2} < p_1 < 1$, we have $\frac{p_1}{1-p_1} > 1$. Hence

$$\begin{aligned} g_1(t,p_1) = th(tp_1)\int_0^1 g_1(t(1-p_1),z)dz \\ = th(tp_1)f^{\tau_1}(t(1-p_1)), \end{aligned}$$

where f^{τ_1} is the density function of τ_1. For $\frac{1}{3} < p_1 \leq \frac{1}{2}$, we have $\frac{1}{2} < \frac{p_1}{1-p_1} \leq 1$ and thus

$$\begin{aligned} g_1(t,p_1) &= th(tp_1)\int_0^{\frac{p_1}{1-p_1}} g_1(t(1-p_1)),z)dz \\ &= th(tp_1)[f^{\tau_1}(t(1-p_1)) - \int_{\frac{p_1}{1-p_1}}^1 g_1(t(1-p_1),z)dz]. \end{aligned}$$

Inductively, $g_1(t,p_1)$ is uniquely determined over $(\frac{1}{k+1},\frac{1}{k}](k=1,2,\ldots)$ and therefore for all $p_1 > 0$. □

2.3 Marginal Distribution

This section is concerned with the derivation of the marginal distributions of $\mathbf{P}(\theta)$ for each $m \geq 1$. We begin with the distribution function of $P_1(\theta)$.

Theorem 2.5. *For any* $0 < p \leq 1$,

$$\mathbb{P}\{P_1(\theta) \leq p\} = 1 + \sum_{n=1}^{\infty} \frac{(-\theta)^n}{n!} \int_{B_p^n} \frac{(1-\sum_{i=1}^n u_i)^{\theta-1}}{u_1 u_2 \cdots u_n} du_1 du_2 \cdots du_n. \tag{2.10}$$

The function $\rho(x) = \mathbb{P}\{P_1(\theta) < 1/x\}$ *for* $x > 0$ *is called the Dickman function.*

Proof. Let $\{\tau_s : s \geq 0\}$ in Theorem 2.4 be the gamma process $\{\gamma_s : s \geq 0\}$ with Lévy measure

$$\Lambda(dx) = \theta x^{-1} e^{-x} dx,\ x \geq 0.$$

It then follows from Theorem 2.2 that $P_1(\theta)$ has the same law as $P_1^{\tau_1}$. By direct calculation,

$$I_n(p,ds) = ds\int_{B_p^n}(u_1\cdots u_n)^{-1}\frac{1}{\Gamma(\theta)}(s\hat{u}_n)^{\theta-1}e^{-s}du_1\cdots du_n.$$

Integrating with respect to s over $[0,\infty)$, it follows that

$$\int_0^\infty I_n(p,ds) = \int_{B_p^n}\frac{(1-\Sigma_{i=1}^n u_i)}{u_1\cdots u_n}du_1\cdots du_n,$$

which, combined with Theorem 2.3, implies the result. □

The joint distribution of $(P_1(\theta),\ldots,P_m(\theta))$ for each $m\geq 2$ is given by the following theorem.

Theorem 2.6. *Let $g_1^\theta(p)$ and $g_m^\theta(p_1,\ldots,p_m)$ denote the density functions of $P_1(\theta)$ and $(P_1(\theta),\ldots,P_m(\theta))$, respectively. Then*

$$pg_1^\theta(p) = \theta(1-p)^{\theta-1}\int_0^{(p/(1-p))\wedge 1} g_1^\theta(x)dx \tag{2.11}$$

$$\begin{aligned} g_m^\theta(p_1,\ldots,p_m) &= \frac{\theta^{m-1}(\hat{p}_{m-1})^{\theta-2}}{p_1p_2\cdots p_{m-1}}g_1^\theta\left(\frac{p_m}{\hat{p}_{m-1}}\right), \\ &= \frac{\theta^m(\hat{p}_m)^{\theta-1}}{p_1\cdots p_m}\mathbb{P}\left\{P_1(\theta)\leq\frac{p_m}{\hat{p}_m}\wedge 1\right\}, \end{aligned} \tag{2.12}$$

where $0<p_m<\cdots<p_1, \Sigma_{k=1}^m p_k<1$.

Proof. For each $m\geq 2$, it follows from Theorem 2.2 that $(P_1(\theta),\ldots,P_m(\theta))$ and $(P_1^{\tau_1},P_2^{\tau_1},\ldots,P_m^{\tau_1})$ have the same distribution, $(P_1^{\tau_1},P_2^{\tau_1},\ldots,P_m^{\tau_1})$ is independent of τ_1, and τ_1 is a *Gamma*$(\theta,1)$ random variable. Hence $g_1(t,p)=\frac{1}{\Gamma(\theta)}t^{\theta-1}e^{-t}g_1^\theta(p)$.

Integrating with respect to t on both sides of (2.8), it follows that

$$\begin{aligned} g_1^\theta(p) &= \frac{\theta}{p}\int^{\frac{p}{1-p}\wedge 1} g_1^\theta(z)dz\int_0^{+\infty}e^{-\frac{tp}{1-p}}\frac{1}{\Gamma(\theta)}t^{\theta-1}e^{-t}\frac{dt}{1-p} \\ &= \frac{\theta}{p}\frac{1}{(1+\frac{p}{1-p})^\theta}\frac{1}{1-p}\int_0^{\frac{p}{1-p}\wedge 1} g_1^\theta(z)dz, \end{aligned}$$

which leads to (2.11).

For $m\geq 2$,

$$g_m(t,p_1,\ldots,p_m) = \frac{1}{\Gamma(\theta)}t^{m-1}e^{-t}g_m^\theta(p_1,\ldots,p_m).$$

Thus

$$
\begin{aligned}
g_m^\theta(p_1,\ldots,p_m) &= \int_0^{+\infty} \frac{t^{m-1}h(tp_1)\cdots h(tp_{m-1})}{\hat{p}_{m-1}} g_1^\theta\left(t\hat{p}_{m-1}, \frac{p_m}{\hat{p}_{m-1}}\right) dt \\
&= g_1^\theta\left(\frac{p_m}{\hat{p}_{m-1}}\right) \int_0^{+\infty} \frac{(\theta t)^{m-1}(tp_1\cdots tp_m)^{-1} e^{-t(p_1+\cdots+p_{m-1})}}{\Gamma(\theta)\hat{p}_{m-1}e^{t\hat{p}_{m-1}}(t\hat{p}_{m-1})^{1-\theta}} dt \\
&= g_1^\theta\left(\frac{p_m}{\hat{p}_{m-1}}\right) \frac{\theta^{m-1}}{\hat{p}_{m-1}^2 p_1\cdots p_{m-1}} \int_0^{+\infty} e^{-\frac{p_1+\cdots+p_{m-1}}{\hat{p}_{m-1}}s} \frac{1}{\Gamma(\theta)} s^{\theta-1} e^{-s} ds \\
&= g_1^\theta\left(\frac{p_m}{\hat{p}_{m-1}}\right) \frac{\theta^{m-1}}{p_1\cdots p_{m-1}} \frac{1}{\hat{p}_{m-1}^2} \frac{1}{(1+\frac{p_1+\cdots+p_{m-1}}{\hat{p}_{m-1}})^\theta} \\
&= \frac{\theta^{m-1}\hat{p}_{m-1}^{\theta-2}}{p_1\cdots p_{m-1}} g_1^\theta\left(\frac{p_m}{\hat{p}_{m-1}}\right).
\end{aligned}
$$

□

2.4 Size-biased Sampling and the GEM Representation

The proof of Theorem 2.1 includes an explicit construction of the Poisson–Dirichlet distribution through the gamma process. The distribution also appears in many different ways. One of these is related to size-biased sampling.

Consider a population of individuals of a countable number of different types labeled $\{1,2,\ldots\}$. Assume that the proportion of type i individuals in the population is p_i. A sample is randomly selected from the population and the type of the selected individual is denoted by $\sigma(1)$. Next remove all individuals of type $\sigma(1)$ from the population and then randomly select the second sample. This is repeated to get more samples. This procedure of sampling is called *size-biased sampling*. Denote the type of the ith selected sample by $\sigma(i)$. Then $(p_{\sigma(1)}, p_{\sigma(2)},\ldots)$ is called a *size-biased permutation* of $(p_1,p_2,\ldots)$.

Theorem 2.7. *Let* $\mathbf{P}(\theta)$ *have the Poisson–Dirichlet distribution with parameter* $\theta > 0$. *Then the size-biased permutation* $(V_1,V_2,\ldots)$ *of* $\mathbf{P}(\theta)$ *is given by*

$$
V_1 = U_1, V_2 = (1-U_1)U_2, V_3 = (1-U_1)(1-U_2)U_3, \ldots \tag{2.13}
$$

where $\{U_n : n \geq 1\}$ *are i.i.d. Beta*$(1,\theta)$ *random variables. Since*

$$
\mathbb{E}\left[1-\sum_{k=1}^n V_k\right] = \mathbb{E}\left[\prod_{k=1}^n (1-U_k)\right] = \left(\frac{\theta}{1+\theta}\right)^n,
$$

it follows that $\sum_{k=1}^\infty V_k = 1$ *with probability one.*

Proof. For each $n \geq 2$, let $(X_1,\ldots,X_n)$ be a symmetric *Dirichlet*$(\alpha,\ldots,\alpha)$ random vector with density function $f(x_1,\ldots,x_n)$. Then X_1 has a *Beta*$(\alpha,(n-1)\alpha)$ distribution with density function $f_1(x)$. Let $(\tilde{X}_1,\ldots,\tilde{X}_n)$ denote the size-biased permutation of $(X_1,\ldots,X_n)$. By definition, for any x in $(0,1)$,

$$\mathbb{P}\{\tilde{X}_1 \in (x, x+\Delta x)\} = nxf_1(x)\Delta x + o(\Delta x),$$

and

$$\begin{aligned}\mathbb{P}\{\tilde{X}_1 \leq x\} &= \mathbb{E}[\mathbb{P}\{\tilde{X}_1 \leq x \mid X_1, \ldots, X_n\}] \\ &= n\frac{\Gamma(n\alpha)}{\Gamma(\alpha)\Gamma((n-1)\alpha)} \int_0^x x_1 x_1^{\alpha-1}(1-x_1)^{(n-1)\alpha-1} dx_1 \\ &= \frac{\Gamma(n\alpha+1)}{\Gamma(\alpha+1)\Gamma((n-1)\alpha)} \int_0^x x_1^{\alpha}(1-x_1)^{(n-1)\alpha-1} dx_1,\end{aligned}$$

which implies that $\tilde{X}_1 = \tilde{U}_1$ has a $Beta(\alpha+1, (n-1)\alpha)$ distribution. By direct calculation,

$$\begin{aligned}&\mathbb{P}\{(\tilde{X}_1, \ldots, X_{\sigma(1)-1}, X_{\sigma(1)+1}, \ldots) \in (x_1, \Delta x_1) \times \cdots \times (x_n, \Delta x_n)\} \\ &\quad \approx nx_1 f(x_1, \ldots, x_n)\Delta x_1 \cdots \Delta x_n,\end{aligned}$$

which shows that the distribution of $(\tilde{X}_1, X_1, \ldots, X_{\sigma(1)-1}, X_{\sigma(1)+1}, \ldots, X_n)$ is the

$$Dirichlet(\alpha+1, \alpha, \ldots, \alpha).$$

Given $\tilde{X}_1$, the remaining $(n-1)$ components divided by $1-\tilde{X}_1$ will thus have the symmetric distribution $Dirichlet(\alpha, \ldots, \alpha)$. Choose $\tilde{U}_2$ in a way similar to that of $\tilde{U}_1$. Then $\tilde{U}_2$ has a

$$Beta(\alpha+1, (n-2)\alpha)$$

distribution. It is clear from the construction that $\tilde{X}_2 = (1-\tilde{U}_1)\tilde{U}_2$. Continuing this process to get all components, we can see that

$$\begin{aligned}\tilde{X}_1 &= \tilde{U}_1, \\ \tilde{X}_k &= (1-\tilde{U}_1)\cdots(1-\tilde{U}_{k-1})\tilde{U}_k, k = 2, \ldots, n,\end{aligned}$$

where $\tilde{U}_1, \ldots, \tilde{U}_n$ are independent and $\tilde{U}_k$ has a $Beta(\alpha+1, (n-k)\alpha)$ distribution.

For fixed $r \geq 1$, let $n \to \infty, \alpha \to 0$ such that $n\alpha \to \theta > 0$. It follows that

$$\begin{aligned}\tilde{U}_k &\to U_k, k = 1, \ldots, r, \\ \tilde{X}_k &\to V_k, k = 1, \ldots, r.\end{aligned}$$

Recall that $\mathbf{P}(\theta)$ is the limit of the order statistics of $(X_1, \ldots, X_n)$ under the same limiting procedure. Since $(\tilde{X}_1, \ldots, \tilde{X}_n)$ is simply a rearrangement of $(X_1, \ldots, X_n)$, it follows that the limit of the order statistics of $(\tilde{X}_1, \ldots, \tilde{X}_n)$ is also $\mathbf{P}(\theta)$. The theorem now follows from Theorem 1 in [35] and the fact that $\sum_{k=1}^{\infty} V_k = 1$.

□

Remark: This theorem provides a very friendly description of the Poisson–Dirichlet distribution through i.i.d. random sequences. The law of $(V_1, V_2, \ldots)$ is called the *GEM distribution* in Ewens [67], named for Griffiths [94] who noted its genetic importance first, and Engen [51] and McCloskey [137] for introducing it in the context of ecology.

2.5 The Ewens Sampling Formula

As is shown in Chapter 1, the Dirichlet distribution characterizes the equilibrium behavior of a population that evolves under the influence of mutation and random sampling. Due to its connection to the Dirichlet distribution, the Poisson–Dirichlet distribution is expected, and will actually be shown in Chapter 5, to describe the equilibrium distribution of certain neutral infinite alleles diffusion models in population genetics. The parameter θ is then the scaled mutation rate of the population. The celebrated *Ewens sampling formula* provides a way of estimating θ assuming the population is selectively neutral.

For any fixed $n \geq 1$, let

$$\mathscr{A}_n = \left\{ (a_1,\ldots,a_n) : a_k \geq 0, k = 1,\ldots,n;\ \sum_{i=1}^{n} i a_i = n \right\}. \tag{2.14}$$

Consider a random sample of size n from a Poisson–Dirichlet population. For any $1 \leq i \leq n$, let $A_i = A_i(n,\theta)$ denote the number of alleles appearing in the sample exactly i times.

Consider the case of $n = 3$. If one type appears in the sample once and another appears twice, then $A_1 = 1,\ A_2 = 1,\ A_3 = 0$. If all three types appear in the sample, then $A_1 = 3,\ A_2 = 0,\ A_3 = 0$. If there is only one type in the sample, then $A_1 = 0,\ A_2 = 0,\ A_3 = 1$.

By definition, $\mathbf{A}_n = (A_1,\ldots,A_n)$ is an $\mathscr{A}_n$-valued random variable. Every element of $\mathscr{A}_n$ will be called an *allelic partition* of the integer n, and $\mathbf{A}_n$ is called the allelic partition of the random sample of size n. The Ewens sampling formula gives the distribution of $\mathbf{A}_n$.

Theorem 2.8. (Ewens sampling formula)

$$ESF(\theta,\mathbf{a}) \equiv \mathbb{P}\{\mathbf{A}_n = (a_1,\ldots,a_n)\} = \frac{n!}{\theta_{(n)}} \prod_{j=1}^{n} \left(\frac{\theta}{j}\right)^{a_j} \frac{1}{a_j!}, \tag{2.15}$$

where $\theta_{(n)} = \theta(\theta+1)\cdots(\theta+n-1)$.

Proof. Let $\{\gamma_s : s \geq 0\}$ be the gamma process with Lévy measure

$$\Lambda(dx) = \theta x^{-1} e^{-x} dx, x > 0,$$

and set $J_i = J_i(\gamma_1), P_i = \frac{J_i}{\gamma_1}, i \geq 1$. Let $(X_1,X_2,\ldots,X_n)$ be a random sample of size n from a population with allele frequency following the Poisson–Dirichlet distribution with parameter θ. Then $X_1,\ldots,X_n$ are (conditionally) independent, and for any k, we have

$$\mathbb{P}\{X_i = k \mid \mathbf{P}(\theta) = (p_1, p_2, \ldots)\} = p_k.$$

Thus the unconditional probability is given by

$$\mathbb{P}(X_i = k) = \mathbb{E}P_k(\theta).$$

For each $(a_1, \ldots, a_n)$ in $\mathscr{A}_n$, the probability of the event $\{A_i = a_i, i = 1, \ldots, n\}$ can be calculated as follows. Let

$$k_{11}, \ldots, k_{1a_1};\ k_{21}, \ldots, k_{2a_2}; \ldots;\ k_{n1}, \ldots, k_{na_n}$$

be the distinct values of $X_1, \ldots, X_n$. The total number of different assignments of $X_1, \ldots, X_n$ is clearly

$$\frac{n!}{(1!)^{a_1} a_1! (2!)^{a_2} a_2! \cdots (n!)^{a_n} a_n!} = \frac{n!}{\prod_{j=1}^n (j!)^{a_j} a_j!}.$$

For each assignment, the conditional probability given $\mathbf{P}(\theta)$ is

$$P_{k_{11}} \cdots P_{k_{1a_1}} P_{k_{21}}^2 \cdots P_{k_{2a_2}}^2 \cdots P_{k_{n1}}^n \cdots P_{k_{na_n}}^n.$$

It then follows that

$$\begin{aligned}
&\mathbb{P}(\mathbf{A}_n = (a_1, \ldots, a_n) \mid \mathbf{P}(\theta)) \\
&= \frac{n!}{\prod_{j=1}^n (j!)^{a_j} a_j!} \sum_{\substack{\text{distinct } k_{11}, \ldots, k_{1a_1}; \\ k_{21}, \ldots, k_{2a_2}; \ldots; k_{n1}, \ldots, k_{na_n}}} P_{k_{11}} \cdots P_{k_{1a_1}} P_{k_{21}}^2 \cdots P_{k_{2a_2}}^2 \cdots P_{k_{n1}}^n \cdots P_{k_{na_n}}^n \\
&= \frac{n!}{\prod_{j=1}^n (j!)^{a_j} a_j!} \frac{1}{\gamma_1^n} \sum_{\substack{\text{distinct } k_{11}, \ldots, k_{1a_1}; \\ k_{21}, \ldots, k_{2a_2}; \ldots; k_{n1}, \ldots, k_{na_n}}} J_{k_{11}} \cdots J_{k_{1a_1}} J_{k21}^2 \cdots J_{k_{2a_2}}^2 \cdots J_{k_{n1}}^n \cdots J_{k_{na_n}}^n.
\end{aligned}$$

Taking into account the independence of γ_1 and $\mathbf{P}(\theta)$, it follows that

$$\begin{aligned}
&\mathbb{E}(\gamma_1^n) \mathbb{P}(\mathbf{A}_n = (a_1, \ldots, a_n)) \\
&= \frac{\Gamma(n+\theta)}{\Gamma(\theta)} \mathbb{P}(\mathbf{A}_n = (a_1, \ldots, a_n)) \\
&= \frac{n!}{\prod_{j=1}^n (j!)^{a_j} a_j!} \mathbb{E}\left(\sum_{\substack{\text{distinct} k_{11}, \ldots, k_{1a_1}; \\ k_{21}, \ldots, k_{2a_2}; \ldots; k_{n1}, \ldots, k_{na_n}}} J_{k_{11}} \cdots J_{k_{1a_1}} J_{k_{21}}^2 \cdots J_{k_{2a_2}}^2 \cdots J_{k_{n1}}^n \cdots J_{k_{na_n}}^n \right) \\
&= \frac{n!}{\prod_{j=1}^n (j!)^{a_j} a_j!} \mathbb{E}\left(\sum_{\text{distinct } Y_{ij} \in \mathscr{J}} f_1(Y_{11}) \cdots f_1(Y_{1a_1}) \cdots f_n(Y_{n1}) \cdots f_n(Y_{na_n}) \right),
\end{aligned}$$

where $\mathscr{J}$ is the set of jump sizes of τ_s over $[0,1]$ and $f_i(x) = x^i$. By Campbell's theorem in Appendix A,

$$\begin{aligned}
&\mathbb{E}\left(\sum_{\text{distinct } Y_{ij} \in \mathscr{J}} f_1(Y_{11})\cdots f_1(Y_{1a_1}) f_2(Y_{21})\cdots f_2(Y_{2a_2})\ldots f_n(Y_{na_n})\right)\\
&= \left(\int_S f_1(x)\Lambda(dx)\right)^{a_1}\left(\int_S f_2(x)\Lambda(dx)\right)^{a_2}\cdots\left(\int_S f_n(x)\Lambda(dx)\right)^{a_n}\\
&= \prod_{j=1}^{n}\left(\theta\int_S x^j x^{-1}e^{-x}dx\right)^{a_j}\\
&= \prod_{j=1}^{n}(\theta\Gamma(j))^{a_j},
\end{aligned}$$

which leads to (2.15).

□

If samples of the same type are put into one group, then there will be at most n different groups. Let $C_i(n)$ denote the size of the ith largest group with $C_i(n) = 0$ for $i > n$. Clearly the sequence $\{C_i(n) : i = 1,2,\ldots\}$ contains the same information about the random sample $X_1,\ldots,X_n$ as $\mathbf{A}_n$ did, and the Poisson–Dirichlet distribution would be recovered from the limit of

$$\left(\frac{C_1(n)}{n}, \frac{C_2(n)}{n}, \ldots\right)$$

as n goes to infinity (cf. [126]). It is more interesting and somewhat unexpected to see that the allelic partition $\mathbf{A}_n$ also has a limit as n approaches infinity, and the limit is a sequence of independent Poisson random variables, as is shown below.

Theorem 2.9. (Arratia, Barbour and Tavaré [6]) *Let $\{\eta_i : i = 1,\ldots\}$ be a sequence of independent Poisson random variables with respective parameters θ/i. Then for every $m \geq 1$,*

$$(A_1(n,\theta),\ldots,A_m(n,\theta)) \to (\eta_1,\ldots,\eta_m),$$

as n goes to infinity, where the convergence is in distribution.

Proof. For each fixed $m \geq 1$, and non-negative $a_1,\ldots,a_m$ satisfying

$$a = a_1 + 2a_2 + \cdots + ma_m \leq n,$$

it follows from direct calculation that

$$\begin{aligned}
&\mathbb{P}\left\{A_i(n,\theta) = a_i,\ i = 1,\ldots,m\right\}\\
&= \sum_{\substack{b_{m+1},\ldots,b_n \geq 0;\\ \sum_{j=m+1}^{n} jb_j = n-a}} \mathbb{P}\{A_i(n,\theta) = a_i,\ i = 1,\ldots,m;\ A_j(n,\theta) = b_j,\ j = m+1,\ldots,n\}
\end{aligned}$$

which equals to

$$\mathbb{P}\left\{A_i(n,\theta)=a_i,\ i=1,\ldots,m;\ \sum_{j=m+1}^{n} jA_j(n,\theta)=n-a\right\}$$
$$=\mathbb{P}\left\{\eta_i=a_i,\ i=1,\ldots,m;\ \sum_{j=m+1}^{n} j\eta_j=n-a\,\middle|\,\sum_{j=1}^{n} j\eta_j=n\right\}$$
$$=\mathbb{P}\{\eta_i=a_i,\ i=1,\ldots,m\}\frac{\mathbb{P}\left\{\sum_{j=m+1}^{n} j\eta_j=n-a\right\}}{\mathbb{P}\left\{\sum_{j=1}^{n} j\eta_j=n\right\}}.$$

Set

$$T_{kn}=(k+1)\eta_{k+1}+\cdots+n\eta_n.$$

Then

$$\mathbb{P}\{T_{0n}=n\}=\mathbb{P}\left\{\sum_{j=1}^{n} j\eta_j=n\right\}$$
$$=\frac{\theta_{(n)}}{n!}\exp\left\{-\theta\sum_{j=1}^{n}\frac{1}{j}\right\}.$$

The generating function of T_{mn} is

$$\mathbb{E}\left[x^{T_{mn}}\right]=\mathbb{E}\left[x^{\sum_{j=m+1}^{n} j\eta_j}\right]$$
$$=\prod_{j=m+1}^{n}\mathbb{E}\left[x^{j\eta_j}\right]$$
$$=\prod_{j=m+1}^{n}\exp\left\{\theta\left(\frac{x^j}{j}-\frac{1}{j}\right)\right\}$$
$$=\exp\left\{-\theta\sum_{j=m+1}^{n}\frac{1}{j}\right\}\cdot\exp\left\{\theta\sum_{j=m+1}^{n}\frac{x^j}{j}\right\}.$$

Set $g(x)=\exp\{-\theta\sum_{j=1}^{m}\frac{x^j}{j}\}$. Then

$$(n-a)!\cdot\mathbb{P}\{T_{mn}=n-a\}\cdot\exp\left\{\theta\sum_{j=m+1}^{n}\frac{1}{j}\right\}$$
$$=\left[\exp\left\{\theta\sum_{j=m+1}^{n}\frac{x^j}{j}\right\}\right]^{(n-a)}\Bigg|_{x=0},$$

where the superscript $(n-a)$ denotes the order of the derivatives. It follows from detailed expansion that

$$
\begin{aligned}
&\left[\exp\left\{\theta \sum_{j=m+1}^{n} \frac{x^j}{j}\right\}\right]^{(n-a)}\Bigg|_{x=0} \\
&= \left[\exp\left\{-\theta \sum_{j=1}^{m} \frac{x^j}{j}\right\} \cdot \exp\{\theta \sum_{j=1}^{\infty} \frac{x^j}{j}\}\right]^{(n-a)}\Bigg|_{x=0} \\
&= \left[g(x) \cdot \exp\{-\theta \log(1-x)\}\right]^{(n-a)}\Bigg|_{x=0} \\
&= \left[(1-x)^{-\theta}\left\{g(1) + g'(1)(x-1) + \frac{g''(1)}{2!}(x-1)^2 + \cdots\right\}\right]^{(n-a)}\Bigg|_{x=0} \\
&= \left[g(1)(1-x)^{-\theta} - g'(1)(1-x)^{-(\theta-1)} + \frac{g''(1)}{2!}(1-x)^{-(\theta-2)} - \cdots\right]^{(n-a)}\Bigg|_{x=0} \\
&= g(1)\theta_{(n-a)} - g'(1)(\theta-1)_{(n-a)} + \frac{g''(1)}{2!}(\theta-2)_{(n-a)} - \cdots .
\end{aligned}
$$

Hence we obtain

$$
\begin{aligned}
&\mathbb{P}\{T_{mn} = n-a\} \\
&= \frac{1}{(n-a)!} \exp\left\{-\theta \sum_{j=m+1}^{n} \frac{1}{j}\right\} \cdot g(1)\theta_{(n-a)} \left[1 - \frac{g'(1)}{g(1)} \frac{(\theta-1)_{(n-a)}}{\theta_{(n-a)}} + o\left(\frac{1}{n^2}\right)\right] \\
&= \frac{\theta_{(n-a)}}{(n-a)!} \exp\left\{-\theta \sum_{j=1}^{n} \frac{1}{j}\right\} \cdot \left[1 + \frac{m\theta(\theta-1)}{\theta+n-a-1} + o\left(\frac{1}{n^2}\right)\right].
\end{aligned}
$$

Therefore,

$$
\begin{aligned}
\frac{\mathbb{P}\{T_{mn} = n-a\}}{\mathbb{P}\{T_{0n} = n\}} &= \frac{\frac{\theta_{(n-a)}}{(n-a)!}\left[1 + \frac{m\theta(\theta-1)}{\theta+n-a-1} + o\left(\frac{1}{n^2}\right)\right]}{\frac{\theta_{(n)}}{n!}} \\
= \quad & \frac{n(n-1)\cdots(n-a+1)}{(\theta+n-a)(\theta+n-a+1)\cdots(\theta+n-1)}\left[1 + \frac{m\theta(\theta-1)}{\theta+n-a-1} + o\left(\frac{1}{n^2}\right)\right] \\
\longrightarrow 1, \quad & (n \to \infty),
\end{aligned}
$$

and

$$
\begin{aligned}
&\mathbb{P}\{A_i(n,\theta) = a_i,\ i = 1, \ldots, m\} \\
&= \mathbb{P}\{\eta_i = a_i,\ i = 1, \ldots, m\} \cdot \frac{\mathbb{P}\{T_{mn} = n-a\}}{\mathbb{P}\{T_{0n} = n\}} \\
&\longrightarrow \mathbb{P}\{\eta_i = a_i,\ i = 1, \ldots, m\} \quad \text{as } n \to \infty.
\end{aligned}
$$

□

On the basis of this result, it is not surprising to have the following representation of the Ewens sampling formula.

Theorem 2.10. *Let $\eta_1, \eta_2, \ldots$ be independent Poisson random variables with*

$$\mathbb{E}[\eta_i] = \frac{\theta}{i}.$$

Then for each $n \geq 2$, and $(a_1, \ldots, a_n)$ in $\mathscr{A}_n$,

$$\mathbb{P}\{\mathbf{A}_n = (a_1, \ldots, a_n)\} = \mathbb{P}\left\{\eta_i = a_i, i = 1, \ldots, n \mid \sum_{j=1}^{n} j\eta_j = n\right\}.$$

Proof. Let $T_n = \sum_{j=1}^{n} j\eta_j$. It follows from the Ewens sampling formula that

$$\sum_{\sum_{j=1}^{n} jb_j = n} \prod_{j=1}^{n} \left(\frac{\theta}{j}\right)^{b_j} \frac{1}{b_j!} = \frac{\theta_{(n)}}{n!}.$$

By direct calculation,

$$\begin{aligned}
&\mathbb{P}\left\{\eta_i = a_i, i = 1, \ldots, n \mid \sum_{j=1}^{n} j\eta_j = n\right\}\\
&= \mathbb{P}\{\eta_i = a_i, i = 1, \ldots, n\}/\mathbb{P}\{T_n = n\}\\
&= \frac{\prod_{i=1}^{n} \frac{(\theta/i)^{a_i}}{a_i!} e^{-\sum_{j=1}^{n} \theta/j}}{\sum_{(b_1,\ldots,b_n)\in \mathbf{A}_n} \prod_{i=1}^{n} \frac{(\theta/i)^{b_i}}{b_i!} e^{-\sum_{j=1}^{n} \theta/j}}\\
&= \frac{n!}{\theta_{(n)}} \prod_{j=1}^{n} \left(\frac{\theta}{j}\right)^{a_j} \frac{1}{a_j!}.
\end{aligned}$$

□

Let $K_n = \sum_{i=1}^{n} A_i$ denote the total number of different alleles in a random sample of size n of a Poisson–Dirichlet population with parameter θ.

Theorem 2.11. *The statistic $K_n/\log n$ is an asymptotically consistent, sufficient estimator of θ and as $n \to \infty$,*

$$\sqrt{\log n}\left(\frac{K_n}{\log n} - \theta\right) \to Z$$

where convergence is in distribution and Z is a normal random variable with mean zero and variance θ.

Proof. For any $1 \leq m \leq n$, let

$$\mathscr{C}_m = \left\{(a_1, \ldots, a_n) \in \mathscr{A}_n : \sum_{i=1}^{n} a_i = m\right\}.$$

Then

$$\begin{aligned}\mathbb{P}\{K_n = m\} &= \mathbb{P}\{\mathbf{A}_n \in \mathscr{C}_m\} \\ &= \sum_{(a_1,\ldots,a_n)\in\mathscr{C}_m} \frac{n!}{\theta_{(n)}} \prod_{j=1}^{n} \left(\frac{\theta}{j}\right)^{a_j} \frac{1}{a_j!} \\ &= |S_n^m| \frac{\theta^m}{\theta_{(n)}},\end{aligned} \tag{2.16}$$

where

$$S_n^m = (-1)^{n-m} n! \sum_{(a_1,\ldots,a_n)\in\mathscr{C}_m} \prod_{j=1}^{n} \frac{1}{j^{a_j} a_j!} \tag{2.17}$$

is the signed Stirling number of the first kind. Later on, we will show that $|S_n^m|$ is the total number of permutations of n numbers into m cycles, and is equal to the coefficient of θ^m in the polynomial $\theta_{(n)}$. What is important now is that $|S_n^m|$ is independent of θ. Given $K_n = m$, we have

$$\mathbb{P}\{\mathbf{A}_n = (a_1,\ldots,a_n) \mid K_n = m\} = \frac{n!}{|S_n^m|} \chi_{\{(a_1,\ldots,a_n)\in\mathscr{C}_m\}} \prod_{j=1}^{n} \frac{1}{j^{a_j} a_j!}$$

which implies that K_n is a sufficient statistic for θ.

Let $\Lambda_n(t) = \frac{1}{\log n} \log \mathbb{E}[e^{tK_n}]$. By Stirling's formula for gamma functions, one has

$$\begin{aligned}\lim_{n\to\infty} \Lambda_n(t) &= \lim_{n\to\infty} \frac{1}{\log n} \log \frac{\Gamma(e^t\theta + n)}{\Gamma(\theta+n)} \frac{\Gamma(\theta)}{\Gamma(e^t\theta)} \\ &= \lim_{n\to\infty} \frac{1}{\log n} \log \frac{(e^t\theta + n)^{(e^t\theta+n)}}{(\theta+n)^{(\theta+n)}} \\ &= \theta(e^t - 1) \equiv \Lambda(t).\end{aligned}$$

Let

$$M_n(t) = \mathbb{E}[e^{tK_n}], N_n(t) = \sum_{i=0}^{n} \frac{e^t\theta}{e^t\theta + i}.$$

Then a direct calculation shows that

$$\begin{aligned}M_n'(t) &= M_n(t)N_n(t), \\ M_n''(t) &= [N_n^2(t) + N_n'(t)]M_n(t), \\ (\log M_n(t))'' &= \frac{M_n''(t)M_n(t) - (M_n'(t))^2}{M_n^2(t)} \\ &= N_n'(t) = \theta e^t \sum_{i=1}^{n} \frac{i}{(e^t\theta + i)^2} \sim (\theta e^t)\log n.\end{aligned}$$

Since $\Lambda''(t) = \theta e^t$, we thus have proved that

$$\Lambda_n''(t) \to \Lambda''(t), \text{uniformly in the neighborhood of zero.}$$

Applying Proposition 1.1 and Theorem 1.2 in [187], we conclude that

$$\lim_{n\to\infty} \frac{K_n}{\log n} = \theta,$$

and

$$\sqrt{\log n}\left(\frac{K_n}{\log n} - \theta\right) \to Z.$$

□

Since K_n is a sufficient statistic for θ, one likelihood function is

$$L_n(\theta, m) = \mathbb{P}\{K_n = m\} = |S_n^m| \frac{\theta^m}{\theta_{(n)}}.$$

Taking the derivative with respect to θ and setting the derivative equal to zero leads to

$$m = \theta \sum_{i=1}^{n} \frac{1}{\theta + i - 1}.$$

For $1 \le m \le n-1$, the solution $\hat{\theta}$ of this equation is the maximum likelihood estimator of θ. The case $m = n$ corresponds to $\theta = \infty$. For a fixed sample size, n, the number of alleles in the sample is an increasing function of θ. Therefore a sample with a larger number of alleles suggests a higher mutation rate.

Assume that we have an unbiased estimator $\hat{g}$ of the function $g(\theta)$. Then by the Rao–Blackwell theorem in statistics, the conditional expectation of $\hat{g}$ given K_n is usually a better estimator than $\hat{g}$.

2.6 Scale-invariant Poisson Process

In addition to the Poisson process representation in Theorem 2.2, the Poisson–Dirichlet distribution can be constructed through another Poisson process: *the scale-invariant Poisson process*.

Let $\eta_1, \eta_2, \ldots$ be the Poisson random variables in Theorem 2.9. For any $n \ge 1$, one can check directly that

$$N_n(\cdot) = \sum_{i=1}^{\infty} \eta_i \delta_{i/n}(\cdot)$$

is a Poisson random measure on $(0, \infty)$ with mean measure

$$\mu_n(\cdot) = \sum_{i=1}^{\infty} \frac{\theta}{i} \delta_{i/n}(\cdot).$$

Since μ_n converges weakly to the measure

$$\mu(dx) = \frac{\theta}{x} dx,\ x > 0,$$

we are led to a Poisson process $\mathcal{F}$ on $(0,\infty)$ with mean measure μ.

Let the points in $\mathcal{F}$ be labeled almost surely as

$$0 < \cdots < \varsigma_2 < \varsigma_2 < 1 < \varsigma_0 < \varsigma_{-1} < \varsigma_{-2} < \cdots < \infty. \tag{2.18}$$

Theorem 2.12. *The Poisson process $\mathcal{F}$ on $(0,\infty)$ with mean measure μ is scale-invariant; i.e., for any $c > 0$, the random set $c\mathcal{F}$ has the same distribution as $\mathcal{F}$.*

Proof. For any $c > 0$, it follows from Theorem A.4 that $c\mathcal{F}$ is a Poisson process with mean measure

$$\mu\left(d\left(\frac{x}{c}\right)\right) = \frac{\theta}{x} dx = \mu(dx).$$

This proves the result.

□

Let

$$\varsigma(\theta) = \sum_{i>0} \varsigma_i,$$

and $B_1, B_2, \ldots$ be a sequence of i.i.d. random variables with common distribution $Beta(\theta, 1)$.

Theorem 2.13. *The random variable $\varsigma(\theta)$ has the same distribution as*

$$\sum_{k=1}^{\infty} \prod_{i=1}^{k} B_i, \tag{2.19}$$

with a density function given by

$$\tilde{g}_\theta(x) = \frac{e^{-\gamma\theta} x^{\theta-1}}{\Gamma(\theta)} \left\{ 1 + \sum_{n=1}^{\infty} \frac{(-\theta)^n}{n!} \int_{G_x^n} \frac{(1-\sum_{i=1}^n y_i)^{\theta-1}}{y_1 \cdots y_n} dy_1 \cdots dy_n \right\}, \tag{2.20}$$

where

$$\gamma = \lim_{n\to\infty} \left\{ \sum_{k=1}^{n} \frac{1}{k} - \log n \right\}$$

is the Euler's constant, and

$$G_x^n = \left\{ y_i > x^{-1}, i = 1, \ldots, n; \sum_{i=1}^{n} y_i < 1 \right\}.$$

Proof. Let $\mathcal{F}_0$ be the homogeneous Poisson process on $\mathbb{R}$ with intensity $\theta > 0$. Consider the map

$$f : \mathbb{R} \longrightarrow (0,\infty),\ x \to e^{-x}.$$

It follows from Theorem A.4 that $f(\digamma_0)$ is a Poisson process with mean measure

$$\theta(\log x)' dx = \mu(dx).$$

Therefore, $f(\digamma_0)$ has the same distribution as $\digamma$. In particular, the set of points $\{\varsigma_i : i > 0\}$ has the same distribution as $f(\digamma_0 \cap (0,\infty))$. Since the points in $\digamma_0 \cap (0,\infty)$ can be represented as $\{W_1, W_1 + W_2, W_1 + W_2 + W_3, \ldots\}$ for a sequence of i.i.d. exponential random variables with mean $1/\theta$, it follows, by choosing $B_i = e^{-W_i}$, that $\varsigma(\theta)$ has the same distribution as $\sum_{k=1}^{\infty} \prod_{i=1}^{k} B_i$. By its construction, B_i follows the $Beta(\theta, 1)$ distribution.

Let

$$g(x) = x \cdot \chi_{(0,1)}(x).$$

Then, it follows from Campbell's theorem, that for any $\lambda \geq 0$

$$\begin{aligned} \mathbb{E}[e^{-\lambda \sum_i g(\varsigma_i)}] &= \mathbb{E}[e^{-\lambda \varsigma(\theta)}] \\ &= \exp\left\{ -\int_0^1 (1 - e^{-\lambda x}) \frac{\theta}{x} dx \right\}. \end{aligned} \tag{2.21}$$

Equation (2.20) now follows from the identity

$$\int_0^x \frac{1 - e^{-y}}{y} dy = \int_x^{\infty} \frac{e^{-y}}{y} dy + \log x + \gamma, \; x > 0,$$

and the fact that $\lambda^{-\theta} (\int_{\lambda}^{\infty} \frac{e^{-y}}{y} dy)^n$ is the Laplace transform of the function

$$\int_{G_x^n} \frac{(1 - \sum_{i=1}^n y_i)^{\theta - 1}}{y_1 \cdots y_n} dy_1 \cdots dy_n.$$

□

Theorem 2.14. (Arratia, Barbour and Tavaré [8]) *For any $\theta > 0$, the conditional distribution of $(\varsigma_1, \varsigma_2, \ldots)$, given $\varsigma(\theta) = 1$, is the Poisson–Dirichlet distribution with parameter θ.*

Proof. It suffices to verify that the conditional marginal distribution of $(\varsigma_1, \varsigma_2, \ldots)$ is the same as the corresponding marginal distribution of $\mathbf{P}(\theta)$. For any $m \geq 1$, let $h(u; p_1, \ldots, p_m)$ denote the conditional probability of $\varsigma(\theta)$, given $\varsigma_i = p_i, i = 1, \ldots, m$. Then the joint density function of $(\varsigma_1, \ldots, \varsigma_m, \varsigma(\theta))$ is given by

$$\exp\left\{ -\left(\int_{p_1}^1 + \cdots + \int_{p_m}^{p_{m-1}} \right) \frac{\theta dx}{x} \right\} \frac{\theta^m}{p_1 \cdots p_m} h(u; p_1, \ldots, p_m). \tag{2.22}$$

The conditional probability of $(\varsigma_1, \ldots, \varsigma_m)$, given $\varsigma(\theta) = 1$, is

$$\frac{p_m^{\theta} \theta^m}{p_1 \cdots p_m} h(1; p_1, \ldots, p_m) / \tilde{g}_{\theta}(1). \tag{2.23}$$

It follows from the scale-invariant property that

$$h(1;p_1,\ldots,p_m)=\frac{1}{p_m}\tilde{g}_\theta\left(\frac{1-\sum_{i=1}^m p_i}{p_m}\right). \tag{2.24}$$

Substituting (2.24) into (2.23), the theorem now follows from (2.12) and the fact that

$$\mathbb{P}\left\{P_1(\theta)\leq\frac{p_m}{\hat{p}_m}\wedge 1\right\}=p_m^{\theta-1}\frac{\tilde{g}_\theta\left(\frac{\hat{p}_m}{p_m}\right)}{\tilde{g}_\theta(1)}.$$

□

2.7 Urn-based Models

In a 1984 paper [103], Hoppe discovered an urn model (called Hoppe's urn) that gives rise to the Ewens sampling formula. This model turns out to be quite useful in various constructions and calculations related to the Poisson–Dirichlet distribution and the Ewens sampling formula. In this section we will discuss several closely related urn-type models.

2.7.1 Hoppe's Urn

Consider an urn that initially contains a black ball of mass θ. Balls are drawn from the urn successively with probabilities proportional to their masses. When the black ball is drawn, it is returned to the urn together with a ball of a new color not previously added with mass one; if a non-black ball is drawn it is returned to the urn with one additional ball of mass one and the same color as that of the ball drawn. Colors are labeled $1,2,3,\ldots$ in the order of appearance. Let X_n be the label of the additional ball returned after the nth drawing.

Theorem 2.15. *For any $i=1,\ldots,n$, let A_i be the number of labels that appear i times in the sequence $\{X_1,\ldots,X_n\}$. Then the distribution of $\mathbf{A}_n=(A_1,\ldots,A_n)$ is given by the Ewens sampling formula; i.e., for each $(a_1,\ldots,a_n)$ in $\mathscr{A}_n$*

$$\mathbb{P}\{\mathbf{A}_n=(a_1,\ldots,a_n)\}=\frac{n!}{\theta_{(n)}}\prod_{j=1}^n\left(\frac{\theta}{j}\right)^{a_j}\frac{1}{a_j!}.$$

Proof. For each $(a_1,\ldots,a_n)$ in $\mathscr{A}_n$, let $k=\sum_{i=1}^n a_i$ be the total number of times that the black ball is selected, which is the same as the total number of different labels. The set $\{n_1,\ldots,n_k\}$ gives the number of balls of each label $i=1,\ldots,k$. Without loss of generality, we assume that these numbers are arranged in decreasing order so that

$$n_1\geq n_2\cdots\geq n_k.$$

Define

$$l = \text{number of distinct integers in } \{n_1,\ldots,n_k\},$$

and

$$b_1 = \text{length of the run of } n_1,\ b_i = \text{length of the } i\text{th run},\ i = 2,\ldots,l.$$

Then one can rewrite the Ewens sampling formula as

$$\mathbb{P}\{\mathbf{A}_n = (a_1,\ldots,a_n)\} = \frac{n!}{\theta_{(n)}} \frac{\theta^k}{\prod_{j=1}^k n_j \prod_{i=1}^l b_i!}. \tag{2.25}$$

Consider a sample path $X_1 = x_1,\ldots,X_n = x_n$ that is compatible with $(a_1,\ldots,a_n)$. It is clear that

$$\mathbb{P}\{X_1 = x_1,\ldots,X_n = x_n\} = \frac{\theta^k \prod_{i=1}^k (n_i - 1)!}{\theta_{(n)}}, \tag{2.26}$$

where θ^k counts for the selection of black balls k times, and each new color i will be selected $n_i - 1$ additional times with a product of successive masses $(n_i - 1)!$. The denominator is the product of successive masses of the balls in the urn of the first n selections.

The total number of paths that are compatible with $(a_1,\ldots,a_n)$ can be counted as the number of permutations of n objects which are divided into k groups with the decreasing order group sizes $\{n_1,\ldots,n_k\}$. The total number of ways of dividing n objects into k groups of sizes $\{n_1,\ldots,n_k\}$ is $\frac{n!}{n_1!\cdots n_k!}$. As the successive runs are characterized by $(b_1,\ldots,b_l)$, the total number of distinct permutations is thus

$$\frac{n!}{n_1!\cdots n_k! \prod_{j=1}^l b_j!}. \tag{2.27}$$

The multiplication of (2.26) and (2.27) leads to (2.25) and the theorem.

□

For each $i = 1,\ldots,n$, set

$$\eta_i = \begin{cases} 1, & \text{if the } i\text{th draw is a black ball,} \\ 0, & \text{else.} \end{cases}$$

Clearly $\eta_1,..,\eta_n$ are independent, and

$$\mathbb{P}\{\eta_i = 1\} = \frac{\theta}{\theta + i - 1}.$$

As a direct application of Theorem 2.15 we obtain the following representation:

$$K_n = \eta_1 + \cdots + \eta_n, \tag{2.28}$$

which provides an alternative proof of Theorem 2.11.

2.7.2 Linear Birth Process with Immigration

Consider a population consisting of immigrants and their descendants. Immigrants enter the population according to a continuous time, pure-birth Markov chain with rate θ, and each immigrant initiates a family, the size of which follows a linear birth process with rate 1. Different families evolve independently. In comparison with Hoppe's urn, the role of the black balls is replaced by immigration. This process provides an elementary way to study the age-ordered samples from infinite alleles models.

We consider a population composed of various numbers of different types (for example mutants, alleles, in a biological context) which is evolving continuously in time. There is an input process $I(t)$ describing how new mutants enter the population and a stochastic structure $x(t)$ with $x(0) = 1$ and the convention $x(t) = 0$ if $t < 0$, prescribing the growth pattern of each mutant population.

Mutants arrive at the times $0 \leq T_1 < T_2 < \cdots$ and initiate lines according to independent versions of $x(t)$. Thus let $\{x_i(t)\}$ be independent copies of $x(t)$ with $x_i(t)$ being initiated by the ith mutant. Then $x_i(t - T_i)$ will be the size at time t of the ith mutant line. The process $N(t)$ represents the total population size at time t:

$$N(t) = \sum_{i=1}^{I(t)} x_i(t - T_i).$$

Assume that:

(i) $I(t)$ is a pure-birth process with rate θ and initial value zero.

(ii) The process $x(t)$ is a pure-birth process starting at $x(0) = 1$ and with infinitesimal birth rate $q_{n,n+1} = n$ for $n \geq 1$.

Then the process $N(t)$ is a linear birth process with immigration. It is a pure-birth process starting at $N(0) = 0$ with infinitesimal rates

$$\rho_n = \lim_{h \to 0} \frac{1}{h} \mathbb{P}\{N(t+h) - N(t) = 1 \mid N(t) = n\}.$$

Obviously $\rho_0 = \theta$ and for small $h > 0$

$$\begin{aligned}
&\mathbb{P}\{N(t+h) - N(t) = 1 \mid N(t) = n\} \\
&= \mathbb{E}\left[\left(\theta + \sum_{i=1}^{I(t)} x_i(t - T_i)\right) h + o(h) \mid N(t) = n\right] \\
&= (n + \theta)h + o(h).
\end{aligned}$$

Thus for $n \geq 1$, $\rho_n = n + \theta$. Let

$$a_i(t) = \sharp\{j : T_j \leq t \text{ and } x_j(t - T_j) = i\}$$

so that

$$(a_1(t), \ldots, a_{N(t)}(t))$$

is the corresponding random allelic partition of $N(t)$. Define

$$\tau_n = \inf\{t \geq 0 : N(t) = n\} \text{ for } n \geq 1.$$

Then

$$\mathbf{A}_n = (A_1, \ldots, A_n) = (a_1(\tau_n), \ldots, a_n(\tau_n))$$

is a random partition of n. Based on our construction we have

$$\mathbb{P}\{\mathbf{A}_{n+1} = (a_1 + 1, \ldots, a_n, 0) \mid \mathbf{A}_n = (a_1, \ldots, a_n)\} = \frac{\theta}{n+\theta}; \tag{2.29}$$

if $a_i \geq 1, 1 \leq i < n$,

$$\mathbb{P}\{\mathbf{A}_{n+1} = (\ldots, a_i - 1, a_{i+1} + 1, \ldots, 0) \mid \mathbf{A}_n = (a_1, \ldots, a_n)\} = \frac{ia_i}{n+\theta}; \tag{2.30}$$

$$\mathbb{P}\{\mathbf{A}_{n+1} = (a_1, \ldots, a_{n-1}, 0, 1) \mid \mathbf{A}_n = (a_1, \ldots, a_n)\} = \frac{n}{n+\theta}, \text{ if } a_n = 1; \tag{2.31}$$

where the change of state in (2.29) corresponds to the introduction of a new mutant, while in (2.30) and (2.31) one of the existing mutant populations is increased by 1. As with Hoppe's urn, we have established the following theorem.

Theorem 2.16. *The distribution of the random partition* $\mathbf{A}_n = (a_1(\tau_n), \ldots, a_n(\tau_n))$ *is given by the Ewens sampling formula.*

Let $y(t)$ be a pure-birth process on $\{0, 1, 2, \ldots\}$ with birth rate λ_n. Then the forward equation for the transition probability $P_{ij}(t)$ has the form

$$P'_{ij}(t) = \lambda_{j-1} P_{i,j-1}(t) - \lambda_j P_{ij}(t), \; t \geq 0. \tag{2.32}$$

Since the process is pure-birth,

$$P_{ij}(t) = 0, \; j < i, \; t \geq 0. \tag{2.33}$$

Thus

$$P'_{ii}(t) = -\lambda_i P_{ii}(t), \; t \geq 0. \tag{2.34}$$

Since $P_{ii}(0) = 1$, we conclude from (2.32) and (2.34),

$$P_{ij}(t) = \lambda_{j-1} \int_0^t e^{-\lambda_j (t-s)} P_{i,j-1}(s) ds, \; j > i, \; t \geq 0, \tag{2.35}$$

and

$$P_{ii}(t) = e^{-\lambda_i t},\ t \geq 0. \tag{2.36}$$

For the process $N(t)$, we have $N(0) = 0$ and $\lambda_i = i + \theta$. Therefore for each i,

$$P_{ii}(t) = e^{-(i+\theta)t},\ t \geq 0, \tag{2.37}$$

and

$$\begin{aligned} P_{i,i+1}(t) &= (i+\theta)\int_0^t e^{-(i+1+\theta)(t-s)} e^{-(i+\theta)s} ds \\ &= (i+\theta) e^{-\theta t} e^{-it}(1-e^{-t}),\ j > i,\ t \geq 0. \end{aligned} \tag{2.38}$$

Choosing $i = 0$ in (2.38) and substituting the latter into (2.35), yields

$$\begin{aligned} P_{0,2}(t) &= (1+\theta)\int_0^t e^{-(2+\theta)(t-s)} \theta e^{-\theta s}(1-e^{-s}) ds \\ &= \binom{\theta+1}{2} e^{-\theta t}(1-e^{-t})^2,\ t \geq 0. \end{aligned} \tag{2.39}$$

By induction we conclude that for each $n \geq 1$,

$$\mathbb{P}\{N(t) = n\} = \binom{\theta+n-1}{n} e^{-\theta t}(1-e^{-t})^n. \tag{2.40}$$

In general, for $n \geq m \geq 1,\ t \geq s \geq 0$

$$\mathbb{P}\{N(t) = n \mid N(s) = m\} = \binom{\theta+n-1}{n-m} e^{-(m+\theta)(t-s)}(1-e^{-(t-s)})^{n-m}. \tag{2.41}$$

Similarly for the linear birth process $x(t)$ and $n \geq m \geq 1,\ t \geq 0$, we have

$$\mathbb{P}\{x(t) = n \mid x(s) = m\} = \binom{n-1}{n-m} e^{-m(t-s)}(1-e^{-(t-s)})^{n-m}. \tag{2.42}$$

In particular, given $x(0) = 1$, the distribution of $x(t)$ is geometric with parameters e^{-t}; i.e.,

$$\mathbb{P}\{x(t) = n \mid x(0) = 1\} = e^{-t}(1-e^{-t})^{n-1}. \tag{2.43}$$

This can be explained intuitively as follows: Note that e^{-t} is the probability that a rate-one Poisson process makes no jumps over $[0,t]$. Consider this as the probability of "success". For $x(t)$ to have a value n, $n-1$ jumps or "failures" need to occur over $[0,t]$. The one lonely "success" will be the starting value $x(0)$.

Theorem 2.17. *Both the processes $e^{-t}x(t)$ and $e^{-t}N(t)$ are non-negative submartingales with respect to their respective natural filtrations. Therefore there are random variables X and Y such that*

$$e^{-t}x(t) \to X,\ e^{-t}N(t) \to Y,\ a.s.,\ t \to \infty. \tag{2.44}$$

Proof. It follows from (2.41) that the expected value of $N(t)$ given $N(s) = m$, is

$$\begin{aligned}
&\mathbb{E}[N(t) \mid N(s) = m] \\
&= \sum_{n=m}^{\infty} (n+\theta) \binom{\theta+n-1}{m+\theta-1} e^{-(m+\theta)(t-s)} (1-e^{-(t-s)})^{n-m} - \theta \\
&= \frac{m+\theta}{e^{-(t-s)}} \sum_{l=m+1}^{\infty} \binom{\theta+l-1}{(m+1)+\theta-1} e^{-(m+1+\theta)(t-s)} (1-e^{-(t-s)})^{l-(m+1)} - \theta \\
&= (m + (1-e^{-(t-s)})\theta) e^{t-s},
\end{aligned}$$

which, combined with the Markov property, implies that $e^{-t}N(t)$ is a non-negative submartingale. A similar argument shows that $e^{-t}x(t)$ is actually a martingale. It follows from (2.40) and (2.43) that

$$\mathbb{E}[e^{-t}N(t)] = (1-e^{-t})\theta,\ \mathbb{E}[e^{-t}x(t)] = 1.$$

By an application of the martingale convergence theorem (cf. Durrett [47]) we conclude that (2.44) holds.

□

Theorem 2.18. *In Theorem 2.17, the random variable X has an exponential distribution with mean one, and the random variable Y has a gamma distribution with parameters $\theta, 1$.*

Proof. Let $p = e^{-t}$. For every real number λ the characteristic function of $px(t)$ is

$$\psi_t(\lambda) = \mathbb{E}[e^{i\lambda px(t)}] = \frac{pe^{ip\lambda}}{1-(1-p)e^{ip\lambda}}.$$

Clearly as t goes to infinity,

$$\psi_t(\lambda) \to \frac{1}{1-i\lambda} = \psi(\lambda),$$

the characteristic function of an exponential random variable with parameter 1. For each $\lambda < 1$, let

$$q = 1-(1-p)e^{p\lambda}.$$

It then follows from (2.40) that

$$\mathbb{E}[e^{\lambda p N(t)}] = \sum_{n=0}^{\infty} (e^{p\lambda})^n \binom{\theta+n-1}{n} e^{-\theta t}(1-e^{-t})^n$$
$$= \sum_{n=0}^{\infty} \binom{\theta+n-1}{n} \frac{e^{-\theta t}}{q^{\theta}} q^{\theta}(1-q)^n \tag{2.45}$$
$$= \left(\frac{1}{e^t-(e^t-1)e^{p\lambda}}\right)^{\theta}$$
$$= \left(\frac{1}{1-\lambda+o(p)}\right)^{\theta} \to \left(\frac{1}{1-\lambda}\right)^{\theta}, \; t\to\infty, \tag{2.46}$$

which implies that Y has a $Gamma(\theta,1)$ distribution.

□

Let $X_1, X_2, ..$ be independent copies of X. Recall that $x_i(t)=0$ for $t<0$. Then the age-ordered family sizes $x_1(t-T_1), x_2(t-T_2), \ldots$ have the following limits.

Theorem 2.19.

$$e^{-t}(x_1(t-T_1), x_2(t-T_2), \ldots) \to (e^{-T_1}X_1, e^{-T_2}X_2, \ldots) \;\; a.s., \; t\to\infty, \tag{2.47}$$

where the convergence is in space $\mathbb{R}_+^{\infty}$.

Proof. It suffices to verify that for every fixed $r \geq 1$,

$$e^{-t}(x_1(t-T_1), \ldots, x_r(t-T_r)) \to (e^{-T_1}X_1, \ldots, e^{-T_r}X_r) \;\; a.s., \; t\to\infty.$$

From (2.44) and the fact that $\chi_{\{T_r\leq t\}}$ converges to one almost surely, we conclude that

$$\begin{aligned} &e^{-t}(x_1(t-T_1), \ldots, x_r(t-T_r)) \\ &= (e^{-T_1}e^{-(t-T_1)}x_1(t-T_1), \ldots, e^{(-T_r)}e^{-(t-T_r)}x_r(t-T_r)) \\ &\to (e^{-T_1}X_1, \ldots, e^{-T_r}X_r) \;\; a.s., \; t\to\infty. \end{aligned}$$

□

Theorem 2.20. *The limit Y in Theorem 2.17 has the following representation:*

$$Y = \sum_{i=1}^{\infty} e^{-T_i}X_i \;\; a.s. \tag{2.48}$$

Proof. For each fixed $r \geq 1$,

$$\begin{aligned} \sum_{i=1}^{r} e^{-T_i}X_i &= \lim_{t\to\infty} \sum_{i=1}^{r} e^{-t}x_i(t-T_i) \\ &\leq \lim_{t\to\infty} e^{-t}N(t) = Y. \end{aligned}$$

Thus with probability one

$$Y \geq \sum_{i=1}^{\infty} e^{-T_i} X_i.$$

This, combined with the fact that

$$\mathbb{E}[Y] = \theta = \sum_{i=1}^{\infty} \left(\frac{\theta}{\theta+1}\right)^i = \mathbb{E}\left[\sum_{i=1}^{\infty} e^{-T_i} X_i\right],$$

implies the theorem. □

Let

$$\begin{aligned} U_i &= \frac{e^{-T_i} X_i}{\sum_{k=i}^{\infty} e^{-T_k} X_k} \\ &= \frac{e^{-(T_i - T_{i-1})} X_i}{\sum_{k=i}^{\infty} e^{-(T_k - T_{i-1})} X_k}, \; i = 1, 2, \ldots, \end{aligned}$$

with $T_0 = 0$. It is clear that $U_1, U_2, \ldots$ are identically distributed. Rewrite U_1 as

$$U_1 = \frac{X_1}{X_1 + \sum_{k=2}^{\infty} e^{-(T_k - T_1)} X_k}.$$

Since $\sum_{k=2}^{\infty} e^{-(T_k - T_1)} X_k$ has the same distribution as Y and is independent of X_1, by Theorem 1.1, U_1 has a $Beta(1, \theta)$ distribution. Since

$$U_2 = \frac{X_2}{\sum_{k=2}^{\infty} e^{-(T_k - T_2)} X_k} = \frac{X_2}{X_2 + \sum_{k=3}^{\infty} e^{-(T_k - T_2)} X_k}$$

is independent of both X_1 and $\sum_{k=2}^{\infty} e^{-(T_k - T_2)} X_k$, it follows that U_1 and U_2 are independent. Using similar arguments, it follows that $U_1, U_2, \ldots$ are independent with common distribution.

Set

$$\tilde{V}_i = \frac{e^{-T_i} X_i}{Y}.$$

Then we have the following theorem.

Theorem 2.21.

$$\tilde{V}_i = (1 - U_1) \cdots (1 - U_{i-1}) U_i. \tag{2.49}$$

Proof. By direct calculation,

$$\begin{aligned} 1 - U_1 &= \frac{\sum_{k=2}^{\infty} e^{-T_k} X_k}{\sum_{k=1}^{\infty} e^{-T_k} X_k}, \\ (1 - U_1) \cdots (1 - U_{i-1}) &= \frac{\sum_{k=i}^{\infty} e^{-T_k} X_k}{\sum_{k=1}^{\infty} e^{-T_k} X_k}. \end{aligned}$$

Thus

$$(1-U_1)\cdots(1-U_{i-1})U_i = \frac{e^{-T_i}X_i}{Y} = \tilde{V}_i.$$

□

Remark: The linear birth process with immigration gives a construction of the GEM representation.

2.7.3 A Model of Joyce and Tavaré

In both Hoppe's urn and the linear birth-with-immigration model, family sizes are studied based on their age orders; but the detailed relations among family members are missing. In [120], Joyce and Tavaré introduced a new representation of the linear birth process with immigration which incorporates the genealogical structure of the population. To describe the model, we start with several concepts of random permutations.

For each integer $n \geq 1$, let π be a permutation of set $\{1,2,\ldots,n\}$, characterized by $\pi(1)\cdots\pi(n)$. Set

$$i_1 = \inf\{i \geq 1 : \pi^i(1) = 1\}.$$

Then the set $\{\pi(1),\ldots,\pi^{i_1-1}(1),1\}$ is called a *cycle* starting at 1. Similarly we can define cycles starting from any integer. Each permutation can be decomposed into several disjoint cycles. Consider the case of $n = 5$ and the permutation 24513. The cycle started at 1 is (241), the cycle started at 3 is (53). We can write the permutation as (241)(53). The *length* of a cycle is the cardinality of the cycle. For each $i = 1,2,\ldots,n$, let $C_i(n)$ denote the number of cycles of length i. Then $(C_1(n),\ldots,C_n(n))$ is a allelic partition of n.

The model of Joyce and Tavaré can be described as follows: consider an urn that initially contains a black ball of mass θ. Balls are drawn from the urn successively with probabilities proportional to their masses. After each ball is drawn, it is returned to the urn with an additional ball of mass 1. The balls added are numbered by the drawing numbers. After the $(n-1)$th drawing, there are $n-1$ additional balls numbered $1,2,\ldots,n-1$ inside the urn. The ball to be added after the nth drawing will be numbered n. If the nth draw is a black ball, the additional ball will start a new cycle; if ball j, where $1 \leq j \leq n-1$, is selected, n is put immediately to the left of j in the cycle where j belongs. We illustrate this through the following example.

$$\begin{aligned}
n &= 1,(1),\\
n &= 2,(1)(2),\\
n &= 3,(31)(2),\\
n &= 4,(31)(42),\\
n &= 5,(351)(42),\\
n &= 6,(3651)(42).
\end{aligned}$$

In this example, the first and the second draws are both black balls. Thus after the second draw there are two cycles. Ball 3 is added after ball 1 is selected, and ball 4 is added after ball 2 is selected. At the 5th draw, ball 1 is selected and at the 6th draw, ball 5 is selected.

There are several basic facts about these cyclical decompositions:

(a) The first cycle is started at 1; removing the numbers from the first cycle, the second cycle is started at the smallest of the remaining numbers and so on.

(b) Inside a cycle, a pair of numbers with the larger one on the left indicates a successive relation with the one on the right being the parent.

(c) A cycle corresponds to a family; the order of the cycles corresponds to the age order; the details within a cycle describe the relations between family members.

(d) The total number of cycles corresponds to total number of families.

In step 6 of the example, we see that there are two families, with family (3651) appearing first. In family (3651), 3 and 5 are the children of 1 and 6 is the child of 5. The advantage of this model is that the probability of any draw that results in k cycles, is

$$\frac{\theta^k 1^{n-k}}{\theta(\theta+1)\cdots(\theta+n-1)}. \tag{2.50}$$

Let T_n^k be the total number of permutations of n objects into k cycles. Then

$$\sum_{i=1}^{n} T_n^k \frac{\theta^k}{\theta_{(n)}} = 1,$$

which implies that T_n^k is the coefficient of θ^k in the polynomial $\theta_{(n)}$. If K denotes the total number of cycles, then

$$\mathbb{P}\{K=k\} = T_n^k \frac{\theta^k}{\theta_{(n)}}. \tag{2.51}$$

Comparing with (2.51) and (2.16), it follows that K has the same distribution as K_n and $T_n^k = |S_n^k|$.

Furthermore, we have

Theorem 2.22. *The distribution of the random partition*

$$\mathbf{C}_n = (C_1(n),\ldots,C_n(n))$$

is given by the Ewens sampling formula.

Proof. For non-negative integers $c_1,..,c_n$ satisfying $\sum_{i=1}^n ic_i = n$, the number of permutations of $1,2,\ldots,n$ with c_i cycles of length i, $i=1,\ldots,n$, can be calculated as follows: first divide the n numbers into $k = c_1+\cdots+c_n$ cycles, the total number of ways being

$$\frac{n!}{\prod_{i=1}^{n}(i!)^{c_i}}. \tag{2.52}$$

Next, for each cycle, the starting point is fixed but the remaining numbers are free to move around which will increase the count by a factor of

$$\prod_{i=1}^{n}((i-1)!)^{c_i}. \tag{2.53}$$

Finally, to discount the repetition of c_i cycles, the total count needs to be divided by

$$\prod_{i=1}^{n} c_i!. \tag{2.54}$$

The theorem is now proved by noting that the probability

$$\mathbb{P}\{C_1(n) = c_1, \ldots, C_n(n) = n\}$$

is just (2.50) multiplied by (2.52) and (2.53) and then divided by (2.54).

□

2.8 The Dirichlet Process

The Poisson–Dirichlet distribution is the distribution of allelic frequencies in descending order and individual type information is lost. Thus it is called *unlabeled.* Let S be a compact metric space, and ν_0 a probability measure on S. The labeled version of the Poisson–Dirichlet distribution is a random measure with mean measure ν_0. If $M_1(S)$ denotes the space of probability measures on S equipped with the usual weak topology, then the labeled version of the Poisson–Dirichlet distribution is the law of

$$\Xi_{\theta,\nu_0} = \sum_{i=1}^{\infty} P_i(\theta)\delta_{\xi_i}, \tag{2.55}$$

where $\mathbf{P}(\theta) = (P_1(\theta), P_2(\theta), \ldots)$ has the Poisson–Dirichlet distribution with parameter θ; δ_x is the Dirac measure at x and is independent of $\mathbf{P}(\theta)$; $\xi_1, \xi_2, \ldots$ are i.i.d. random variables with common distribution ν_0. The random measure Ξ_{θ,ν_0} is called the *Dirichlet process*, and its law is denoted by Π_{θ,ν_0}. Occasionally we may refer to Π_{θ,ν_0} as the Dirichlet process for notational convenience. Later on, we will see that there are also corresponding labeled and unlabeled dynamical models.

Let $\varsigma_1 \geq \varsigma_2 \geq \cdots$ be the random points of the nonhomogeneous Poisson process F on $(0,\infty)$ with intensity measure $\mu(dx) = \theta x^{-1}e^{-x}dx$. Then it follows from the Poisson process representation in Section 2.2 that $(\frac{\varsigma_1}{\sum_{i=1}^{\infty}\varsigma_i}, \frac{\varsigma_2}{\sum_{i=1}^{\infty}\varsigma_i}, \ldots,)$ has the Poisson–Dirichlet distribution with parameter θ.

Theorem 2.23. (Labeling theorem) *Assume that each point of* $\{\varsigma_1, \varsigma_2, \ldots,\}$ *is labeled independently with labels* $1, \ldots, m$. *The probability of any particular point getting label* i *is denoted by* p_i. *For any* $i = 1, \ldots, m$, *let* $\varsigma_1^{(i)} \geq \varsigma_2^{(i)} \geq \cdots$ *be the points with label* i *in descending order. Then* $\varsigma_1^{(i)} \geq \varsigma_2^{(i)} \geq \cdots$ *are the points of a nonhomogeneous Poisson process on* $(0, \infty)$ *with intensity measure* $\theta p_i x^{-1} e^{-x} dx$. *Furthermore, let*

$$\sigma_i = \sum_{k=1}^{\infty} \varsigma_j^{(i)}, \sigma = \sum_{i=1}^{m} \sigma_i.$$

Then $(\frac{\sigma_1}{\sigma}, \ldots, \frac{\sigma_m}{\sigma})$ *has the Dirichlet*$(\theta p_1, \ldots, \theta p_m)$ *distribution, for each* $1 \leq i \leq m$; $(\frac{\varsigma_1^{(i)}}{\sigma_i}, \frac{\varsigma_2^{(i)}}{\sigma_i}, \ldots,)$ *is independent of* σ_i *and has the Poisson–Dirichlet distribution with parameter* θp_i.

Proof. For each ς in $\mathcal{F}$, let x_ς be a discrete random variable taking value i with probability distribution $\nu(i) = \mathbb{P}\{x_\varsigma = i \mid \varsigma\} = p_i$ for $i = 1, \ldots, m$. Assuming that $\{x_\varsigma : \varsigma \in \mathcal{F}\}$ are independent given $\mathcal{F}$, then by Theorem A.5, $\{(\varsigma, x_\varsigma) : \varsigma \in \mathcal{F}\}$ is a Poisson process with state space $(0, \infty) \times \{1, \ldots, m\}$ and mean measure $\mu \times \nu$. The result now follows from Theorem A.3.

□

Our next result explores the connection between the Dirichlet process and the Dirichlet distribution.

Theorem 2.24. *Assume that* ν_0 *is supported on a finite subset* $\{s_1, \ldots, s_m\}$ *of* S *with* $\gamma_i = \nu_0(\{s_i\}), i = 1, \ldots, m$. *Let* $(X_1, \ldots, X_m)$ *have the Dirichlet*$(\theta\gamma_1, \ldots, \theta\gamma_m)$ *distribution. Then* Π_{θ,ν_0} *is the law of* $\sum_{i=1}^{m} X_i \delta_{s_i}$.

Proof. Let ξ be distributed according to ν_0 and $\xi_1, \xi_2, \ldots,$ independent copies of ξ. Note that Π_{θ,ν_0} is the law of

$$\sum_{j=1}^{\infty} \frac{\varsigma_j}{\sigma} \delta_{\xi_j}.$$

Reorganizing the summation according to the labels in $\{s_1, \ldots, s_m\}$, one gets

$$\sum_{j=1}^{\infty} \frac{\varsigma_j}{\sigma} \delta_{\xi_j} = \sum_{i=1}^{m} \sum_{j:\xi_j = s_i} \frac{\varsigma_j^{(i)}}{\sigma} \delta_{s_i} = \sum_{i=1}^{m} \frac{\sigma_i}{\sigma} \delta_{s_i},$$

which, combined with Theorem 2.23, implies the result.

□

Theorem 2.25. *For any* ν_1, ν_2 *in* $M_1(S)$ *and any* $\theta_1, \theta_2 > 0$, *let* Ξ_{θ_1,ν_1} *and* Ξ_{θ_2,ν_2} *be independent and have respective laws* Π_{θ_1,ν_1} *and* Π_{θ_2,ν_2}. *Let* β *be a Beta*(θ_1, θ_2) *random variable and assume that* β, Ξ_{θ_1,ν_1}, *and* Ξ_{θ_2,ν_2} *are independent. Then the law of* $\beta \Xi_{\theta_1,\nu_1} + (1-\beta)\Xi_{\theta_2,\nu_2}$ *is* $\Pi_{\theta_1+\theta_2, \frac{\theta_1\nu_1+\theta_2\nu_2}{\theta_1+\theta_2}}$.

Proof. Let $z_1 \geq z_2 \geq \cdots$ be the random points of a nonhomogeneous Poisson process on $(0, \infty)$ with intensity measure $(\theta_1 + \theta_2)x^{-1}e^{-x}dx$. Now label each such point independently by 1 or 2 with respective probabilities, $\frac{\theta_1}{\theta_1+\theta_2}$ and $\frac{\theta_2}{\theta_1+\theta_2}$. For $i=1,2$, let $z_1^{(i)} \geq z_2^{(i)} \geq \cdots$ be the points labeled i, in descending order. It follows from Theorem 2.23 that $z_1^{(i)} \geq z_2^{(i)} \geq \cdots$ are the random points of a nonhomogeneous Poisson process on $(0,\infty)$ with intensity measure $\theta_i x^{-1}e^{-x}dx$.

For $i=1,2$, let $\{\xi_j^{(i)} : j \geq 1\}$ be i.i.d. with common distribution ν_i. Then applying Theorem 2.23 again, one has that for

$$\tilde{\sigma} = \sum_{k=1}^{\infty} z_k,\ \tilde{\sigma}_i = \sum_{k=1}^{\infty} z_k^{(i)},$$

$\frac{\tilde{\sigma}_1}{\tilde{\sigma}}$ has the same law as β, and Ξ_{θ_i,ν_i} has the same law as $\sum_{j=1}^{\infty} \frac{z_j^{(i)}}{\tilde{\sigma}_i}\delta_{\xi_j^{(i)}}$. Hence

$$\beta \Xi_{\theta_1,\nu_1} + (1-\beta)\Xi_{\theta_2,\nu_2} \stackrel{d}{=} \frac{\tilde{\sigma}_1}{\tilde{\sigma}} \sum_{j=1}^{\infty} \frac{z_j^{(1)}}{\tilde{\sigma}_1}\delta_{\xi_j^{(1)}} + \frac{\tilde{\sigma}_2}{\tilde{\sigma}} \sum_{j=1}^{\infty} \frac{z_j^{(2)}}{\tilde{\sigma}_2}\delta_{\xi_j^{(2)}}, \tag{2.56}$$

where $\stackrel{d}{=}$ means equal in distribution.

To finish the proof we need to write down the exact relation between $z_1 \geq z_2 \geq \cdots$ and $z_1^{(i)} \geq z_2^{(i)} \geq \cdots$ and, using this relation, construct an i.i.d. sequence $\{\xi_k : k \geq 1\}$ from $\{\xi_j^{(i)}\}$ with common distribution $\frac{\theta_1\nu_1+\theta_2\nu_2}{\theta_1+\theta_2}$.

Let $\{x_k : k \geq 1\}$ be an i.i.d. sequence of random variables with

$$\mathbb{P}\{x_1 = 1\} = \frac{\theta_1}{\theta_1+\theta_2} = 1 - \mathbb{P}\{x_1 = 2\}.$$

Assume that $\{x_k : k \geq 1\}$, $\{\xi_j^{(1)} : j \geq 1\}$,$\{\xi_j^{(2)} : j \geq 1\}$, $\{z_j^{(1)} : j \geq 1\}$, and $\{z_j^{(2)} : j \geq 1\}$ are all independent. Then the labeling procedure for the first three goes as follows:

$$z_1 = z_1^{(x_1)},$$
$$z_2 = \begin{cases} z_2^{(x_2)}, & \text{if } x_2 = x_1 \\ z_1^{(x_2)}, & \text{if } x_2 \neq x_1, \end{cases}$$

and

$$z_3 = \begin{cases} z_3^{(x_3)}, & \text{if } x_3 = x_2 = x_1 \\ z_1^{(x_3)}, & \text{if } x_3 \neq x_2 = x_1 \\ z_2^{(x_3)}, & \text{if } x_3 = x_2 \neq x_1 \\ z_2^{(x_3)}, & \text{if } x_3 = x_1 \neq x_2. \end{cases}$$

In general, one has

$$z_k = z_j^{(x_k)}, \text{ if } j = \#\{1 \leq i \leq k,\ x_i = x_k\}. \tag{2.57}$$

Using this relation, we construct the sequence $\{\xi_k : k \geq 1\}$

$$\xi_k = \xi_j^{(x_k)}, \text{ if } j = \#\{1 \leq i \leq k,\ x_i = x_k\}. \tag{2.58}$$

Now we show that $\{\xi_k : k \geq 1\}$ is i.i.d. with common distribution $\frac{\theta_1 \nu_1 + \theta_2 \nu_2}{\theta_1 + \theta_2}$. For each fixed k,

$$\begin{aligned}
\mathbb{P}\{\xi_k \leq u\} &= \sum_{j=1}^{k} \mathbb{P}\{\xi_j^{(x_k)} \leq u, j = \#\{1 \leq i \leq k,\ x_i = x_k\},\ x_k = 1 \text{ or } 2\} \\
&= \mathbb{P}\{\xi_1^{(1)} \leq u,\ x_1 = 1\} + \mathbb{P}\{\xi_1^{(2)} \leq u,\ x_1 = 2\} \\
&= \frac{\theta_1 \nu_1 + \theta_2 \nu_2}{\theta_1 + \theta_2}((-\infty, u]).
\end{aligned}$$

Thus the sequence $\{\xi_k : k \geq 1\}$ has common distribution $\frac{\theta_1 \nu_1 + \theta_2 \nu_2}{\theta_1 + \theta_2}$. For any $k < l$,

$$\begin{aligned}
\mathbb{P}\{\xi_k \leq u, \xi_l \leq v\} = \sum_{j=1}^{k} \sum_{i=1}^{l} \mathbb{P}\{&\xi_j^{(x_k)} \leq u, \xi_i^{(x_l)} \leq v, \\
&j = \#\{1 \leq r \leq k,\ x_r = x_k\}, i = \#\{1 \leq r \leq l,\ x_r = x_l\}\} \\
= I_1 + I_2 + I_3 + I_4,&
\end{aligned} \tag{2.59}$$

where

$$\begin{aligned}
I_1 &= \sum_{j=1}^{k} \sum_{i=1}^{l} \mathbb{P}\{\xi_j^{(1)} \leq u, \xi_i^{(1)} \leq v\} \\
&\times \mathbb{P}\{j = \#\{1 \leq r \leq k,\ x_r = x_k = 1\}, i = \#\{1 \leq r \leq l,\ x_r = x_l = 1\}\} \\
&= \nu_1((-\infty, u])\nu_1((-\infty, v]) \\
&\times \sum_{j=1}^{k} \sum_{i=j+1}^{l} \mathbb{P}\{j = \#\{1 \leq r \leq k,\ x_r = x_k = 1\}, i - j = \#\{k < r \leq l,\ x_r = x_l = 1\}\}, \\
&= \nu_1((-\infty, u])\nu_1((-\infty, v])\mathbb{P}\{x_k = 1, x_l = 1\}, \\
I_2 &= \sum_{j=1}^{k} \sum_{i=1}^{l} \mathbb{P}\{\xi_j^{(1)} \leq u, \xi_i^{(1)} \leq v\} \\
&\times \mathbb{P}\{j = \#\{1 \leq r \leq k,\ x_r = x_k = 1\}, i = \#\{1 \leq r \leq l,\ x_r = x_l = 2\}\} \\
&= \nu_1((-\infty, u])\nu_2((-\infty, v]) \\
&\times \sum_{j=1}^{k} \sum_{i=k-j+1}^{l} \mathbb{P}\{j = \#\{1 \leq r \leq k,\ x_r = x_k = 1\}, \\
&\qquad i - (k - j) = \#\{k < r \leq l,\ x_r = x_l = 2\}\} \\
&= \nu_1((-\infty, u])\nu_2((-\infty, v])\mathbb{P}\{x_k = 1, x_l = 2\},
\end{aligned}$$

$$\begin{aligned}
I_3 &= \sum_{j=1}^{k}\sum_{i=1}^{l}\mathbb{P}\{\xi_j^{(1)} \le u, \xi_i^{(1)} \le v\} \\
&\quad \times \mathbb{P}\{j = \#\{1 \le r \le k,\ x_r = x_k = 2\}, i = \#\{1 \le r \le l,\ x_r = x_l = 1\}\} \\
&= \nu_2((-\infty,u])\nu_1((-\infty,v]) \\
&\quad \times \sum_{j=1}^{k}\sum_{i=k-j+1}^{l}\mathbb{P}\{j = \#\{1 \le r \le k,\ x_r = x_k = 2\}, \\
&\qquad\qquad i-(k-j) = \#\{k < r \le l,\ x_r = x_l = 1\}\} \\
&= \nu_2((-\infty,u])\nu_1((-\infty,v])\mathbb{P}\{x_k = 2, x_l = 1\}, \\
I_4 &= \sum_{j=1}^{k}\sum_{i=1}^{l}\mathbb{P}\{\xi_j^{(1)} \le u, \xi_i^{(1)} \le v\} \\
&\quad \times \mathbb{P}\{j = \#\{1 \le r \le k,\ x_r = x_k = 2\}, i = \#\{1 \le r \le l,\ x_r = x_l = 2\}\} \\
&= \nu_2((-\infty,u])\nu_2((-\infty,v]) \\
&\quad \times \sum_{j=1}^{k}\sum_{i=j+1}^{l}\mathbb{P}\{j = \#\{1 \le r \le k,\ x_r = x_k = 2\}, i-j = \#\{k < r \le l,\ x_r = x_l = 1\}\} \\
&= \nu_2((-\infty,u])\nu_2((-\infty,v])\mathbb{P}\{x_k = 2, x_l = 2\}.
\end{aligned}$$

The independence of ξ_k and ξ_l now follows from (2.59). Putting (2.57) and (2.58) together, we obtain that

$$\frac{\tilde{\sigma}_1}{\tilde{\sigma}}\sum_{j=1}^{\infty}\frac{z_j^{(1)}}{\tilde{\sigma}_1}\delta_{\xi_j^{(1)}} + \frac{\tilde{\sigma}_2}{\tilde{\sigma}}\sum_{j=1}^{\infty}\frac{z_j^{(2)}}{\tilde{\sigma}_2}\delta_{\xi_j^{(2)}} = \sum_{k=1}^{\infty}\frac{z_k}{\tilde{\sigma}}\delta_{\xi_k}. \tag{2.60}$$

This proves the theorem.

□

Before moving on to a new chapter, we describe an urn-type construction for the Dirichlet process.

Example 2.2 (Ethier and Kurtz [63]). *For $n \ge 1$, let S^n denote the n-fold product of S and $E = S \cup S^2 \cup \cdots$. Construct a Markov chain $\{\mathbf{X}(m) : m \ge 1\}$ on E as follows:*

$$\mathbf{X}(0) = (\xi), \mathbf{X}(1) = (\xi, \xi),$$

where ξ is an S-valued random variable with distribution ν_0. For $n \ge 2$,

$$\begin{aligned}
\mathbb{P}\{\mathbf{X}(m+1) = (x_1,\ldots,x_{j-1},\xi_j,x_{j+1},\ldots,x_n) \mid \mathbf{X}(m) = (x_1,\ldots,x_n)\} \\
= \frac{\theta}{n(n+\theta-1)},
\end{aligned} \tag{2.61}$$

and

$$\mathbb{P}\{\mathbf{X}(m+1) = (x_1,\ldots,x_n,x_j) \mid \mathbf{X}(m) = (x_1,\ldots,x_n)\} = \frac{n-1}{n(n+\theta-1)}, \tag{2.62}$$

where $\xi_j, j = 1,2,\ldots$, *are i.i.d. random variables with common distribution* ν_0.

Define

$$\tau_n = \inf\{m \geq 2 : \mathbf{X}(m) \in S^n\}, \tag{2.63}$$

and

$$\zeta_n(x_1,\ldots,x_n) = \frac{1}{n}\sum_{i=1}^{n} \delta_{x_i}. \tag{2.64}$$

Assume that ν_0 *is diffuse on S; i.e.,* $\nu_0(\{x\}) = 0$ *for all x in S. Then*

$$\zeta_n(\mathbf{X}(\tau_{n+1}-1)) \to \zeta,$$

where the convergence is almost surely in $M_1(S)$, *and the law of the random measure* ζ *is* Π_{θ,ν_0}.

Remark: Let us return to Hoppe's urn and assume that there are n balls inside the urn. Take a ball from the urn at random. Then the chance for a particular ball to be chosen is $1/n$. The color of the selected ball is black with probability $\theta/(n-1+\theta)$, and a new color with probability $(n-1)/(n-1+\theta)$. The similarities of these probabilities to the transition probabilities in the above Markov chain model and the fact that the Dirichlet process does not depend on the particular ordering of its atoms explain why the result is expected naturally.

2.9 Notes

The Poisson–Dirichlet distribution was introduced by Kingman in [125]. It arises as the stationary distribution of the ranked allele frequencies in the infinitely many alleles model in [181]. The proof of Theorem 2.1 comes from [125] and [126]. The connections to the limiting distribution of the ranked relative cycle length in a random permutation were shown in [163], [176] and [177]. The result in the example on prime factorizations of integers was first proved in [15]. A different proof was given in [34] using the GEM distribution. In [83], the Poisson–Dirichlet distribution was used to describe the distribution of random Belyi surfaces. A brief survey on the Poisson–Dirichlet distribution can be found in [102].

The material in Section 2.2 is from [145]. Dickman [31] obtained the asymptotic distribution of the largest prime factor in a large integer, which corresponds to $\theta = 1$ in Theorem 2.5. The current form of Theorem 2.5 first appeared in [95]. The marginal distribution in Theorem 2.6 is given in [181]. The case of $\theta = 1$ was derived in Vershik and Schmidt [176]. The proofs in Section 2.3 follow [145].

The GEM distribution was introduced in [137] and [51] in the context of ecology. Its importance in genetics was recognized by Griffiths[94]. The result in Theorem 2.7 originated in [142]. The proof here is from Donnelly and Joyce [35], which seems to be the first paper to give a published proof of the result. Further development involving a subprobability limit can be found in [88].

The Ewens sampling formula was discovered by Ewens in [66]. A derivation of the formula using a genealogical argument was carried out in [122]. It was also derived in [3] by sampling from a Dirichlet prior. The proof here follows the treatment in [130]. Theorem 2.9 in the case of $\theta = 1$ first appeared in [90]. The general case and the proof here are from [6]. Further discussions on approximations to the Ewens sampling formula can be found in [9]. Extensive applications in population genetics can be found in [67]. A general logarithmic structure including the Ewens sampling formula was studied in [7] and [9]. Theorem 2.11 first appeared in [90] in the case of θ equal to 1. The general case was obtained in [180]. A generalization of Theorem 2.11 to a functional central limit theorem was obtained in [26] for θ equal to one and in [101] for the general case. A nice alternative proof of this functional central limit theorem was obtained in [37] based on a Poisson embedding using a model in [121]. The associated large deviation principles were obtained in [75] and [70]. More properties of the Stirling numbers of the first kind can be found in [1]. Sampling formulae involving selection can be found in [99], [54], and [107].

The scale-invariant Poisson process representation was established in [8]. The approach taken in Section 2.6 is from [9]. Another remarkable property of the scale-invariant Poisson process, the scale-invariant spacing lemma, can be found in [5] and [10].

The urn schemes in Section 2.7 have roots in the paper of Blackwell and MacQueen [16]. Hoppe's urn was first introduced in [103], which also included the proof of Theorem 2.15. The interplay between the genealogical structure of the infinitely-many alleles model, the Poisson–Dirichlet distribution, GEM distribution, size-biased permutation, the Ewens sampling formula and Hoppe's urn, were further explored in [33] and [104]. Section 2.7.2 is from [171]. The earliest reference for the equality (2.41) seems to be [123]. The model and results in Section 2.7.3 come from [120].

The existence of the Dirichlet process in Section 2.8 was established in [80]. The representation (2.55) was discovered in [162]. Theorem 2.23 is called the coloring theorem in [130]. The proofs of Theorem 2.24 and Theorem 2.25 are from [59] with more details included. The Markov chain urn model in Section 2.8 was first introduced in [58]. The study of the asymptotic behavior of the urn model can be found in [39].

There are extensive studies of the Poisson–Dirichlet distribution and related structures from the viewpoint of Bayesian statistics. Detail information can be found in [108], [109], [111], [112], [113], and the references therein. The references [173] and [174] include studies of the multiplicative property of the gamma random measure.

Chapter 3
The Two-Parameter Poisson–Dirichlet Distribution

The representation (2.13) is a particular case of the general residual allocation model (RAM), where $\{U_n\}_{n\geq 1}$ are simply $[0,1]$-valued independent random variables. In this chapter, we study another particular class of RAM. The law of the ranked sequence of this RAM is called the two-parameter Poisson–Dirichlet distribution, a natural generalization of the Poisson–Dirichlet distribution. Our focus is on several two-parameter generalizations of the corresponding properties of the Poisson–Dirichlet distribution.

3.1 Definition

For $0 \leq \alpha < 1$ and $\theta > -\alpha$, let $U_k, k = 1, 2, \ldots$, be a sequence of independent random variables such that U_k has the $Beta(1-\alpha, \theta + k\alpha)$ distribution. Set

$$V_1^{\alpha,\theta} = U_1,\ V_n^{\alpha,\theta} = (1-U_1)\cdots(1-U_{n-1})U_n,\ n \geq 2. \tag{3.1}$$

Then with probability one,

$$\sum_{k=1}^{\infty} V_k^{\alpha,\theta} = 1.$$

Definition 3.1. The law of $(V_1^{\alpha,\theta}, V_2^{\alpha,\theta}, \ldots)$ is called the *two-parameter GEM distribution*, denoted by $GEM(\alpha,\theta)$. Let $\mathbf{P}(\alpha,\theta) = (P_1(\alpha,\theta), P_2(\alpha,\theta), \ldots)$ denote $(V_1^{\alpha,\theta}, V_2^{\alpha,\theta}, \ldots)$ sorted in descending order. Then the law of $\mathbf{P}(\alpha,\theta)$ is called the two-parameter Poisson–Dirichlet distribution, and is denoted by $PD(\alpha,\theta)$.

Proposition 3.1 (Perman, Pitman, and Yor [146]). *The law of the size-biased permutation of* $\mathbf{P}(\alpha,\theta)$ *is the two-parameter GEM distribution.*

Consider the following general RAM with $\{\tilde{U}_n : n \geq 1\}$ being a sequence of independent random variables taking values in $[0,1]$, and

S. Feng, *The Poisson–Dirichlet Distribution and Related Topics*,
Probability and its Applications, DOI 10.1007/978-3-642-11194-5_3,

$$\tilde{V}_1 = \tilde{U}_1,\ \tilde{V}_n = (1-\tilde{U}_1)\cdots(1-\tilde{U}_{n-1})\tilde{U}_n,\ n \geq 2.$$

Let H_{sbp} denote the map of size-biased permutation. The next proposition gives a characterization of the $GEM(\alpha, \theta)$ distribution.

Proposition 3.2 (Pitman [151]). *The law of* $(\tilde{V}_1, \tilde{V}_2, \ldots)$ *is the same as the law of* $H_{sbp}(\tilde{V}_1, \tilde{V}_2, \ldots)$ *iff* $\tilde{U}_i$ *is a* $Beta(1-\alpha, \theta + i\alpha)$ *random variable for* $i \geq 1$*; i.e.,* $(\tilde{V}_1, \tilde{V}_2, \ldots)$ *has the two-parameter* $GEM(\alpha, \theta)$ *distribution.*

Proposition 3.1, combined with Proposition 3.2, shows that the two-parameter Poisson–Dirichlet distribution characterizes the ranked random discrete distribution $\mathbf{P} = (P_1, P_2, \ldots)$ satisfying

$$P_i > 0, i = 1, 2, \ldots$$
$$H_{sbp}(\mathbf{P}) = H_{sbp}(H_{sbp}(\mathbf{P})) \text{ in law.}$$

The Dirichlet process also has a two-parameter generalization.

Definition 3.2. Let $\xi_k, k = 1, \ldots$ be a sequence of i.i.d. random variables with common distribution ν_0 on $[0,1]$. Set

$$\Xi_{\theta,\alpha,\nu_0} = \sum_{k=1}^{\infty} P_k(\alpha,\theta)\delta_{\xi_k}. \tag{3.2}$$

The random measure $\Xi_{\theta,\alpha,\nu_0}$ is called the *two-parameter Dirichlet process*. For notational convenience, this term is also used for the law of $\Xi_{\theta,\alpha,\nu_0}$, denoted by $Dirichlet(\theta, \alpha, \nu_0)$ or $\Pi_{\alpha,\theta,\nu_0}$.

3.2 Marginal Distributions

The marginal distributions of the two-parameter Poisson–Dirichlet distribution can be derived in a manner similar to those of the Poisson–Dirichlet distribution. For this we would need results on the subordinator representation of the two-parameter Poisson–Dirichlet distribution and a change-of-measure formula.

Definition 3.3. For any $0 < \alpha < 1$, the suborbinator $\{\rho_s : s \geq 0\}$ is called a *stable subordinator* with index α if its Lévy measure is

$$\Lambda_\alpha(dx) = c_\alpha x^{-(1+\alpha)}dx,\ x > 0,$$

for some constant $c_\alpha > 0$. For convenience, we will choose $c_\alpha = \frac{\alpha}{\Gamma(1-\alpha)}$ in the sequel.

Let $\mathbb{E}_{\alpha,\theta}$ denote the expectation with respect to $PD(\alpha, \theta)$. The following proposition provides a subordinator representation for $PD(\alpha, 0)$, and establishes a change-of-measure formula between $PD(\alpha, \theta)$ and $PD(\alpha, 0)$.

Proposition 3.3 (Perman, Pitman, and Yor [146]). *For any $t > 0$, let $J_1(\rho_t) \geq J_2(\rho_t) \geq \cdots$ be the ranked values of the jump sizes of the stable subordinator with index α over the time interval $[0,t]$. Then the following hold:*

(1) *The law of $(\frac{J_1(\rho_t)}{\rho_t}, \frac{J_2(\rho_t)}{\rho_t}, \ldots)$ is $PD(\alpha,0)$.*
(2) *For any non-negative, measurable function f on $\mathbb{R}_+^\infty$,*

$$\mathbb{E}_{\alpha,\theta}[f(p_1,p_2,\ldots)] = C_{\alpha,\theta}\mathbb{E}_{\alpha,0}\left[\rho_1^{-\theta} f\left(\frac{J_1(\rho_1)}{\rho_1}, \frac{J_2(\rho_1)}{\rho_1}, \ldots\right)\right], \tag{3.3}$$

where

$$C_{\alpha,\theta} = \frac{\Gamma(\theta+1)}{\Gamma(\frac{\theta}{\alpha}+1)}.$$

As an application of the representation in (1), we get:

Theorem 3.4. *Let $Z_n = \Lambda_\alpha(J_n(\rho_1),+\infty)$ and $\psi(x) = \Lambda_\alpha(x,+\infty)$. Then:*

(1) *$Z_1 < Z_2 < \cdots$ are the points of a homogeneous Poisson process on $(0,\infty)$ with intensity 1, and $Z_n = Y_1 + \cdots + Y_n$, where $\{Y_n\}$ are i.i.d. exponential random variables with parameter 1;*

(2) $\frac{J_n(\rho_1)}{\rho_1} = \frac{Z_n^{-\frac{1}{\alpha}}}{\sum_{i=1}^\infty Z_i^{-\frac{1}{\alpha}}}$;

(3) $\lim_{n\to+\infty} n(\frac{J_n(\rho_1)}{\rho_1})^\alpha = \frac{\rho_1^{-\alpha}}{\Gamma(1-\alpha)}$ *a.s.*

Proof. (1) By Theorem A.4, $\{\psi(J_n(\rho_1)) : n \geq 1\}$ are the points of a Poisson process with mean measure $\mu(\cdot) = \Lambda_\alpha(\psi^{-1}(\cdot))$. By direct calculation,

$$\mu(dz) = \frac{d\Lambda_\alpha(\psi^{-1}(z))}{d\psi^{-1}(z)}\frac{d\psi^{-1}(z)}{dz}dz = dz.$$

It follows by definition that for any $u > 0$,

$$\mathbb{P}(Z_1 > u) = \mathbb{P}(N((0,u]) = 0) = e^{-\mu((0,u))} = e^{-u}.$$

Thus $Y_1 = Z_1, Y_2 = Z_2 - Z_1, Y_3 = Z_3 - Z_2, \ldots$ are i.i.d. exponential random variables with parameter 1.

(2) Since $\Lambda_\alpha(x,+\infty) = \frac{x^{-\alpha}}{\Gamma(1-\alpha)}$, it follows that

$$Z_n = \frac{J_n(\rho_1)^{-\alpha}}{\Gamma(1-\alpha)}.$$

Thus

$$\frac{J_n(\rho_1)}{\rho_1} = \frac{J_n(\rho_1)}{\sum_{i=1}^{+\infty} J_i(\rho_1)} = \frac{Z_n^{-\frac{1}{\alpha}}}{\sum_{i=1}^{+\infty} Z_i^{-\frac{1}{\alpha}}}.$$

(3) By definition, $\frac{J_n(\rho_1)^\alpha}{\rho_1^\alpha} = \frac{1}{Z_n\Gamma(1-\alpha)\rho_1^\alpha}$ or, equivalently, $Z_n = \frac{1}{\Gamma(1-\alpha)\rho_1^\alpha P_n^\alpha}$. By (1) and the law of large numbers for the i.i.d. sequence $\{Y_n\}$, we obtain that

$$\lim_{n\to+\infty}\frac{Z_n}{n} = 1, \tag{3.4}$$

and

$$\lim_{n\to\infty} n\left(\frac{J_n(\rho_1)}{\rho_1}\right)^\alpha \to \Gamma(1-\alpha)^{-1}\rho_1^{-\alpha} \ \ a.s. \tag{3.5}$$

□

The change-of-measure formula (3.3), yields the following characterization of the two-parameter Poisson–Dirichlet distribution.

Theorem 3.5. *Fix $0 < \alpha < 1, \theta > 0$. Let g be a non-negative, measurable function defined on $(0,\infty)$ such that*

$$\int_0^\infty \frac{du}{u^{\alpha+1}}|1-g(u)| < \infty, \tag{3.6}$$

and

$$\int_0^\infty \frac{du}{u^{\alpha+1}}(1-g(u)) > 0. \tag{3.7}$$

Then, for any $\lambda > 0$,

$$\int_0^\infty du\, e^{-\lambda u}\frac{u^{\theta-1}}{\Gamma(\theta)}\mathbb{E}_{\alpha,\theta}\left[\prod_{n=1}^\infty g(uP_n(\alpha,\theta))\right] = \left(\frac{\Gamma(1-\alpha)}{\alpha H_g(\alpha,\lambda)}\right)^{\theta/\alpha}, \tag{3.8}$$

where

$$H_g(\alpha,\lambda) = \int_0^\infty \frac{du}{u^{\alpha+1}}(1-e^{-\lambda u}g(u)).$$

Proof. By (3.3) and a change of variable from u/ρ_1 to v, we get

$$\begin{aligned}
&\int_0^\infty du\, e^{-\lambda u}\frac{u^{\theta-1}}{\Gamma(\theta)}\mathbb{E}_{\alpha,\theta}\left[\prod_{n=1}^\infty g(uP_n(\alpha,\theta))\right]\\
&= \frac{C_{\alpha,\theta}}{\Gamma(\theta)}\int_0^\infty du\, u^{\theta-1}\mathbb{E}_{\alpha,0}\left[\rho_1^{-\theta}\prod_{n=1}^\infty\left(g\left(u\frac{J_n(\rho_1)}{\rho_1}\right)e^{-\lambda u\frac{J_n(\rho_1)}{\rho_1}}\right)\right]\\
&= \frac{C_{\alpha,\theta}}{\Gamma(\theta)}\int_0^\infty dv\, v^{\theta-1}\mathbb{E}\left[\prod_{n=1}^\infty\left(g(vJ_n(\rho_1))e^{-\lambda vJ_n(\rho_1)}\right)\right].
\end{aligned} \tag{3.9}$$

By condition (3.6), we can apply Campbell's theorem so that the last expression in (3.9) becomes

$$
\begin{aligned}
&\frac{C_{\alpha,\theta}}{\Gamma(\theta)}\int_0^\infty dv\left[v^{\theta-1}\exp\left\{-c_\alpha\int_0^\infty(1-g(vx)e^{-\lambda vx})\frac{dx}{x^{\alpha+1}}\right\}\right]\\
&=\frac{C_{\alpha,\theta}}{\Gamma(\theta)}\int_0^\infty dv\left(v^{\theta-1}e^{-c_\alpha H_g(\alpha,\lambda)v^\alpha}\right)\\
&=\left(\frac{\Gamma(1-\alpha)}{\alpha H_g(\alpha,\lambda)}\right)^{\theta/\alpha}.
\end{aligned}
\tag{3.10}
$$

Putting together (3.9) and (3.10), we get (3.8).

□

Now we are ready to derive the marginal distributions for $PD(\alpha,\theta)$. For any $\beta > -\alpha$, let $(P_1(\alpha,\beta),P_2(\alpha,\beta),\ldots)$ be distributed as $PD(\alpha,\beta)$ and let $g_{\alpha,\beta}(p)$ denote the distribution function of $P_1(\alpha,\beta)$.

Theorem 3.6. *Assume that α is strictly positive. Then for each $n\geq 1$, the joint density function $\varphi_n^{\alpha,\theta}(p_1,\ldots,p_n)$ of $(P_1(\alpha,\theta),\ldots,P_n(\alpha,\theta))$ is given by*

$$
\varphi_n^{\alpha,\theta}(p_1,\ldots,p_n)=\frac{C_{\alpha,\theta}c_\alpha^n}{C_{\alpha,\theta+n\alpha}}\frac{\hat{p}_n^{\theta+n\alpha-1}}{(p_1\cdots p_n)^{(1+\alpha)}}g_{\alpha,\theta+n\alpha}\left(\frac{p_n}{\hat{p}_n}\right). \tag{3.11}
$$

Proof. Part (1) of Proposition 3.3, combined with Perman's formula for $h(x)=c_\alpha x^{-(\alpha+1)}$, implies that the joint density function of $(\rho_1,P_1^{\rho_1},\ldots,P_n^{\rho_1})$ is

$$
\begin{aligned}
\phi_n(t,p_1,\ldots,p_n)&=\frac{t^{n-1}c_\alpha^{n-1}(tp_1\cdots tp_{n-1})^{-(\alpha+1)}}{\hat{p}_{n-1}}\cdot\phi_1\left(t\hat{p}_{n-1},\frac{p_n}{\hat{p}_{n-1}}\right)\\
&=\frac{c_\alpha^{n-1}(p_1\cdots p_{n-1})^{(-\alpha+1)}t^{-(n-1)\alpha}}{\hat{p}_{n-1}}\cdot\phi_1\left(t\hat{p}_{n-1},\frac{p_n}{\hat{p}_{n-1}}\right).
\end{aligned}
$$

Since

$$
\phi_1(t,p)=th(tp)\int_0^{\frac{p}{1-p}\wedge 1}\phi_1(t(1-p),u)du,
$$

it follows that

$$
\phi_1\left(t\hat{p}_{n-1},\frac{p_n}{\hat{p}_{n-1}}\right)=c_\alpha t\hat{p}_{n-1}(tp_n)^{-(\alpha+1)}\cdot\int_0^{\frac{p_n}{\hat{p}_n}\wedge 1}\phi_1(t\hat{p}_n,u)du,
$$

and

$$
\phi_n(t,p_1,\ldots,p_n)=c_\alpha^n(p_1\cdots p_n)^{-(\alpha+1)}t^{-n\alpha}\cdot\int_0^{\frac{p_n}{\hat{p}_n}\wedge 1}\phi_1(t\hat{p}_n,u)du.
$$

Taking into account the change-of-measure formula in Proposition 3.3, we get that the density function of $(P_1(\alpha,\theta),\ldots,P_n(\alpha,\theta))$ is

$$\begin{aligned}
&\varphi_n^{\alpha,\theta}(p_1,\ldots,p_n)\\
&= C_{\alpha,\theta}c_\alpha^n(p_1\cdots p_n)^{-(\alpha+1)}\cdot\int_0^{\frac{p_n}{\hat{p}_n}\wedge 1}du\int_0^{+\infty}t^{-(n\alpha+\theta)}\phi_1(t\hat{p}_n,u)dt\\
&= C_{\alpha,\theta}c_\alpha^n(p_1\cdots p_n)^{-(\alpha+1)}\hat{p}_n^{n\alpha+\theta-1}\int_0^{\frac{p_n}{\hat{p}_n}\wedge 1}du\int_0^{+\infty}s^{-(n\alpha+\theta)}\phi_1(s,u)ds\\
&= \frac{C_{\alpha,\theta}}{C_{\alpha,\theta+n\alpha}}c_\alpha^n(p_1\cdots p_n)^{-(\alpha+1)}\hat{p}_n^{n\alpha+\theta-1}\int_0^{\frac{p_n}{\hat{p}_n}\wedge 1}\varphi_1^{\alpha,\theta+n\alpha}(u)du\\
&= \frac{C_{\alpha,\theta}}{C_{\alpha,\theta+n\alpha}}c_\alpha^n\frac{\hat{p}_n^{n\alpha+\theta-1}}{(p_1\cdots p_n)^{\alpha+1}}g_{\alpha,\theta+n\alpha}\left(\frac{p_n}{\hat{p}_n}\right),
\end{aligned}$$

where, in the third equality, we used the fact that

$$\varphi_1^{\alpha,\theta+n\alpha}(v) = C_{\alpha,\theta+n\alpha}\int_0^{+\infty}s^{-(\theta+n\alpha)}\phi_1(s,v)ds,$$

which itself follows from the change-of-measure formula with $\theta+n\alpha$ in place of α.

□

Remark: Clearly the independent sequence of $Beta(1-\alpha,\theta+n\alpha)$ random variables $U_n(\alpha,\theta)$ converges in law to the i.i.d. sequence of $Beta(1,\theta)$ random variables $U_n(0,\theta)$ when α converges to zero. The continuity of the GEM representation and the map of ordering imply that for each $n\geq 1$, $g_{\alpha,\theta+n\alpha}(p)$ converges to the distribution function of $P_1(\theta)$ in Theorem 2.5. By direct calculation, we have

$$\lim_{\alpha\to 0}\frac{C_{\alpha,\theta}}{C_{\alpha,\theta+n\alpha}}c_\alpha^n = 1.$$

Thus by taking the limit of α going to zero, we get Theorem 2.6 from Theorem 3.6.

3.3 The Pitman Sampling Formula

Due to the similarity between the one-parameter GEM distribution and the two-parameter GEM distribution, it is natural to expect a sampling formula in the two-parameter setting that is similar to the Ewens sampling formula. This expectation is realized and the resulting formula is called the two-parameter Ewens sampling formula or the *Pitman sampling formula*.

In order to derive the formula, we start with a subordinator representation for the two-parameter Poisson–Dirichlet distribution.

Proposition 3.7 (Pitman and Yor [156]). *Assume that* $0<\alpha<1, \theta>0$.

(1) *Let* $\{\sigma_t : t\geq 0\}$ *and* $\{\gamma_t : t\geq 0\}$ *be two independent subordinators with respective Lévy measures* $\alpha C x^{-(1+\alpha)}e^{-x}dx$ *and* $\theta x^{-1}e^{-x}dx$ *for some* $C>0$. *Set*

$$\sigma_{\alpha,\theta} = \sigma((C\Gamma(1-\alpha))^{-1}\gamma_{1/\alpha}),$$

where, for notational convenience, we write $\sigma(t)$ for σ_t. Let $J_1(\alpha,\theta) \geq J_2(\alpha,\theta) \geq \ldots$ denote the ranked jump sizes of $\{\sigma_t : t \geq 0\}$ over the random interval

$$[0,(C\Gamma(1-\alpha))^{-1}\gamma_{1/\alpha}].$$

Then $\sigma_{\alpha,\theta}$ and

$$\left(\frac{J_1(\alpha,\theta)}{\sigma_{\alpha,\theta}}, \frac{J_2(\alpha,\theta)}{\sigma_{\alpha,\theta}}, \ldots\right)$$

are independent, and have respective distributions $Gamma(\theta,1)$ and $PD(\alpha,\theta)$.

(2) *Suppose*

$$(P_1(0,\theta),\ldots) \text{ and } (P_1(\alpha,0),\ldots)$$

have respective distributions

$$PD(0,\theta) \text{ and } PD(\alpha,0).$$

Independent of $(P_1(0,\theta),\ldots)$, let

$$(P_1^m(\alpha,0),P_2^m(\alpha,0),\ldots),\ m=1,2,\ldots,$$

be a sequence of i.i.d. copies of $(P_1(\alpha,0),\ldots)$. Then $\{P_m(0,\theta)P_n^m(\alpha,0) : n,m = 1,2,\ldots\}$, ranked in descending order, follows the $PD(\alpha,\theta)$ distribution.

Theorem 3.8. *For each $n \geq 1$, let $\mathscr{A}_n$ denote the space of allelic partitions defined in (2.14). For any $0<\alpha<1$, $\theta>-\alpha$, and $\mathbf{a}=(a_1,\ldots,a_n)$ in $\mathscr{A}_n$, the law of the random allelic partition $\mathbf{A}_n$ of a random sample of size n from a $PD(\alpha,\theta)$ population is given by the Pitman sampling formula*

$$\begin{aligned} PSF(\alpha,\theta,\mathbf{a}) &= \mathbb{P}\{\mathbf{A}_n=(a_1,a_2,\ldots,a_n)\} \\ &= \frac{n!}{\theta_{(n)}}\prod_{l=0}^{k-1}(\theta+l\alpha)\prod_{j=1}^{n}\frac{((1-\alpha)_{(j-1)})^{a_j}}{(j!)^{a_j}(a_j!)}, \end{aligned} \tag{3.12}$$

where $k=\sum_{j=1}^n a_j$.

Proof. We divide the proof into two cases: $\theta>0$ and $-\alpha<\theta\leq 0$. First consider the case, $\theta>0$. Due to the subordinator representation in Proposition 3.7, the result is obtained by applying Theorem A.6. This is similar to the proof of the Ewens sampling formula in Theorem 2.8. Let $X_1,\ldots,X_n$ be a random sample of size $n\geq 1$ from the $PD(\alpha,\theta)$ population. All distinct values of the sample are denoted by $l_{ij}, i=1,\ldots,n; j=1,\ldots,a_i$. Set

$$C(a_1,\ldots,a_n)=\frac{n!}{\prod_{j=1}^n (j!)^{a_j}(a_j!)}.$$

Then

$$PSF(\alpha,\theta,\mathbf{a}) = C(a_1,\ldots,a_n)\mathbb{E}_{\alpha,\theta}\left[\sum_{\substack{\text{distinct } l_{11},\ldots,l_{1a_1};\\ l_{21},\ldots,l_{2a_2};\ldots;l_{n1},\ldots,l_{na_n}}} \prod_{i=1}^{n}\prod_{j=1}^{a_i}\left(\frac{J_{l_{ij}}(\alpha,\theta)}{\sigma_{\alpha,\theta}}\right)^i\right].$$

This, combined with the facts that $\sigma_{\alpha,\theta}$ is a $Gamma(\theta,1)$ random variable, and is independent of $(\frac{J_1(\alpha,\theta)}{\sigma_{\alpha,\theta}}, \frac{J_2(\alpha,\theta)}{\sigma_{\alpha,\theta}},\ldots)$, implies that

$$\begin{aligned}
&PSF(\alpha,\theta,\mathbf{a})\\
&= \frac{C(a_1,\ldots,a_n)\Gamma(\theta)}{\Gamma(\theta+n)}\mathbb{E}\left[\sum_{\substack{\text{distinct } l_{11},\ldots,l_{1a_1};\\ l_{21},\ldots,l_{2a_2};\ldots;l_{n1},\ldots,l_{na_n}}} \prod_{i=1}^{n}\prod_{j=1}^{a_i} J_{l_{ij}}^i(\alpha,\theta)\right] \qquad (3.13)\\
&= \frac{C(a_1,\ldots,a_n)\Gamma(\theta)}{\Gamma(\theta+n)}\mathbb{E}\left[\mathbb{E}\left[\sum_{\substack{\text{distinct } l_{11},\ldots,l_{1a_1};\\ l_{21},\ldots,l_{2a_2};\ldots;l_{n1},\ldots,l_{na_n}}} \prod_{i=1}^{n}\prod_{j=1}^{a_i} J_{l_{ij}}^i(\alpha,\theta)|\gamma_{1/\alpha}\right]\right].
\end{aligned}$$

Applying Theorem A.6 to the conditional expectation in (3.13), it follows that

$$\begin{aligned}
&PSF(\alpha,\theta,\mathbf{a})\\
&= \frac{C(a_1,\ldots,a_n)\Gamma(\theta)}{\Gamma(\theta+n)}\mathbb{E}\left[\prod_{i=1}^{n}\prod_{j=1}^{a_i}\mathbb{E}\left[\sum_{m=1}^{\infty} J_m^i(\alpha,\theta)|\gamma_{1/\alpha}\right]\right]\\
&= \frac{C(a_1,\ldots,a_n)\Gamma(\theta)}{\Gamma(\theta+n)}\mathbb{E}\left[\prod_{i=1}^{n}\left(\mathbb{E}\left[\sum_{m=1}^{\infty} J_m^i(\alpha,\theta)|\gamma_{1/\alpha}\right]\right)^{a_i}\right] \qquad (3.14)\\
&= \frac{C(a_1,\ldots,a_n)\Gamma(\theta)}{\Gamma(\theta+n)}\mathbb{E}\left[\left(\frac{\gamma_{1/\alpha}}{C\Gamma(1-\alpha)}\right)^k\right]\prod_{i=1}^{n}\left(\alpha C\int_0^{\infty} x^{i-1-\alpha}e^{-x}dx\right)^{a_i}\\
&= \frac{C(a_1,\ldots,a_n)\Gamma(\theta)}{\Gamma(\theta+n)}\left(\frac{\alpha}{\Gamma(1-\alpha)}\right)^k\frac{\Gamma(\frac{\theta}{\alpha}+k)}{\Gamma(\frac{\theta}{\alpha})}\prod_{i=1}^{n}\Gamma(i-\alpha)^{a_i},
\end{aligned}$$

which leads to (3.12).

Next, consider the case $-\alpha < \theta \leq 0$. Since $\theta+\alpha > 0$, formula (3.12) holds for

$$PSF(\alpha,\theta+\alpha,\mathbf{a}).$$

Since the function

$$f_{\mathbf{a}}(x_1,x_2,\ldots) = \sum_{\substack{\text{distinct } l_{11},\ldots,l_{1a_1};\\ l_{21},\ldots,l_{2a_2};\ldots;l_{n1},\ldots,l_{na_n}}} \prod_{i=1}^{n}\prod_{j=1}^{a_i} x_{l_{ij}}^i$$

is symmetric, the sampling formula from the $PD(\alpha,\theta)$ population is the same as the sampling formula from the $GEM(\alpha,\theta)$ population. Thus

$$PSF(\alpha,\theta,\mathbf{a}) = C(a_1,\ldots,a_n)\mathbb{E}[f_{\mathbf{a}}(V_1^{\alpha,\theta},V_2^{\alpha,\theta},\ldots)]. \tag{3.15}$$

Decomposing the expectation $\mathbb{E}[f_{\mathbf{a}}(V_1^{\alpha,\theta},V_2^{\alpha,\theta},\ldots)]$ in terms of the powers of $V_1^{\alpha,\theta}$, it follows that

$$\begin{aligned}
&PSF(\alpha,\theta,\mathbf{a})\\
&= C(a_1,\ldots,a_n)\sum_{l=0}^{n}\chi_{\{\mathbf{a}^l\in\mathscr{A}_n\}}a_l\frac{\Gamma(\theta+1)}{\Gamma(1-\alpha)\Gamma(\theta+\alpha)}\\
&\times\frac{\Gamma(l+1-\alpha)\Gamma(\theta+n-l+\alpha)}{\Gamma(\theta+n+1)}\mathbb{E}[f_{\mathbf{a}^l}(V_1^{\alpha,\theta+\alpha},V_2^{\alpha,\theta+\alpha},\ldots)],
\end{aligned} \tag{3.16}$$

where $a_0=1, \mathbf{a}^l=(\ldots,a_l-1,\ldots)$ and the coefficient a_l in the summation accounts for the fact that $V_1^{\alpha,\theta}$ could correspond to the proportion of any of the a_l families of size l. By the sampling formula (3.12), we have that for $l=0$,

$$\mathbb{E}[f_{\mathbf{a}^l}(V_1^{\alpha,\theta+\alpha},V_2^{\alpha,\theta+\alpha},\ldots)] = \frac{\prod_{j=1}^{n}((1-\alpha)_{(j-1)})^{a_j}}{(\theta+\alpha)_{(n)}}\prod_{l=0}^{k-1}(\theta+\alpha+l\alpha), \tag{3.17}$$

and for $l>0$

$$\begin{aligned}
&\mathbb{E}[f_{\mathbf{a}^l}(V_1^{\alpha,\theta+\alpha},V_2^{\alpha,\theta+\alpha},\ldots)]\\
&= \frac{(1-\alpha)_{l-1}^{a_l-1}\prod_{j=1,j\neq l}^{n}((1-\alpha)_{(j-1)})^{a_j}}{(\theta+\alpha)_{(n-l)}}\prod_{l=0}^{k-2}(\theta+\alpha+l\alpha).
\end{aligned} \tag{3.18}$$

Putting (3.16)–(3.18) together, it follows that

$$\begin{aligned}
PSF(\alpha,\theta,\mathbf{a}) &= C(a_1,\ldots,a_n)\left[\frac{(\theta+\alpha)_{(n)}}{(\theta+1)_{(n)}}\frac{\prod_{j=1}^{n}((1-\alpha)_{(j-1)})^{a_j}}{(\theta+\alpha)_{(n)}}\prod_{l=0}^{k-1}(\theta+\alpha+l\alpha)\right.\\
&+\sum_{l=1}^{n}\chi_{\{\mathbf{a}^l\in\mathscr{A}_n\}}a_l\frac{(1-\alpha)_{(l)}(\theta+\alpha)_{(n-l)}}{(\theta+1)_{(n)}}\frac{(1-\alpha)_{l-1}^{a_l-1}\prod_{j=1,j\neq l}^{n}((1-\alpha)_{(j-1)})^{a_j}}{(\theta+\alpha)_{(n)}}\\
&\left.\times\prod_{l=0}^{k-2}(\theta+\alpha+l\alpha)\right] \qquad (3.19)\\
&= C(a_1,\ldots,a_n)\frac{\prod_{l=1}^{k-1}(\theta+l\alpha)}{(\theta+1)_{(n)}}\left[\theta+k\alpha+\sum_{l=1}^{n}(la_l-a_l\alpha)\right]\\
&= C(a_1,\ldots,a_n)\frac{\prod_{l=0}^{k-1}(\theta+l\alpha)}{(\theta)_{(n)}},
\end{aligned}$$

which is just (3.12). □

3.4 Urn-type Construction

In this section, we consider an urn construction for the Pitman sampling formula. Even though it is not clear how this formula arises in a genetic setting, we will use some genetic terminologies in our discussion due to its similarity to the Ewens sampling formula.

A random sample of size n could arise from a random sample of size $n+1$ with one element being removed at random. Equivalently, a random sample of size $n+1$ can be obtained by adding one element to a random sample of size n. Thus it is natural to expect that the two-parameter sampling formula has certain consistent structures. Given that a random sample $X_1,\ldots,X_n$ has an allelic partition $\mathbf{a}=(a_1,\ldots,a_n)$ or equivalently, $(X_1,\ldots,X_n)=(a_1,\ldots,a_n)$, the value of X_{n+1} could be one of $X_1,\ldots,X_n$ or a completely new value. If X_{n+1} takes on a value that is not among the first n samples, then the allelic partition of $X_1,\ldots,X_{n+1}$ is $(a_1+1,a_2,\ldots,a_n,0)$. By (3.12),

$$\begin{aligned} p_{n+1}&(a_1+1,a_2,\ldots,a_n,0)\\ &=\frac{(n+1)!}{(a_1+1)\theta_{(n+1)}}\prod_{l=0}^{k}(\theta+l\alpha)\prod_{j=1}^{n}\frac{((1-\alpha)_{(j-1)})^{a}_{j}}{(j!)^{a_j}a_j!} \qquad (3.20)\\ &=\frac{n+1}{a_1+1}\frac{\theta+k\alpha}{n+\theta}p_n(a_1,a_2,\ldots,a_n). \end{aligned}$$

Note that moving X_{n+1} around in the sample will not alter the allelic partition $(a_1+1,a_2,\ldots,a_n,0)$. Thus X_{n+1} could appear as the last appearance is only one out of $n+1$ possibilities. Since any of the a_1+1 single appearance values could appear in the last sample, it follows that

$$\mathbb{P}\{X_{n+1}\text{ takes on a new value}|(X_1,\ldots,X_n)=\mathbf{a}\}=\frac{\theta+k\alpha}{\theta+n}. \qquad (3.21)$$

Similarly if X_{n+1} takes on a value that appears in $X_1,\ldots,X_n$ r times for some $1\le r\le n$ (denote this event by $C(n,r)$), then the allelic partition of $X_1,\ldots,X_{n+1}$ is $(\ldots,a_r-1,a_{r+1}+1,\ldots,a_{n+1})$, and

$$\mathbb{P}\{C(n,r)|(X_1,\ldots,X_n)=(\ldots,a_r-1,a_{r+1}+1,\ldots,a_{n+1})\}=\frac{a_r(r-\alpha)}{\theta+n}. \qquad (3.22)$$

Conversely, starting with the initial condition,

$$\mathbb{P}\{X_1=a_1\}=1,$$

the two-parameter sampling formula (3.12) is uniquely determined by the conditional probabilities in (3.21) and (3.22). This suggests an urn structure for the two-parameter sampling formula that is similar to the model in Section 2.7.2.

Consider again a population composed of immigrants and their descendants. Each new immigrant starts a new family. Individuals in the same family are considered to be the same type. Let $I(t)$ be the immigration process describing how immigrants enter the population and let $x(t)$ be a linear growth process, with $x(0) = 1$, describing the growth pattern of each family.

Immigrants arrive at the times $0 \leq T_1 < T_2 < \cdots$ and initiate families evolving as independent versions of $x(t)$. If $\{x_i(t)\}$ are independent copies of $x(t)$ with $x_i(t)$ being initiated by the ith type, then $x_i(t - T_i)$ will be the size at time t of the type i family. The process $N(t)$ represents the total population size at time t:

$$N(t) = \sum_{i=1}^{I(t)} x_i(t - T_i).$$

For $N(t) \geq 1$, the families induce a random partition $\Pi(t)$ of the integer $N(t)$. Let $A_j(t)$ denote the number of families of size j at time t so that

$$\Pi(t) = (A_1(t), \ldots, A_{N(t)}(t))$$

is the corresponding random allelic partition.

The transition mechanisms of these processes are as follows.

(i) $I(t)$ is a pure-birth process with $I(0) = 0$ and the birth rate

$$r_i = \lim_{h \to 0} \frac{1}{h} \mathbb{P}\{I(t+h) - I(t) = 1 | I(t) = i\}$$

where

$$r_0 = \beta_0, \quad r_i = \alpha i + \theta, \quad \text{for } i \geq 1$$

with $\beta_0 > 0, \theta > -\alpha$, and $0 < \alpha < 1$.

(ii) The process $x(t)$ is also a pure-birth process but starting at $x(0) = 1$ and with infinitesimal birth rate $\ell_n = n - \alpha$ for $n \geq 1$.

(iii) The cumulative process $N(t)$ is then a pure-birth starting at $N(0) = 0$ with rates determined by the processes $\{I(t), x_1(t - T_1), \ldots, x_{I(t)}(t - T_{I(t)})\}$. Let

$$\rho_n = \lim_{h \to 0} \frac{1}{h} \mathbb{P}\{N(t+h) - N(t) = 1 \mid N(t) = n\}.$$

Obviously $\rho_0 = \beta_0$ and for small $h > 0$

$$\begin{aligned}
&\mathbb{P}\{N(t+h) - N(t) = 1 \mid N(t) = n\} \\
&= \mathbb{E}\left[\left(\alpha I(t) + \theta + \sum_{i=1}^{I(t)} (x_i(t - T_i) - \alpha)\right) h + o(h) \mid N(t) = n\right] \\
&= (n + \theta) h + o(h).
\end{aligned}$$

Thus for $n \geq 1$, $\rho_n = n + \theta$.

Define

$$\tau_n = \inf\{t \geq 0 : N(t) = n\} \text{ for } n \geq 1.$$

Then

$$\Pi_n = \Pi(\tau_n) = (A_1(\tau_n), \ldots, A_n(\tau_n))$$

will be a random partition of n.

For any $(a_1, \ldots, a_n)$ in $\mathscr{A}_n$ with $k = \sum_{i=1}^n a_i$, it follows from our construction that

$$\mathbb{P}\{\Pi_{n+1} = (a_1 + 1, \ldots, a_n, 0) \mid \Pi_n = (a_1, \ldots, a_n)\} = \frac{k\alpha + \theta}{n + \theta}. \tag{3.23}$$

If $a_i > 1$ for some $1 \leq i < n$, then

$$\begin{aligned} &\mathbb{P}\{\Pi_{n+1} = (a_1, \ldots, a_i - 1, a_{i+1} + 1, \ldots, a_n, 0) \mid \Pi_n = (a_1, \ldots, a_n)\} \\ &= \frac{a_i(i - \alpha)}{n + \theta}. \end{aligned} \tag{3.24}$$

Finally for $a_n = 1$, we have

$$\mathbb{P}\{\Pi_{n+1} = (0, \ldots, 0, 0, 1) \mid \Pi_n = (0, \ldots, 1)\} = \frac{n - \alpha}{n + \theta}. \tag{3.25}$$

Comparing (3.23)-(3.25) with the conditional probabilities in (3.21) and (3.22), we obtain:

Theorem 3.9. *The distribution of the random partition $\Pi_n(\tau_n)$ is given by the Pitman sampling formula.*

The processes $I(t)$, $x(t)$, and $N(t)$ are special cases of linear birth process with immigration, generically denoted by $Y(t)$. This is a time homogeneous Markov chain with infinitesimal rates

$$\lambda_n = \lim_{h \to 0} \frac{1}{h} \mathbb{P}[Y(t+h) - Y(t) = 1 \mid Y(t) = n] = \lambda n + c, \tag{3.26}$$

where $\lambda > 0, c > 0$.

By an argument similar to that used in Theorem 2.17 and Theorem 2.18, we have

$$\mathbb{P}[Y(t) = n \mid Y(0) = 0] = \binom{\frac{c}{\lambda} + n - 1}{n} e^{-ct}(1 - e^{-\lambda t})^n, \quad n = 0, 1, \ldots, \tag{3.27}$$

and

$$\lim_{t \to \infty} e^{-\lambda t} Y(t) = W_{\frac{c}{\lambda}} \ a.s., \tag{3.28}$$

where W_d is a $Gamma(d, 1)$ random variable.

Starting with $Y(0) = 1$, the corresponding results for $n \geq 1$ become

$$\mathbb{P}[Y(t)=n \mid Y(0)=1] = \binom{\frac{c}{\lambda}+n-1}{n-1} e^{-(\lambda+c)t}(1-e^{-\lambda t})^{n-1}, \tag{3.29}$$

and

$$\lim_{t\to\infty} e^{-\lambda t} Y(t) = W_{1+\frac{c}{\lambda}} \ \ a.s. \tag{3.30}$$

Specializing to $I(t)$, $x(t)$, and $N(t)$, we obtain the following results:

(i) For $I(t)$, one has

$$I(0)=0,\ \lambda_0=\beta_0,\ \lambda_n=\alpha n+\theta,\ n\geq 1.$$

Let $I^*(t)=I(t+T_1)$. Then $I^*(t)$ is a linear birth process with immigration starting with $I^*(0)=1$ and with infinitesimal rates $\lambda_n^*=\alpha n+\theta$. Thus

$$\lim_{t\to\infty} e^{-\alpha t} I(t) = \lim_{t\to\infty} e^{-\alpha T_1} e^{-\alpha(t-T_1)} I^*(t-T_1) = e^{-\alpha T_1} W_{\theta/\alpha+1} \ \ a.s., \tag{3.31}$$

where T_1 and $W_{\theta/\alpha}$ are independent, and T_1 has an exponential distribution with mean β_0^{-1}.

(ii) For $x(t)$, one has $x(0)=1,\ \lambda_n=n-\alpha,\ n\geq 1$, and

$$\lim_{t\to\infty} e^{-t} x(t) = W_{1-\alpha} \ \ a.s. \tag{3.32}$$

(iii) For $N(t)$, one has $N(0)=0,\ \lambda_0=\beta_0,\ \lambda_n=n+\beta-\alpha,\ n\geq 1$ and

$$\lim_{t\to\infty} e^{-t} N(t) = e^{-T_1} W_{1+\theta} \ \ a.s. \tag{3.33}$$

With these results in hand, we are ready to prove the following theorem.

Theorem 3.10. *For $0<\alpha<1$, $\theta>-\alpha$, let*

$$K_n^{\alpha,\theta} = \sum_{i=1}^n A_i(\tau_n). \tag{3.34}$$

Then

$$\lim_{n\to\infty} \frac{K_n^{\alpha,\theta}}{n^\alpha} = S_{\alpha,\theta} \ \ a.s. \tag{3.35}$$

Proof. By applying (3.31) and (3.33), we have

$$\lim_{t\to\infty} \frac{I(t)}{(N(t))^\alpha} = \lim_{t\to\infty} \frac{e^{-\alpha t} I(t)}{[e^{-t}N(t)]^\alpha} = \frac{W_{\theta/\alpha}}{(W_{1+\theta})^\alpha} \ \ a.s., \tag{3.36}$$

which gives (3.35). Here $W_{\theta/\alpha}$ and $W_{1+\theta}$ are correlated. □

Remarks:
(a) More information about the distribution of $S_{\alpha,\theta}$ can be found in [155], where its connection to the Mittag–Leffler distribution is explored.
(b) A general scheme, called the *Chinese restaurant process*, can be adapted to give a unified treatment for the models in Section 2.7 and Section 3.4. Our presentation reflects the historical development of the subject.

3.5 Notes

The most fundamental reference for the two-parameter Poisson–Dirichlet distribution is Pitman and Yor [156], which is also the source for the proof of Theorem 3.4 and Theorem 3.5. The connection to the Blackwell–MacQueen urn scheme [16] is included in the survey paper of Pitman [152]. The proof of Theorem 3.6 presented here appears in [73]. A different proof is found in Handa [100] using the GEM representation and the theory of point processes. The residual allocation model was first introduced in [98].

The Pitman sampling formula in Theorem 3.8 first appeared in Pitman [148], and was further studied in [149]. The first part of the proof of Theorem 3.8 can be found in Carlton [17], which also includes detailed calculations of moments and parameter estimation for the two-parameter Poisson–Dirichlet distribution. The proof for the second part seems to be new. Further discussions on sampling formulae can be found in [89].

The models in Section 3.4, motivated by the work in [121], are from [75]. Pitman [155] includes other urn-type models and related references. The conditional structures for the two-parameter sampling formulae are obtained in [149]. Theorem 3.10 is due to Pitman [150] who obtained his results by moment calculations and martingale convergence techniques. The proof here is from [75]. The Chinese restaurant process first appeared in [2]. Further development and background can be found in [155].

We have focused on the two-parameter Poisson–Dirichlet distribution with infinitely many types. For a population with m types, the proper ranges of α and θ for the two-parameter Poisson–Dirichlet distribution are $\alpha = -\kappa, \theta = m\kappa$ for some $\kappa > 0$.

In [29] and the references therein, connections can be found between the two-parameter Poisson–Dirichlet distribution and models in physics, including mean-field spin glasses, random map models, fragmentation, and returns of a random walk to the origin. Applications in macroeconomics and finance can be found in [4]. Reference [115] includes results in Bayesian statistics. Additional distributional results on subordinators are included in [116].

Chapter 4
The Coalescent

Consider a population that evolves under the influence of mutation and random sampling. In any generation, the gene of each individual is either a copy of a gene of its parent or is a mutant. It is thus possible to trace the lines of descent of a gene back through previous generations. By tracing back in time, the genealogy of a sample from the population emerges. The coalescent is a mathematical model that describes the ancestry of a finite sample of individuals, genes or DNA sequences taken from a large population. In this chapter, we introduce Kingman's coalescent, derive its distribution, and study the embedded pure-death Markov chain. All calculations are based on the Wright–Fisher model.

4.1 Kingman's n-Coalescent

For each $n \geq 1$, and $1 \leq k \leq n$, let $S(n,k)$ denote the total number of ways of partitioning n elements into k nonempty sets. The set $\{S(n,k) : n \geq 1,\ 1 \leq k \leq n\}$ is the collection of the Stirling numbers of the second kind. They are different from the unsigned Stirling numbers of the first kind $T_n^k = |S_n^k|$, the total number of permutations of n objects into k cycles. We list several known facts about $S(n,k)$.

$$\begin{aligned}
S(n,1) &= S(n,n) = 1,\\
S(n,2) &= 2^{n-1} - 1,\ S(n,n-1) = \binom{n}{2},\ n \geq 2,\\
S(n,k) &= S(n-1,k-1) + kS(n-1,k),\\
S(n,k) &= \frac{1}{k!}\sum_{i=1}^{k}(-1)^{k-i}\binom{k}{i} i^n.
\end{aligned}$$

Consider the Wright–Fisher model with a large population size $2N$ and no mutation. Take a sample of size n from the population at the present time. Since the only

S. Feng, *The Poisson–Dirichlet Distribution and Related Topics*,
Probability and its Applications, DOI 10.1007/978-3-642-11194-5_4,

influence on the population is random genetic drift, any two individuals in the sample will share a most recent common ancestor (MRCA) sometime in the past. Going further back in time, more individuals will share an MRCA. Eventually all individuals in the sample will coalesce at a single individual which we call the ancestor of the sample.

Consider the present time as $t = 0$. The time $t = m$ will be the time of going back m generations in the past. Starting at time zero and going back one generation, we can ask the following question: What is the probability P_{nk} that the number of ancestors of the current sample is k in the previous generation? Clearly k has to be less than or equal to n; i.e., $P_{nk} = 0$ for $k > n$. Let $Z_0 = n$, Z_1 be the total number of ancestors, one generation back, and Z_m denote the total number of ancestors, m generations back. Then Z_m is a Markov chain with transition probability matrix $(P_{ij})_{1\le i,j\le n}$.

The transition probability P_{nk} can be calculated as follows. First, the sample is partitioned into k groups. The total number of ways is given by $S(n,k)$. Next for each fixed partition, we calculate the probability. Since the ancestor in the previous generation of each individual in the sample can be any of the $2N$ individuals, the total number of choices is $(2N)^n$. There are $2N$ choices for the first ancestor, $2N-1$ choices for the second ancestor, and so on. The probability of each partition is $(2N)(2N-1)\cdots(2N-k+1)/(2N)^n$. Thus

$$
\begin{aligned}
P_{nk} &= S(n,k)\frac{(2N)(2N-1)\cdots(2N-k+1)}{(2N)^n}\\
&= \begin{cases} \frac{\binom{n}{2}}{2N} + o(\frac{1}{(2N)^2}), & k = n-1\\ 1 - \frac{\binom{n}{2}}{2N} + o(\frac{1}{(2N)^2}), & k = n\\ o(\frac{1}{(2N)^2}), & \text{else.} \end{cases}
\end{aligned}
$$

If $2N$ is much bigger than n, then by ignoring higher order terms in $\frac{1}{2N}$, the number of ancestors will be $n-1$ or n with respective approximate probabilities $\binom{n}{2}\frac{1}{2N}$ and $1-\binom{n}{2}\frac{1}{2N}$. Similar arguments can be applied to any P_{ij} with $1 \le j \le i \le n$.

Starting at i, let σ_i^{2N} denote the number of steps that the Markov chain Z_t spent at state i. Then, for $m \ge 1$,

$$
\mathbb{P}\{\sigma_i^{2N} = m\} \approx \frac{\binom{i}{2}}{2N}\left(1 - \frac{\binom{i}{2}}{2N}\right)^{m-1}.
$$

Using the same scaling as in diffusion approximations, we can see that for any $t > 0$

$$\mathbb{P}\{\frac{\sigma_i^{2N}}{2N} \leq t\} = \sum_{k=1}^{[2Nt]} \frac{\binom{i}{2}}{2N}\left(1 - \frac{\binom{i}{2}}{2N}\right)^{k-1}$$
$$\approx 1 - e^{-\binom{i}{2}t}.$$

In other words, the macroscopic time $\frac{\sigma_i^{2N}}{2N}$ approaches an exponential random variable with parameter $\binom{i}{2}$. This brings us to a process, called the coalescent.

To describe the coalescent, consider the current time as zero. For each $n \geq 1$, let $E_n = \{1, 2, \ldots, n\}$ and $\mathscr{E}_n$ denote the collection of equivalence relations of E_n. Each element of $\mathscr{E}_n$ is thus a subset of $E_n \times E_n$. For example, in the case of $n = 3$, the set

$$\{(1,1), (2,2), (3,3), (1,3), (3,1)\}$$

defines an equivalence relation that results in two equivalence classes $\{1,3\}$ and $\{2\}$. The set $\mathscr{E}_n$ is clearly finite and its elements will be denoted by η, ξ, etc.

The equivalence relations that are of interest to us here are defined through the ancestral structures. Two individuals are equivalent if they have the same ancestor at some time t in the past. For ξ, η in $\mathscr{E}_n$, we write $\xi \prec \eta$ if η is obtained from ξ by combining exactly two equivalence classes of ξ into one. For distinct ξ, η in $\mathscr{E}_n$, set

$$q_{\xi\eta} = \begin{cases} 1, & \xi \prec \eta \\ 0, & \text{else.} \end{cases}$$

Let $|\xi|$ be the number of equivalence classes induced by ξ. Define

$$q_\xi := -q_{\xi,\xi} = \binom{|\xi|}{2}.$$

Definition 4.1. Kingman's n-coalescent is a $\mathscr{E}_n$-valued, continuous-time, Markov chain X_t with infinitesimal matrix $(q_{\xi\eta})$ starting at $X_0 = \{(i,i) : i = 1, \ldots, n\}$.

For each $2 \leq i \leq n$, denote by τ_i, the waiting time between the $(n-i)$th jump and the $(n-i+1)$th jump of X_t. Then $\tau_2, \ldots, \tau_n$ are independent and τ_i is an exponential random variable with parameter $\binom{i}{2}$. Set $T_0 = 0$ and

$$T_m = \tau_n + \cdots + \tau_{n+1-m}, 1 \leq m \leq n-1$$

and

$$D_t = |X_t|.$$

Then D_t is a pure-death process starting at n with intensity

$$\lim_{h \to 0} h^{-1}\mathbb{P}\{D_{t+h} = k-1 \mid D_t = k\} = \binom{k}{2}.$$

The marginal distribution of D_t is given by

$$\mathbb{P}\{D_t = k\} = \mathbb{P}\{T_{n-k} \leq t\} - \mathbb{P}\{T_{n-k+1} \leq t\}.$$

Introduce a discrete time process

$$Y_k = X_{T_{n-k}},\ 1 \le k \le n.$$

By direct calculation, for $\xi \prec \eta$

$$\mathbb{P}\{Y_{k-1} = \eta \mid Y_k = \xi\} = \frac{1}{\binom{|\xi|}{2}}.$$

It is clear from the construction that

$$X_t = Y_{D_t},$$

and the processes Y_k and D_t are independent.

Theorem 4.1. *For each η in $\mathscr{E}_n$ with $|\eta| = k$,*

$$\mathbb{P}\{Y_k = \eta\} = \frac{(n-k)!k!(k-1)!}{n!(n-1)!}\prod_{i=1}^{k}(a_i!), \tag{4.1}$$

where $a_1,\ldots,a_k$ are the sizes of the equivalence classes in η.

Proof. We will prove the result by induction. Let $P_k(\eta) = \mathbb{P}\{Y_k = \eta\}$. Then

$$\begin{aligned} P_k(\eta) &= \sum_{\xi \prec \eta} \mathbb{P}\{Y_{k+1} = \xi, Y_k = \eta\} \\ &= \frac{1}{\binom{k+1}{2}} \sum_{\xi \prec \eta} P_{k+1}(\xi). \end{aligned}$$

For $k = n$, we have

$$\eta = \{(i,i) : i = 1,\ldots,n\},\ \ a_1 = \cdots = a_n = 1.$$

Since $P_n(\eta) = 1$, (4.1) holds in this case. Assume the formula holds for $k+1$ with $2 \le k < n$. If the sizes of equivalence classes of η are $a_1,..,a_k$, then for any $\xi \prec \eta$, the sizes $b_1,\ldots,b_{k+1}$ of equivalence classes of ξ will be of the form $a_1,\ldots,a_{r-1},\lambda,a_r-\lambda,\ldots,a_k$ for some $1 \le r \le k$ and $1 \le \lambda < a_r$. For each fixed r and λ, there are $\frac{1}{2}\binom{a_r}{\lambda}$ of such ξ. Here the factor $1/2$ is due to the double counting of the pair $\lambda, a_r - \lambda$ in $\binom{a_r}{\lambda}$; i.e., the same ξ is counted twice: once by choosing λ elements and once by choosing the other $a_r - \lambda$ elements. From the recursion above, we obtain

$$
\begin{aligned}
P_k(\eta) &= \frac{1}{\binom{k+1}{2}} \sum_{\xi \prec \eta} \frac{(n-k-1)!(k+1)!k!}{n!(n-1)!} \prod_{i=1}^{k+1} b_i! \\
&= \frac{1}{\binom{k+1}{2}} \sum_{r=1}^{k} \sum_{\lambda=1}^{a_r-1} \frac{(n-k-1)!(k+1)!k!}{n!(n-1)!} a_1! \cdots \lambda!(a_r-\lambda)! \cdots a_k! \frac{1}{2} \binom{a_r}{\lambda} \\
&= \frac{(n-k-1)!k!(k-1)!}{n!(n-1)!} \prod_{i=1}^{k} a_i! \sum_{r=1}^{k} (a_r - 1).
\end{aligned}
$$

Since $\sum_{r=1}^{k}(a_r-1) = n-k$, it follows that

$$
P_k(\eta) = \frac{(n-k)!k!(k-1)!}{n!(n-1)!} \prod_{i=1}^{k} a_i!.
$$

Thus the formula (4.1) also holds for k, which implies the theorem.

□

The random variable T_{n-1} is the total time needed for the n individuals to reach the MRCA. The mean and the variance of T_{n-1} can be calculated explicitly:

$$
\begin{aligned}
\mathbb{E}[T_{n-1}] &= \sum_{k=2}^{n} \mathbb{E}[\tau_k] \qquad (4.2) \\
&= 2 \sum_{k=2}^{n} \left[\frac{1}{k(k-1)} \right] = 2\left(1 - \frac{1}{n}\right), \\
Var[T_{n-1}] &= \sum_{k=2}^{n} Var[\tau_k] \\
&= 4 \sum_{k=2}^{n} \frac{1}{(k(k-1))^2} \qquad (4.3) \\
&= 4 \sum_{k=2}^{n} \left[\frac{1}{(k-1)^2} + \frac{1}{k^2} - \frac{2}{k(k-1)} \right] \\
&= 8 \sum_{k=2}^{n} \frac{1}{k^2} - 4\left(1 - \frac{1}{n}\right)^2.
\end{aligned}
$$

The fact that $\mathbb{E}[\tau_2] = 1 \approx \frac{1}{2}\mathbb{E}[T_{n-1}]$, indicates the significant influence of the most ancient coalescence time. An alternate time measure, taking into account the sample sizes, is given by

$$
\tilde{T}_{n-1} = \sum_{k=2}^{n} k\tau_k. \qquad (4.4)
$$

The corresponding mean and variance are

$$\mathbb{E}[\tilde{T}_{n-1}] = 2\sum_{k=2}^{n}\left[\frac{1}{k-1}\right] \approx \log n, \tag{4.5}$$

$$Var[\tilde{T}_{n-1}] = 4\sum_{k=2}^{n}\frac{1}{(k-1)^2}. \tag{4.6}$$

4.2 The Coalescent

Letting n converge to infinity, one would expect to get a pure-death process $\mathbf{D}_t$ with entrance boundary ∞, and transition intensity

$$\mathbb{P}\{\mathbf{D}_{t+h} = k-1 \mid \mathbf{D}_t = k\} = \binom{k}{2}h + o(h).$$

The rigorous construction of this process will be carried out in Section 4.3. Assuming the existence of the pure-death process, one can then construct a Markov process $\mathbf{X}_t$ on $\mathscr{E}$, which is the collection of all equivalence relations on $\{1,2,\ldots\}$, such that $|\mathbf{X}_t|$ is a pure-death Markov process with transition probability $P_{\xi\eta} = 1/\binom{|\xi|}{2}, \xi \prec \eta$. For $n \geq 1$, define a map $\rho_n : \mathscr{E} \to \mathscr{E}_n$ such that $\rho_n(\xi) = \{(i,j) \in \xi : 1 \leq i,j \leq n\}$. Then $\rho_n(\mathbf{X}_t)$ is the n-coalescent. Thus all n-coalescents can be constructed on a single probability space. The process $\mathbf{X}_t$ is called a *coalescent*.

The coalescent can be generalized to include mutation. This is done by considering the number of ancestors at a time t in the past which have not experienced mutation. Consider the infinitely many alleles model ([184]) where each mutation leads to a new allelic type. Take a sample of size n at time zero. Going back in time, two individuals are equivalent if they have a common ancestor and, for the most recent one, no mutation was experienced along the direct lines of descent from that common ancestor to the two individuals. If at this time an individual has an ancestor that is a mutant gene, the equivalence class containing the individual will be excluded from thereon. Assuming the rate of mutation is $\theta/2$ along each line of descent. Thus going back into the past, one equivalence class will disappear, due to either a mutation or a coalescing event. The coalescent will now involve a pure-death process $\mathbf{D}_t$ with transition intensity

$$\mathbb{P}\{\mathbf{D}_{t+h} = k-1 \mid \mathbf{D}_t = k\} = \left(\binom{k}{2} + \frac{1}{2}k\theta\right)h + o(h).$$

The embedded chain will experience a coalescing event with probability $\frac{\binom{k}{2}}{\binom{k}{2}+\frac{1}{2}k\theta} = \frac{k-1}{\theta+k-1}$ and a mutation with probability $\frac{\theta}{\theta+k-1}$.

On the basis of the probabilities $\frac{k-1}{\theta+k-1}$ and $\frac{\theta}{\theta+k-1}$, we can see that the coalescent with mutation corresponds to running Hoppe's urn backward in time. Therefore the Ewens sampling formula can be derived from coalescent arguments.

Starting with n individuals, the total time $T_{n,\theta}$ needed to reach the MRCA is now given by

$$T_{n,\theta} = \sum_{k=1}^{n} \tau_{k,\theta},$$

where $\tau_{k,\theta}$ is an exponential random variable with parameter $\binom{k}{2} + \frac{1}{2}k\theta$. Note that $\tau_{1,\theta}$ is a new component due to mutation. The mean and variance are

$$\begin{aligned}
\mathbb{E}[T_{n,\theta}] &= \sum_{k=1}^{n} \mathbb{E}[\tau_{k,\theta}] \\
&= 2\sum_{k=1}^{n} \frac{1}{k(\theta+k-1)}, \\
Var[T_{n,\theta}] &= \sum_{k=1}^{n} Var[\tau_{k,\theta}] \\
&= 4\sum_{k=1}^{n} \frac{1}{(k(\theta+k-1))^2}.
\end{aligned}$$

Depending on the value of θ, the total time could be longer or shorter, compared with the mutation-free case.

Similarly, by taking into account the sample sizes, we can introduce the new time measure

$$\tilde{T}_{n,\theta} = \sum_{k=1}^{n} k\tau_{k,\theta},$$

which is the total length of all the branches in the genealogy. The mean and variance of $\tilde{T}_{n,\theta}$ are given by

$$\begin{aligned}
\mathbb{E}[\tilde{T}_{n,\theta}] &= 2\sum_{k=1}^{n} \frac{1}{\theta+k-1}, \\
Var[\tilde{T}_{n-1}] &= 4\sum_{k=1}^{n} \frac{1}{(\theta+k-1)^2}.
\end{aligned}$$

4.3 The Pure-death Markov Chain

In this section, we give a rigorous construction of the process $\{\mathbf{D}_t : t \geq 0\}$ in Section 4.2, and derive some important properties that will be used later.

Let $\mathbb{N} = \{0,1,2,\ldots\}$ and $\bar{\mathbb{N}} = \mathbb{N} \bigcup \{\infty\}$. Set

$$\lambda_n = \frac{n(n-1+\theta)}{2}, \; n \geq 0,$$
$$\mathbf{C} = \{f \in C(\bar{\mathbb{N}}) : \lim_{n\to\infty} \lambda_n(f(n-1) - f(n)) \text{ exists and is finite}\},$$
$$\Omega f(n) = \begin{cases} \lambda_n(f(n-1) - f(n)), & n \in \mathbb{N} \\ \lim_{n\to\infty} \lambda_n(f(n-1) - f(n)), & n = \infty, \end{cases}$$

where the topology on $\bar{\mathbb{N}}$ is the one-point compactification of $\mathbb{N}$.

Theorem 4.2. *The linear operator Ω with domain $\mathbf{C}$ is the generator of a strongly continuous, conservative, contraction semigroup (Feller semigroup) $\{T(t)\}$ defined on the space $C(\bar{\mathbb{N}})$.*

Proof. Let

$$\mathbf{C}_0 = \{f \in C(\bar{\mathbb{N}}) : f(m) = f(\infty) \text{ for all sufficiently large enough } m\}.$$

Clearly $\mathbf{C}_0$ is a subset of $\mathbf{C}$. For every function g in $C(\bar{\mathbb{N}})$, one has that

$$\lim_{n\to\infty} g(n) = g(\infty).$$

For each $m \geq 1$, set

$$g_m(n) = \begin{cases} g(n), & n \leq m \\ g(m), & n > m. \end{cases}$$

Then

$$\sup_{n\in\bar{\mathbb{N}}} |g_m(n) - g(n)| \to 0, \; m \to \infty.$$

Hence $\mathbf{C}_0$, and thus $\mathbf{C}$ is dense in $C(\bar{\mathbb{N}})$.

For each $\lambda > 0$ and f in $\mathbf{C}$, denote $\lambda f - \Omega f$ by g. Next we show that

$$\lambda \|f\| \leq \|g\|, \tag{4.7}$$

where

$$\|f\| = \sup_{n\in\bar{\mathbb{N}}} |f(n)|.$$

By definition,

$$\lambda f(0) = g(0) \leq \|g\|.$$

Assume that for any $m \geq 0$,

$$\lambda |f(n)| \leq \|g\|, \; n \leq m.$$

If $f(m+1) \geq f(m)$, then $\|g\| \geq g(m+1) \geq \lambda f(m+1)$. The case of $f(m+1) < f(m)$ is clear. Hence (4.7) holds and the operator Ω is dissipative.

Finally we will show that for any g in $C(\bar{\mathbb{N}})$, and $\lambda > 0$, there exists an f in $\mathbf{C}$ such that

$$\lambda f - \Omega f = g.$$

This can be done recursively. First, let $f(0) = \lambda^{-1}g(0)$. Assuming that $f(n)$ has been defined, we then have

$$f(n+1) = (\lambda + \lambda_n)^{-1}(\lambda_n f(n) + g(n+1)).$$

It follows from (4.7) that for any $n \geq 1$

$$|f(n-1) - f(n)| \leq \lambda_n^{-1}(\|\lambda f\| + \|g\|) \leq 2\lambda_n^{-1}\|g\|. \tag{4.8}$$

Since $\sum_{n=1}^{\infty} \lambda_n^{-1} < \infty$, $\{f(n)\}$ is a Cauchy sequence. Rewriting $\Omega f(n)$ as $\lambda f(n) - g(n)$, reveals that the sequence $\{\Omega f(n)\}$ is also Cauchy. Hence $\Omega f(n)$ has a finite limits as n converges to infinity. Thus f is in $\mathbf{C}$.

Since Ω is clearly conservative, the theorem follows from the Hille–Yosida theorem.

□

The Markov process, associated with the semigroup $\{T(t)\}$ in Theorem 4.2, is just the pure-death Markov chain $\{\mathbf{D}_t : t \geq 0\}$.

Theorem 4.3. *For each n in $\mathbb{N}$ and $t > 0$, let*

$$d_n^{\theta}(t) = T(t)(\chi_{\{n\}})(\infty) = \mathbb{P}\{\mathbf{D}_t = n\}. \tag{4.9}$$

Then

$$d_n^{\theta}(t) = \begin{cases} 1 - \sum_{k=1}^{\infty} \frac{(2k-1+\theta)}{k!}(-1)^{k-1}\theta_{(k-1)}e^{-\lambda_k t}, & n = 0 \\ \sum_{k=n}^{\infty} \frac{(2k-1+\theta)}{k!}(-1)^{k-n}\binom{k}{n}(n+\theta)_{(k-1)}e^{-\lambda_k t}, & n > 0, \end{cases} \tag{4.10}$$

where the series are clearly absolutely convergent.

Proof. For any $m \geq 1$ and $t > 0$, define

$$d_{mn}^{\theta}(t) = T(t)(\chi_{\{n\}})(m). \tag{4.11}$$

Clearly $d_{mn}^{\theta}(t) = 0$ for $n > m$. Since $T(t)(\chi_{\{n\}})(\cdot)$ is in $C(\bar{\mathbb{N}})$, it follows that

$$\lim_{m \to \infty} d_{mn}^{\theta}(t) = d_n^{\theta}(t). \tag{4.12}$$

Next we fix $m \geq 1$ and set

$$\mathbf{Q} = \begin{pmatrix} -\lambda_0 & 0 & 0 & \cdots & 0 & 0 \\ \lambda_1 & -\lambda_1 & 0 & \cdots & 0 & 0 \\ \cdots & \cdots & \cdots & \cdots & \cdots & \cdots \\ 0 & 0 & 0 & \cdots & \lambda_m & -\lambda_m \end{pmatrix}.$$

The semigroup generated by $\mathbf{Q}$ is then the same as $T(t)$ restricted to

$$C(\{0,1,\ldots,m\}).$$

Let $\mathbf{T}(t) = (d_{ij}^{\theta}(t))_{0\le i,j\le m}$. Then one has

$$\mathbf{T}(t) = e^{\mathbf{Q}t}. \tag{4.13}$$

Next we look for the spectral representation of $\mathbf{Q}$. It is clear that the eigenvalues of $\mathbf{Q}$ are $-\lambda_0, -\lambda_1, \ldots, -\lambda_m$ and each has multiplicity one. Let $\mathbf{x} = (x_0,\ldots,x_m)$ be a left eigenvector corresponding to eigenvalue $-\lambda_i$. If $i = 0$, we can choose $\mathbf{x} = (1,0,\ldots,0)$. For $i \ge 1$, $\mathbf{x}$ solves the following system of equations:

$$\begin{aligned}
&(\lambda_1)x_1 = (-\lambda_i)x_0,\\
&(-\lambda_1)x_1 + \lambda_2 x_2 = (-\lambda_i)x_1,\\
&(-\lambda_k)x_k + \lambda_{k+1}x_{k+1} = (-\lambda_i)x_k,\ k \ge 2,
\end{aligned}$$

which implies

$$x_k = 0,\ k > i, \tag{4.14}$$

$$x_k = (-1)\frac{\lambda_{k+1}}{\lambda_i - \lambda_k}x_{k+1},\ k < i; \tag{4.15}$$

and by choosing $x_i = 1$, it follows that

$$\begin{aligned}
x_k &= (-1)^{i-k}\frac{\lambda_{k+1}\cdots\lambda_i}{(\lambda_i-\lambda_k)\cdots(\lambda_i-\lambda_{i-1})}\\
&= (-1)^{i-k}\binom{i}{k}\frac{(k+\theta)\cdots(i+\theta-1)}{(i+\theta+k-1)\cdots(i+\theta+i-2)}\\
&= (-1)^{i-k}\binom{i}{k}\frac{(k+\theta)_{(i-1)}}{(i+\theta)_{(i-1)}}.
\end{aligned} \tag{4.16}$$

Similarly, let $\mathbf{y} = (y_0,\ldots,y_m)'$ be a right eigenvector corresponding to the eigenvalue $-\lambda_i$. If $i = 0$, we can choose $\mathbf{y} = (1,\ldots,1)'$. For $i \ge 1$, $\mathbf{y}$ solves the following system of equations:

$$\begin{aligned}
&y_0 = 0,\\
&\lambda_k y_{k-1} - \lambda_k y_k = -\lambda_i y_k,\ k \ge 1,
\end{aligned}$$

which implies, by choosing $y_i = 1$, that

$$\begin{aligned}
&y_k = 0,\ k < i,\\
&y_k = \frac{\lambda_k\cdots\lambda_{i+1}}{(\lambda_k-\lambda_i)\cdots(\lambda_{i+1}-\lambda_i)} = \binom{k}{i}\frac{(i+\theta)_{(i)}}{(k+\theta)_{(i)}},\ k > i.
\end{aligned}$$

For each i in $\{0,1,\ldots,m\}$, let

$$u_{ij} = \begin{cases} \delta_{0j}, & i=0 \\ 0, & j>i>0 \\ (-1)^{i-j}\binom{i}{j}\frac{(j+\theta)_{(i-1)}}{(i+\theta)_{(i-1)}}, & j\le i,\ i>0, \end{cases}$$

and

$$v_{ij} = \begin{cases} 1, & i=0 \\ 0, & j<i \\ \binom{j}{i}\frac{(i+\theta)_{(i)}}{(j+\theta)_{(i)}}, & j\ge i>0, \end{cases}$$

$$\mathbf{u}_i = (u_{i0},\ldots,u_{im}),\ \mathbf{v}_i = \begin{pmatrix} v_{i0} \\ v_{i1} \\ \vdots \\ v_{im} \end{pmatrix},$$

$$\mathbf{U} = \begin{pmatrix} \mathbf{u}_0 \\ \mathbf{u}_1 \\ \vdots \\ \mathbf{u}_m \end{pmatrix},\ \mathbf{V} = (\mathbf{v}_0,\ldots,\mathbf{v}_m).$$

Then $\mathbf{u}_i$ and $\mathbf{v}_i$ are the respective left and right eigenvectors of $\mathbf{Q}$ corresponding to the eigenvalue $-\lambda_i$. By definition, for $i<j$, we have $\mathbf{u}_i\cdot\mathbf{v}_j = 0$ and $\mathbf{u}_i\cdot\mathbf{v}_i = 1$. For $i>j$,

$$\begin{aligned}\mathbf{u}_i\cdot\mathbf{v}_j &= \sum_{k=j}^{i} u_{ik}v_{jk} \\ &= C_{ij}A_{ij},\end{aligned}$$

where

$$\begin{aligned} C_{ij} &= \binom{i}{j}\frac{(\theta+j)_{(j)}}{(\theta+i)_{(i-1)}}, \\ A_{ij} &= \sum_{k=j}^{i}(-1)^{i-k}\binom{i-j}{i-k}\frac{(\theta+k)_{(i-1)}}{(\theta+k)_{(j)}}. \end{aligned}$$

Let $h(t) = (t-1)^{i-j}t^{\theta+i+j-2}$. Then it is easy to see that

$$A_{ij} = \frac{d^{i-j-1}h(t)}{dt^{i-j-1}}|_{t=1} = 0, \tag{4.17}$$

which implies that

$$\mathbf{u}_i\cdot\mathbf{v}_j = \delta_{ij}. \tag{4.18}$$

Let $\mathbf{I}$ be the $(m+1)\times(m+1)$ identity matrix and

$$\Lambda = \begin{pmatrix} -\lambda_0 & 0 & 0 & \cdots & 0 & 0 \\ 0 & -\lambda_1 & 0 & \cdots & 0 & 0 \\ \cdots & \cdots & \cdots & \cdots & \cdots & \cdots \\ 0 & 0 & 0 & \cdots & 0 & -\lambda_m \end{pmatrix}$$

and

$$e^{\Lambda t} = \begin{pmatrix} e^{-\lambda_0 t} & 0 & 0 & \cdots & 0 & 0 \\ 0 & e^{-\lambda_1 t} & 0 & \cdots & 0 & 0 \\ \cdots & \cdots & \cdots & \cdots & \cdots & \cdots \\ 0 & 0 & 0 & \cdots & 0 & e^{-\lambda_m t} \end{pmatrix}.$$

It follows from (4.18) that

$$\mathbf{UV} = \mathbf{VU} = \mathbf{I},$$
$$\mathbf{UQV} = \Lambda,$$
$$\mathbf{Q} = \mathbf{V}\Lambda\mathbf{U}.$$

These combined with (4.13) imply that

$$\mathbf{T}(t) = \mathbf{V}e^{\Lambda t}\mathbf{U}. \tag{4.19}$$

Hence

$$\begin{aligned} d^{\theta}_{mn}(t) &= \sum_{k=0}^{m} v_{km}u_{kn}e^{-\lambda_k t} \\ &= \begin{cases} 1+\sum_{k=1}^{m}(-1)^k\binom{m}{k}\frac{(\theta+k)_{(k)}}{(\theta+m)_{(k)}}\frac{(\theta)_{(k-1)}}{(\theta+k)_{(k-1)}}e^{-\lambda_k t}, & n=0 \\ \sum_{k=n}^{m}(-1)^{k-n}\binom{m}{k}\binom{k}{n}\frac{(\theta+k)_{(k)}}{(\theta+m)_{(k)}}\frac{(\theta+n)_{(k-1)}}{(\theta+k)_{(k-1)}}e^{-\lambda_k t}, & n\geq 1 \end{cases} \\ &= \begin{cases} 1+\sum_{k=1}^{m}\frac{(m)_{[k]}}{(\theta+m)_{(k)}}(-1)^k\frac{2k-1+\theta}{k!}(\theta+n)_{(k-1)}e^{-\lambda_k t}, & n=0 \\ \sum_{k=n}^{m}\frac{(m)_{[k]}}{(\theta+m)_{(k)}}(-1)^{k-n}\frac{2k-1+\theta}{k!}\binom{k}{n}(\theta+n)_{(k-1)}e^{-\lambda_k t}, & n\geq 1. \end{cases} \end{aligned} \tag{4.20}$$

which, by letting m approach infinity, implies (4.10). □

Corollary 4.4 *For any $t>0$ and any $r\geq 1$,*

$$\mathbb{P}\{\mathbf{D}_t \in \mathbb{N}\} = 1, \tag{4.21}$$

and

$$\mathbb{E}[\mathbf{D}^r(t)] < \infty. \tag{4.22}$$

Proof. Fix $t > 0$ and $r \geq 1$. It follows from Theorem 4.3 that

$$\begin{aligned}
\mathbb{P}\{\mathbf{D}_t \in \mathbb{N}\} &= \sum_{n=0}^{\infty} d_n^\theta(t) \\
&= \sum_{n=0}^{\infty}\sum_{k=n}^{\infty} \frac{(2k-1+\theta)}{k!}(-1)^{k-n}\binom{k}{n}(n+\theta)_{(k-1)}e^{-\lambda_k t} \\
&= \sum_{k=0}^{\infty}\sum_{n=0}^{k} \frac{(2k-1+\theta)}{k!}(-1)^{k-n}\binom{k}{n}(n+\theta)_{(k-1)}e^{-\lambda_k t} \\
&= 1 + \sum_{k=0}^{\infty}\left[\sum_{n=0}^{k}(-1)^{k-n}\binom{k}{n}(n+\theta)_{(k-1)}\right]\frac{(2k-1+\theta)}{k!}e^{-\lambda_k t} \\
&= 1 + \sum_{k=0}^{\infty} A_{k0}\frac{(2k-1+\theta)}{k!}e^{-\lambda_k t},
\end{aligned}$$

where the interchange of summation is justified by the absolute convergence of the series. The result (4.21) now follows from (4.17).
From (4.21), we have

$$\mathbb{E}[\mathbf{D}^r(t)] = \sum_{n=1}^{\infty} n^r d_n^\theta(t). \tag{4.23}$$

For any $n \geq 1$ and $k \geq n$, let

$$B_{nk} = \frac{(2k-1+\theta)}{k!}\binom{k}{n}(n+\theta)_{(k-1)}e^{-\lambda_k t}.$$

By direct calculation, there exists $k_0 \geq 1$ such that

$$\begin{aligned}
\frac{B_{n(k+1)}}{B_{nk}} &= e^{-(\lambda_{k+1}-\lambda_k)t}\,\frac{2k+1+\theta}{2k-1+\theta}\,\frac{n+\theta+k-1}{k+1-n} \\
&\leq (2k+1+\theta)e^{-kt} < \frac{1}{2},\ k \geq k_0,
\end{aligned}$$

which, combined with (4.10), implies that for $n \geq k_0$

$$d_n^\theta(t) \leq \sum_{k=n}^{\infty} B_{nk} \leq B_{nn}. \tag{4.24}$$

Set

$$E_n = n^r B_{nn} = \frac{n^r}{n!}(n+\theta)_{(n)}e^{-\lambda_n t}.$$

It follows by direct calculation that for n large enough, we have

$$
\begin{aligned}
\frac{E_{n+1}}{E_n} &= e^{-(\lambda_{n+1}-\lambda_n)}\left(1+\frac{1}{n}\right)^r \frac{(2n+\theta+1)(2n+\theta)}{(n+1)(n+\theta)} \\
&\le (2+\theta)^{r+2} e^{-(\lambda_{n+1}-\lambda_n)} < 1.
\end{aligned}
$$

Hence

$$
\sum_{n=1}^{\infty} E_n < \infty
$$

which, combined with (4.23) and (4.24), implies (4.21).

□

4.4 Notes

The coalescent was introduced by Kingman in [128] and [129]. The study on the theory of lines of descent in Griffiths [93] has the coalescent's ingredients. In Watterson [184], the coalescent is analyzed in the context of infinite alleles models. Further studies on the coalescent for the infinite alleles model are found in [42], [43], and [36]. Other developments of coalescent theory include Hudson [105], Tajima [169], Tavaré [170], and Neuhauser and Krone [140].

In recent years, coalescent processes allowing multiple collisions have been studied extensively. For details, see [153], [158], [161], [138], [27], and the references therein. A more comprehensive account of coalescent theory can be found in a recent book by Wakeley [178].

The treatments in Sections 4.1 and 4.2 follow Kingman [128]. The result in Theorem 4.3 is due to Tavaré [170] and a more detailed version of his proof is provided here. Theorem 4.2 and Corollary 4.4 are based on results in Ethier and Griffiths [59].

Chapter 5
Stochastic Dynamics

The backward-looking analysis presented in the coalescent represents a major part in the current theory of population genetics. Traditionally, population genetics focuses mainly on the impact of various evolutionary forces on future generations of the population. In this chapter, we turn to this forward-looking viewpoint and study two stochastic models: the infinitely-many-neutral-alleles model, and the Fleming–Viot process with parent-independent mutation. The evolutionary forces in both models are mutation and random sampling. We establish the reversibility of both models, with the Poisson–Dirichlet distribution and the Dirichlet process as the respective reversible measures. Explicit representations involving the coalescent, are obtained for the transition probability functions for each of the two models. These reflect the natural relation between the forward-looking and backward-looking viewpoints. The two-parameter counterpart of the infinitely-many-neutral-alleles model is discussed briefly, and the relation between the Fleming–Viot process and a continuous branching process with immigration is investigated. The latter can be viewed as a dynamical analog of the beta–gamma relation, derived in Chapter 1. For basic terms and results on semigroup and generators, we refer to [62].

5.1 Infinitely-many-neutral-alleles Model

For each $K \geq 2$, let

$$\Delta_K = \left\{ (x_1, \ldots, x_K) \in [0,1]^K : \sum_{i=1}^{K} x_i = 1 \right\}. \tag{5.1}$$

Recall from Chapter 1 that the K-allele Wright–Fisher diffusion with symmetric mutation is a Δ_K-valued diffusion process $\mathbf{x}_K(t) = (x_1(t), \ldots, x_K(t))$ generated by

S. Feng, *The Poisson–Dirichlet Distribution and Related Topics*,
Probability and its Applications, DOI 10.1007/978-3-642-11194-5_5,

$$L_K f(\mathbf{x}) = \frac{1}{2}\sum_{i,j=1}^{K} a_{ij}(\mathbf{x})\frac{\partial^2 f}{\partial x_i \partial x_j} + \sum_{i=1}^{K} b_i(\mathbf{x})\frac{\partial f}{\partial x_i},$$

$$a_{ij}(\mathbf{x}) = x_i(\delta_{ij} - x_j),\ b_i(\mathbf{x}) = \frac{K\theta}{2(K-1)}\left(\frac{1}{K} - x_i\right),$$

with domain

$$\mathscr{D}(L_K) = \{f \in C(\Delta_K) : f = g|_{\Delta_K} \text{ for some } g \in C^2(\mathbb{R}^K)\}.$$

The operator L_K is closable in $C(\Delta_K)$, and for simplicity, L_K will also be used to denote the closure. This process has many interesting properties. In particular, we will show in the next theorem that it is reversible with reversible measure $\Pi^K = Dirichlet(\frac{\theta}{K-1},\ldots,\frac{\theta}{K-1})$.

Theorem 5.1. *The diffusion process $\mathbf{x}_K(t) = (x_1(t),\ldots,x_K(t))$ is reversible and the reversible measure is Π^K; i.e., for any f,g in $\mathscr{D}(L_K)$,*

$$\int_{\Delta_K} f L_K g \, d\Pi^K = \int_{\Delta_K} g L_K f \, d\Pi^K. \tag{5.2}$$

Proof. For any $K \geq 1$, let $\mathbb{N}^K = \mathbb{N} \times \cdots \times \mathbb{N}$. The generic element of $\mathbb{N}$ is denoted by $\mathbf{n} = (n_1,\ldots,n_K)$. Define

$$\mathscr{D}_0(L_K) = \{f_{\mathbf{n}} : f_{\mathbf{n}}(\mathbf{x}) = \mathbf{x}^{\mathbf{n}} = x_1^{n_1}\cdots x_K^{n_K}, \mathbf{x} \in \Delta_K, \mathbf{n} \in \mathbb{N}^K\},$$

where $f_{\mathbf{0}} \equiv 1$. Clearly $\mathscr{D}_0(L_K)$ is a subset of $\mathscr{D}(L_K)$. Furthermore, it is a core for L_K. Thus by linearity, it suffices to verify (5.2) for any f,g in $\mathscr{D}_0(L_K)$. For any $\mathbf{n},\mathbf{m}$ in $\mathbb{N}^K$, choose $f = f_{\mathbf{n}}, g = f_{\mathbf{m}}$ in (5.2) and set

$$|\mathbf{n}| = n_1 + \cdots + n_K,\ |\mathbf{m}| = m_1 + \cdots + m_K,$$
$$e_i = (0,\ldots,\underset{i\text{th}}{1},\ldots,0) \in \Delta_K,\ i = 1,\ldots,K.$$

Clearly (5.2) holds for $\mathbf{n} = \mathbf{m} = \mathbf{0}$. Assume that $\mathbf{m} \neq \mathbf{0}$ and set $a = \frac{\theta}{K-1}$. Then

$$\begin{aligned} L_K f_{\mathbf{m}}(\mathbf{x}) &= \frac{1}{2}\sum_{i=1}^{K} m_i(m_i-1) f_{\mathbf{m}-e_i}(\mathbf{x}) - \frac{1}{2}[|\mathbf{m}|^2 - |\mathbf{m}|] f_{\mathbf{m}}(\mathbf{x}) \\ &+ \frac{a}{2}\sum_{i=1}^{K} m_i f_{\mathbf{m}-e_i}(\mathbf{x}) - \frac{Ka}{2}|\mathbf{m}| f_{\mathbf{m}}(\mathbf{x}) \\ &= \frac{1}{2}\sum_{i=1}^{K} m_i(m_i+a-1) f_{\mathbf{m}-e_i}(\mathbf{x}) - \frac{1}{2}|\mathbf{m}|(|\mathbf{m}|+Ka-1) f_{\mathbf{m}}(\mathbf{x}), \end{aligned}$$

and

$$\begin{aligned}\int_{\Delta_K} f L_K g \, d\Pi^K &= \int_{\Delta_K} \left\{ \frac{1}{2} \sum_{i=1}^{K} m_i(m_i + a - 1) f_{\mathbf{m}+\mathbf{n}-e_i}(\mathbf{x}) \right. \\ &\quad \left. - \frac{1}{2} |\mathbf{m}|(|\mathbf{m}| + Ka - 1) f_{\mathbf{m}+\mathbf{n}}(\mathbf{x}) \right\} d\Pi^K \\ &= \frac{1}{2} \left\{ \sum_{i=1}^{K} \frac{m_i(m_i + a - 1)}{m_i + n_i + \frac{\theta}{K-1} - 1} - \frac{|\mathbf{m}|(|\mathbf{m}| + Ka - 1)}{|\mathbf{m}| + |\mathbf{n}| + \frac{K\theta}{K-1} - 1} \right\} \\ &\quad \times \frac{\Gamma(m_1 + n_1 + a) \cdots \Gamma(m_K + n_K + a)}{\Gamma(|\mathbf{m}| + |\mathbf{n}| + Ka - 1)} \frac{\Gamma(Ka)}{\Gamma(a) \cdots \Gamma(a)} \\ &= \frac{1}{2} \left\{ \frac{|\mathbf{m}||\mathbf{n}|}{|\mathbf{m}| + |\mathbf{n}| + \frac{K\theta}{K-1} - 1} - \sum_{i=1}^{K} \frac{m_i n_i}{m_i + n_i + \frac{\theta}{K-1} - 1} \right\} \\ &\quad \times \frac{\Gamma(m_1 + n_1 + a) \cdots \Gamma(m_K + n_K + a)}{\Gamma(|\mathbf{m}| + |\mathbf{n}| + Ka - 1)} \frac{\Gamma(Ka)}{\Gamma(a) \cdots \Gamma(a)},\end{aligned}$$

which is symmetric with respect to **n** and **m**. Hence we have (5.2) and the theorem. □

Let ∇_K, ∇_∞ and ∇ be defined as in (2.1). For notational convenience, we use $\mathbf{x}, \mathbf{y}$, and $\mathbf{z}$ to denote the generic elements of $\Delta_K, \nabla_K, \nabla_\infty$, and ∇ in the sequel. Set $\sigma_K(\mathbf{x}) = (x_{(1)}, \ldots, x_{(K)})$, the decreasing order statistics of $\mathbf{x}$. Then clearly $\nabla_K = \sigma_K(\Delta_K)$.

When K approaches infinity, Ethier and Kurtz [61] show that $\sigma_K(\mathbf{x}_K(t))$ converges in distribution to an infinite dimensional diffusion process $\mathbf{x}(t)$ in ∇ with generator

$$Lf(\mathbf{x}) = \frac{1}{2} \sum_{i,j=1}^{\infty} a_{ij}(\mathbf{x}) \frac{\partial^2 f}{\partial x_i \partial x_j} - \frac{\theta}{2} \sum_{i=1}^{\infty} x_i \frac{\partial f}{\partial x_i}. \tag{5.3}$$

The domain of the generator L is

$$\mathscr{D}(L) = span\{1, \varphi_2, \varphi_3, \ldots\} \subset C(\nabla), \tag{5.4}$$

where

$$\varphi_n(\mathbf{x}) = \sum_{i=1}^{\infty} x_i^n, \tag{5.5}$$

is defined on ∇_∞ and extends continuously to ∇. The process $\mathbf{x}(t)$ is called the *infinitely-many-neutral-alleles model*. It can be shown that starting from any point in ∇, the process $\mathbf{x}(t)$ stays in ∇_∞ for positive t with probability one.

Lemma 5.1. *For each $r \geq 1$, and $\mathbf{n} = (n_1, \ldots n_r)$, let*

$$g_{\mathbf{n}}(\mathbf{x}) = \varphi_{n_1}(\mathbf{x}) \cdots \varphi_{n_r}(\mathbf{x}),$$

and $g_{\mathbf{n}}^l(\mathbf{x})$ denote $g_{\mathbf{n}}(\mathbf{x})$ without the factor φ_{n_l}. Similarly $g_{\mathbf{n}}(\mathbf{x})$ without factors $\varphi_{n_l}, \varphi_{n_m}$ is denoted by $g_{\mathbf{n}}^{lm}(\mathbf{x})$. Then

$$Lg_{\mathbf{n}} = \sum_{l=1}^{r} \binom{n_l}{2} \varphi_{n_l-1} g_{\mathbf{n}}^l + \sum_{1 \le l < m \le r} n_l n_m \varphi_{n_l+n_m-1} g_{\mathbf{n}}^{lm} - \frac{|\mathbf{n}|}{2}(|\mathbf{n}| + \theta - 1) g_{\mathbf{n}}, \tag{5.6}$$

where $\varphi_1 \equiv 1$.

Proof. Write

$$L = L^1 - L^2 - L^3,$$

with

$$\begin{aligned} L^1 &= \frac{1}{2} \sum_{i=1}^{\infty} x_i \frac{\partial^2}{\partial x_i^2}, \\ L^2 &= \frac{1}{2} \sum_{i,j=1}^{\infty} x_i x_j \frac{\partial^2 f}{\partial x_i \partial x_j}, \\ L^3 &= \frac{\theta}{2} \sum_{i=1}^{\infty} x_i \frac{\partial}{\partial x_i}. \end{aligned}$$

By direct calculation,

$$\begin{aligned} \frac{\partial g_{\mathbf{n}}}{\partial x_i} &= \sum_{l=1}^{r} g_{\mathbf{n}}^l \frac{\partial \varphi_{n_l}}{\partial x_i}, \\ \frac{\partial^2 g_{\mathbf{n}}}{\partial x_i \partial x_j} &= \sum_{l=1}^{r} g_{\mathbf{n}}^l \frac{\partial^2 \varphi_{n_l}}{\partial x_i^2} \\ &\quad + \sum_{1 \le l \ne m \le r} g_{\mathbf{n}}^{lm} \frac{\partial \varphi_{n_l}}{\partial x_i} \frac{\partial \varphi_{n_m}}{\partial x_j} \end{aligned}$$

which implies

$$L^1 g_{\mathbf{n}} = \sum_{l=1}^{r} \binom{n_l}{2} \varphi_{n_l-1} g_{\mathbf{n}}^l + \frac{1}{2} \sum_{1 \le l \ne m \le r} n_l n_m \varphi_{n_l+n_m-1} g_{\mathbf{n}}^{lm}, \tag{5.7}$$

$$L^2 g_{\mathbf{n}} = \left[\sum_{l=1}^{r} \binom{n_l}{2} + \frac{1}{2} \sum_{1 \le l \ne m \le r} \right] g_{\mathbf{n}} = \binom{|\mathbf{n}|}{2} g_{\mathbf{n}}, \tag{5.8}$$

$$L^3 g_{\mathbf{n}} = \frac{\theta |\mathbf{n}|}{2} g_{\mathbf{n}}. \tag{5.9}$$

The equality (5.6) now follows from (5.7)–(5.8).

□

Theorem 5.2. *The Poisson–Dirichlet distribution* Π_θ *with parameter* θ *is the unique stationary and reversible distribution of the diffusion process* $\mathbf{x}(\cdot)$.

Proof. We first show that Π_θ is stationary and reversible for $\mathbf{x}(\cdot)$. Set

$$\rho_K : \nabla_K \to \nabla_\infty, (x_1,\ldots,x_K) \to (x_1,\ldots,x_K,0,\ldots).$$

Let $\mathbf{X}_K = (X_1,\ldots,X_k)$ have the *Dirichlet*$(\frac{\theta}{K-1},\ldots,\frac{\theta}{K-1})$ distribution and $\Pi^{\theta,K}$ denote the law of $\rho_K(\sigma_K(\mathbf{X}_K))$. Then it follows from Theorem 2.1 that the sequence $\Pi^{\theta,K}$ converges weakly to Π_θ. For any f in $C(\nabla)$, let $\pi_K(f) = f \circ \rho_K \circ \sigma_K$. Then π_K maps $\mathscr{D}(L)$ into $\mathscr{D}(L_K)$. For any f,g in $\mathscr{D}(L)$, by direct calculation,

$$\begin{aligned}
\int_\nabla fLg\,d\Pi_\theta &= \lim_{K\to\infty}\int_{\nabla_\infty} fLg\,d\Pi^{\theta,K}\\
&= \lim_{K\to\infty}\int_{\Delta_K} \pi_K(f)\pi_K(Lg)\,d\Pi^{\theta,K}\\
&= \lim_{K\to\infty}\int_{\Delta_K} \pi_K(f)L_K\pi_K(g)\,d\Pi^{\theta,K}\\
&= \lim_{K\to\infty}\int_{\Delta_K} \pi_K(g)L_K\pi_K(f)\,d\Pi^{\theta,K}\\
&= \int_\nabla gLf\,d\Pi_\theta,
\end{aligned}$$

where the fourth equality follows from Theorem 5.1.

Assume that Π is another stationary distribution. Then for any $r \geq 1$, and any $\mathbf{n} = (n_1,\ldots,n_r)$ satisfying $n_i \geq 2$,

$$\int_\nabla Lg_{\mathbf{n}}\,d\Pi = 0. \tag{5.10}$$

This, combined with (5.6), implies that $\mathbb{E}^\Pi[g_{\mathbf{n}}]$ can be calculated from $g_{\mathbf{m}}$ such that $|\mathbf{m}| < |\mathbf{n}|$. It follows from (5.6) and (5.10) that

$$L\varphi_2 = 1-(1+\theta)\varphi_2$$

and

$$\mathbb{E}^\Pi[\varphi_2] = \frac{1}{\theta+1}.$$

Induction on $\mathbf{n}$ shows that $\mathbb{E}^\Pi[g_{\mathbf{n}}] = \mathbb{E}^{\Pi_\theta}[g_{\mathbf{n}}]$ for all $\mathbf{n}$. Hence $\Pi = \Pi_\theta$.

□

5.2 A Fleming–Viot Process

In this section, we will study a labeled version of the infinitely-many-neutral-alleles model. This process turns out to be a special Fleming–Viot (FV) process.

The FV process is a probability-valued stochastic process describing the evolution of the distribution of genotypes in a population under the influence of mutation and sampling with replacement.

Let S be a compact metric space, $C(S)$ the set of continuous functions on S, and $M_1(S)$ the space of all probability measures on S equipped with the weak topology. For each μ in $M_1(S)$ and ϕ in $C(S)$, let $\langle \mu, \phi \rangle = \int_S \phi(x)\mu(dx)$.

Let A be the generator of a Markov process on S with domain $D(A)$. Define

$$\begin{aligned}\mathscr{D} = \{F : F(\mu) = f(\langle \mu, \phi_1 \rangle, \ldots, \langle \mu, \phi_k \rangle), f \in C_b^2(\mathbb{R}^k), \phi_i \in D(A),\\ k = 1, 2, \ldots; i = 1, \ldots, k, \mu \in M_1(S)\},\end{aligned} \tag{5.11}$$

where $C_b^2(\mathbb{R})$ denotes the set of all functions on $\mathbb{R}$ with bounded continuous second order derivatives. Then the generator of the FV process with sampling rate 1 has the form

$$\begin{aligned}\mathscr{L}F(\mu) &= \int_S \left(A \frac{\delta F(\mu)}{\delta \mu(x)} \right) \mu(dx)\\ &\quad + \frac{1}{2} \int_S \int_S \left(\frac{\delta^2 F(\mu)}{\delta \mu(x) \delta \mu(y)} \right) Q(\mu; dx, dy)\end{aligned} \tag{5.12}$$

where

$$\frac{\delta F(\mu)}{\delta \mu(x)} = \lim_{\varepsilon \to 0+} \frac{F(\mu + \varepsilon \delta_x) - F(\mu)}{\varepsilon}, \tag{5.13}$$

$$\frac{\delta^2 F(\mu)}{\delta \mu(x) \delta \mu(y)} = \lim_{\varepsilon_1 \to 0+, \varepsilon_2 \to 0+} \frac{F(\mu + \varepsilon_1 \delta_x + \varepsilon_2 \delta_y) - F(\mu)}{\varepsilon_1 \varepsilon_2}, \tag{5.14}$$

$$Q(\mu; dx, dy) = \mu(dx)\delta_x(dy) - \mu(dx)\mu(dy),$$

and δ_x stands for the Dirac measure at $x \in S$. For $F(\mu) = f(\langle \mu, \phi \rangle)$, one has

$$\frac{\delta F(\mu)}{\delta \mu(x)} = f'(\langle \mu, \phi \rangle)\phi(x),$$

$$\frac{\delta^2 F(\mu)}{\delta \mu(x) \delta \mu(y)} = f''(\langle \mu, \phi \rangle)\phi(x)\phi(y).$$

Nonprobability measures are used in the definition of $\delta F(\mu)/\delta \mu(x)$ even though F is only defined on $M_1(S)$. But F is clearly well defined for any finite measure. Another way to define the derivative is to use a convex combination and let

$$\frac{\delta F(\mu)}{\delta \mu(x)} = \lim_{\varepsilon \to 0+} \frac{F((1-\varepsilon)\mu + \varepsilon \delta_x) - F(\mu)}{\varepsilon}.$$

This only uses F on $M_1(E)$. Although it will result in a different derivative, it will not alter $\mathscr{L}F(\mu)$. For this reason, we will stick with (5.13) and (5.14).

The domain of $\mathscr{L}$ is $\mathscr{D}$. The space S is called the type space or the space of alleles, A is known as the mutation operator, and the last term describes the *continuous random sampling*.

Let

$$\mathscr{M} = \{F_{\phi_1,\ldots,\phi_n} : F_{\phi_1,\ldots,\phi_n}(\mu) = \langle \mu, \phi_1 \rangle \cdots \langle \mu, \phi_n \rangle, \\ \phi_i \in D(A), 1 \leq i \leq n, n = 1, 2, \ldots\}. \tag{5.15}$$

Each element in $\mathscr{M}$ is called a *monomial* and n is the order of monomial $F_{\phi_1,\ldots,\phi_n}$. $\mathscr{M}$ is a core for $\mathscr{L}$.

Let ν_0 be a diffuse probability measure in $M_1(S)$, and let Π_{θ,ν_0} be the labeled Poisson–Dirichlet distribution or Dirichlet process. By introducing labels to the infinitely-many-neutral-alleles model, we get an FV process with mutation operator

$$A\phi(x) = \frac{\theta}{2} \int_S (\phi(y) - \phi(x)) \nu_0(dy), \ D(A) = C(S). \tag{5.16}$$

We will call this process an FV process with the *parent-independent mutation.*

For any $n \geq 1, \ 1 \leq k \leq n$, define

$$\sigma(n,k) = \{\mathbf{c} = (c_1, \ldots, c_k) : \min c_1 < \cdots < \min c_k\}, \tag{5.17}$$

where each $\mathbf{c}$ is a partition of $\{1, \ldots, n\}$ with k non-empty sets. The number of elements in c_i is denoted by $|c_i|, \ i = 1, \ldots, k$.

Lemma 5.2. *For any $n \geq 1$, and $\phi_1, \ldots, \phi_n$ in $C(S)$,*

$$\int_{M_1(S)} \langle \mu, \phi_1 \rangle \cdots \langle \mu, \phi_n \rangle \Pi_{\theta,\nu_0}(d\mu) \tag{5.18}$$
$$= \sum_{k=1}^{n} \sum_{\mathbf{c} \in \sigma(n,k)} (|c_1| - 1)! \cdots (|c_k| - 1)! \frac{\theta^k}{\theta_{(n)}} \prod_{i=1}^{k} \left\langle \nu_0, \prod_{j \in c_i} \phi_j \right\rangle.$$

Proof. Let $(Y_1, Y_2, \ldots)$ have law Π_θ and $\xi_1, \xi_2, \ldots,$ be independent and identically distributed with common distribution ν_0. Then by direct calculation,

$$\int_{M_1(S)} \langle \mu, \phi_1 \rangle \cdots \langle \mu, \phi_n \rangle \Pi_{\theta,\nu_0}(d\mu) \tag{5.19}$$
$$= \sum_{i_1,\ldots,i_n=1}^{\infty} \mathbb{E}\left[\prod_{l=1}^{n} Y_{i_l} \phi_l(\xi_{i_l})\right]$$
$$= \sum_{k=1}^{n} \sum_{\mathbf{c} \in \sigma(n,k)} \prod_{i=1}^{k} \langle \nu_0, \prod_{j \in c_i} \phi_j \rangle \sum_{n_1,\ldots,n_k \text{ distinct}} \mathbb{E}[Y_{n_1}^{|c_1|} \cdots Y_{n_k}^{|c_k|}].$$

It follows from the Poisson process representation of the Poisson–Dirichlet distribution in Section 2.1, that

$$\sum_{n_1,\dots,n_k \text{ distinct}} \mathbb{E}[Y_{n_1}^{|c_1|}\cdots Y_{n_k}^{|c_k|}]$$

$$= (\mathbb{E}[\tau_\theta^n])^{-1} \sum_{n_1,\dots,n_k \text{ distinct}} \mathbb{E}[X_{n_1}^{|c_1|}\cdots X_{n_k}^{|c_k|}], \tag{5.20}$$

$$= \frac{\Gamma(\theta)}{\Gamma(\theta+n)} \prod_{i=1}^{k} \mathbb{E}[\sum_{j=1}^{\infty} X_j^{|c_i|}]$$

$$= (|c_1|-1)!\cdots(|c_k|-1)!\frac{\theta^k}{\theta_{(n)}},$$

where $\{X_i : i = 1,\dots\}$ is the Poisson process with intensity measure $\theta x^{-1}e^{-x}dx$, and the last equality follows from equation (A.2). The result (5.18) now follows from (5.19) and (5.20).

□

Theorem 5.3. *The FV process with parent-independent mutation has a unique stationary distribution Π_{θ,ν_0}.*

Proof. Fix a $F_{\phi_1,\dots,\phi_n}$ in $\mathscr{M}$. Write

$$F^i_{\phi_1,\dots,\phi_n}(\mu) = \prod_{l\neq i}\langle \mu,\phi_l\rangle, \tag{5.21}$$

$$F^{ij}_{\phi_1,\dots,\phi_n}(\mu) = \prod_{l\neq i,j}\langle \mu,\phi_l\rangle, \tag{5.22}$$

$$F^{(i,j)}_{\phi_1,\dots,\phi_n}(\mu) = \langle \mu,\phi_i\phi_j\rangle F^{ij}_{\phi_1,\dots,\phi_n}(\mu). \tag{5.23}$$

Then by direct calculation

$$\frac{\delta F_{\phi_1,\dots,\phi_n}(\mu)}{\delta\mu(x)} = \sum_{i=1}^{n}\phi_i(x)F^i_{\phi_1,\dots,\phi_n}, \tag{5.24}$$

$$\frac{\delta^2 F_{\phi_1,\dots,\phi_n}(\mu)}{\delta\mu(x)\delta\mu(y)} = \sum_{1\le i\neq j\le n}\phi_i(x)\phi_j(y)F^{ij}_{\phi_1,\dots,\phi_n}, \tag{5.25}$$

and

$$\mathscr{L}F_{\phi_1,\dots,\phi_n}(\mu) = \sum_{1\le i<j\le n} F^{(i,j)}_{\phi_1,\dots,\phi_n}(\mu) + \frac{\theta}{2}\sum_{i=1}^{n}\langle \nu_0,\phi_i\rangle F^i_{\phi_1,\dots,\phi_n}(\mu) - \frac{n(n-1+\theta)}{2}F_{\phi_1,\dots,\phi_n}(\mu). \tag{5.26}$$

If Π is a stationary distribution, then

$$\int_{M_1(S)} \mathscr{L}F_{\phi_1,\dots,\phi_n}(\mu)\Pi(d\mu) = 0,$$

or equivalently

$$\int_{M_1(S)}\left\{\sum_{1\le i<j\le n} F^{(i,j)}_{\phi_1,\ldots,\phi_n}(\mu)+\frac{\theta}{2}\sum_{i=1}^{n}\langle\nu_0,\phi_i\rangle F^{i}_{\phi_1,\ldots,\phi_n}(\mu)\right\}\Pi(d\mu)$$
$$=\frac{n(n-1+\theta)}{2}\int_{M_1(S)} F_{\phi_1,\ldots,\phi_n}(\mu)\Pi(d\mu).$$

Thus $\int_{M_1(S)} F_{\phi_1,\ldots,\phi_n}(\mu)\Pi(d\mu)$ is determined recursively from the moments of the lower order monomials. For $n=1$, we have

$$\int_{M_1(S)} F_{\phi_1}(\mu)\Pi(d\mu)=\langle\nu_0,\phi_1\rangle=\int_{M_1(S)} F_{\phi_1}(\mu)\Pi_{\theta,\nu_0}(d\mu),$$

which implies that $\Pi=\Pi_{\theta,\nu_0}$. This gives uniqueness. Next we show that Π_{θ,ν_0} is indeed a stationary distribution.

It follows from Lemma 5.2 that

$$\int_{M_1(S)}\frac{n(n-1+\theta)}{2}F_{\phi_1,\ldots,\phi_n}(\mu)\Pi_{\theta,\nu_0}(d\mu) \tag{5.27}$$
$$=\frac{n(n-1+\theta)}{2}\sum_{k=1}^{n}\sum_{\mathbf{c}\in\sigma(n,k)}(|c_1|-1)!\cdots(|c_k|-1)!\frac{\theta^k}{\theta_{(n)}}\prod_{i=1}^{k}\left\langle\nu_0,\prod_{j\in c_i}\phi_j\right\rangle,$$

$$\int_{M_1(S)} F^{(i,j)}_{\phi_1,\ldots,\phi_n}(\mu)\Pi_{\theta,\nu_0}(d\mu) \tag{5.28}$$
$$=\sum_{k=1}^{n-1}\sum_{\mathbf{b}\in\sigma(n-1,k)}(|b_1|-1)!\cdots(|b_k|-1)!\frac{\theta^k}{\theta_{(n-1)}}\prod_{r=1}^{k}\left\langle\nu_0,\prod_{l\in b_r}h_l\right\rangle$$

$$\langle\nu_0,\phi_i\rangle\int_{M_1(S)} F^{i}_{\phi_1,\ldots,\phi_n}(\mu)\Pi_{\theta,\nu_0}(d\mu) \tag{5.29}$$
$$=\sum_{k=1}^{n-1}\sum_{\mathbf{d}\in\sigma(n-1,k)}(|d_1|-1)!\cdots(|d_k|-1)!\frac{\theta^k}{\theta_{(n-1)}}\langle\nu_0,\phi_i\rangle\prod_{r=1}^{k}\left\langle\nu_0,\prod_{l\in d_r}h_l\right\rangle,$$

where for every $\mathbf{b}$ in $\sigma(n-1,k)$, the term $\prod_{r=1}^{k}\langle\nu_0,\prod_{l\in b_r}h_l\rangle$ in (5.28) corresponds to $\prod_{r=1}^{k}\langle\nu_0,\prod_{l\in c_r}\phi_l\rangle$ with $\mathbf{c}$ in $\sigma(n,k)$ and certain $|c_l|\ge 2$; similarly, the term $\langle\nu_0,\phi_i\rangle\prod_{r=1}^{k}\langle\nu_0,\prod_{l\in d_r}h_l\rangle$ in (5.29), corresponds to a term $\prod_{r=1}^{k+1}\langle\nu_0,\prod_{l\in c_r}\phi_l\rangle$ with $\mathbf{c}$ in $\sigma(n,k)$ and certain $|c_l|=1$. Thus we can write

$$\int_{M_1(S)}\left\{\sum_{1\le i<j\le n} F^{(i,j)}_{\phi_1,\ldots,\phi_n}(\mu)+\frac{\theta}{2}\sum_{i=1}^{n}\langle\nu_0,\phi_i\rangle F^{i}_{\phi_1,\ldots,\phi_n}(\mu)\right\}\Pi_{\theta,\nu_0}(d\mu)$$
$$=\sum_{k=1}^{n}\sum_{\mathbf{c}\in\sigma(n,k)}B_{\mathbf{c}}\prod_{i=1}^{k}\left\langle\nu_0,\prod_{j\in c_i}\phi_j\right\rangle.$$

It remains to make explicit the expression of $B_{\mathbf{c}}$. In order to get contributions from

$$\int_{M_1(S)} \sum_{1\le i<j\le n} F^{(i,j)}_{\phi_1,\ldots,\phi_n}(\mu)\Pi_{\theta,\nu_0}(d\mu),$$

there is at least one $1 \le i \le k$ such that $|c_i| \ge 2$. For each i satisfying $|c_i| \ge 2$, there are $\binom{|c_i|}{2}$ number of ways to obtain element $\mathbf{b}$ in $\sigma(n-1,k)$ by combining two elements in c_i into one. The coefficients of the $\mathbf{b}$'s so obtained are the same. Thus the total contributions will be

$$\sum_{i:|c_i|\ge 2} \binom{|c_i|}{2} \frac{1}{|c_i|-1}(|c_1|-1)!\cdots(|c_k|-1)!\frac{\theta^k}{\theta_{(n-1)}}. \tag{5.30}$$

To have contributions from

$$\int_{M_1(S)} \frac{\theta}{2}\sum_{i=1}^{n}\langle \nu_0,\phi_i\rangle F^{i}_{\phi_1,\ldots,\phi_n}(\mu)\Pi_{\theta,\nu_0}(d\mu),$$

there is at least one $1 \le i \le k$ such that $|c_i| = 1$. One element $\mathbf{b}$ in $\sigma(n-1,k-1)$ is obtained by removing c_i from $\mathbf{c}$. Hence the total contribution from this term is

$$\frac{\theta}{2}\sum_{i:|c_i|=1}(|c_1|-1)!\cdots(|c_k|-1)!\frac{\theta^{k-1}}{\theta_{(n-1)}}. \tag{5.31}$$

Putting (5.30) and (5.31) together, one has

$$\begin{aligned} B_{\mathbf{c}} &= (|c_1|-1)!\cdots(|c_k|-1)!\left[\sum_{i:|c_i|\ge 2}\frac{|c_i|}{2}\frac{\theta^k}{\theta_{(n-1)}} + \frac{\theta}{2}\sum_{i:|c_i|=1}\frac{\theta^{k-1}}{\theta_{(n-1)}}\right] \\ &= \frac{n(n-1+\theta)}{2}(|c_1|-1)!\cdots(|c_k|-1)!\frac{\theta^k}{\theta_{(n)}}, \end{aligned}$$

which is the same as the coefficient of $\prod_{i=1}^{k}\langle \nu_0,\prod_{j\in c_i}\phi_j\rangle$ in

$$\frac{n(n-1+\theta)}{2}\int_{M_1(S)} F_{\phi_1,\ldots,\phi_n}(\mu)\Pi(d\mu).$$

Hence Π_{θ,ν_0} is a stationary distribution.

□

Theorem 5.4. *The stationary distribution Π_{θ,ν_0} is reversible for the FV process with parent-independent mutation.*

Proof. To prove the reversibility, it suffices to verify that

$$\int_{M_1(S)} F(\mu)\mathscr{L}G(\mu)\Pi_{\theta,\nu_0}(d\mu) = \int_{M_1(S)} G(\mu)\mathscr{L}F(\mu)\Pi_{\theta,\nu_0}(d\mu) \tag{5.32}$$

for any F, G in $\mathscr{M}$. Since $\langle\mu,1\rangle$ can be included in both F and G if necessary, we can assume that

$$F(\mu) = \langle \mu, \phi_1 \rangle \cdots \langle \mu, \phi_n \rangle, \; G(\mu) = \langle \mu, \psi_1 \rangle \cdots \langle \mu, \psi_n \rangle,$$

where $\phi_1, \ldots, \phi_n, \psi_1, \ldots, \psi_n$ are in $C(S)$. Let

$$\begin{aligned} \phi &= (\phi_1, \ldots, \phi_n), \psi = (\psi_1, \ldots, \psi_n), \\ H(\phi, \psi) &= \sum_{1 \le i < j \le n} \int_{M_1(S)} \prod_{l=1}^{n} \langle \mu, \phi_l \rangle \langle \mu, \psi_i \psi_j \rangle \prod_{l \neq i,j} \langle \mu, \psi_l \rangle \Pi_{\theta, \nu_0}(d\mu) \\ &\quad + \frac{\theta}{2} \sum_{i=1}^{n} \langle \nu_0, \psi_i \rangle \int_{M_1(S)} \prod_{l=1}^{n} \langle \mu, \phi_l \rangle \prod_{l \neq i} \langle \mu, \psi_l \rangle \Pi_{\theta, \nu_0}(d\mu). \end{aligned}$$

It follows from (5.26), that (5.32) is equivalent to $H(\phi, \psi) = H(\psi, \phi)$. To deal with $2n$ functions $(\phi_1, \ldots, \phi_n, \psi_1, \ldots, \psi_n)$, we are led to consider partitions of $\{1, \ldots, 2n\}$ into non-empty sets. Let $\varsigma(n,k)$ denote the set of partitions of $\{1, \ldots, n\}$ into k sets. Here $1 \le k \le 2n$, and the empty set is allowed. Then for every $1 \le k \le 2n$ and $\{\mathbf{c}\}$ in $\sigma(2n,k)$ there is a unique way of finding $\mathbf{b}$ and $\mathbf{d}$ in $\varsigma(n,k)$ such that

$$b_i \bigcup d_i \neq \emptyset, \; c_i = b_i \bigcup \{n + l : l \in d_i\}, i = 1, \ldots, k. \tag{5.33}$$

The integration with respect to Π_{θ, ν_0} in $H(\phi, \psi)$ involves only monomials of order $2n - 1$. Thus

$$\begin{aligned} H(\phi, \psi) &= \frac{1}{2} \sum_{k=1}^{2n} \sum_{\mathbf{b}, \mathbf{d} \in \varsigma(n,k)} h(\mathbf{b}, \mathbf{d}) \\ &\quad \times \prod_{i=1}^{k} (|b_i| + |d_i| - 1)! \frac{\theta^k}{\theta_{(2n-1)}} \prod_{i=1}^{k} \left\langle \nu_0, \prod_{j \in b_i} \phi_j \prod_{j \in d_i} \psi_j \right\rangle. \end{aligned} \tag{5.34}$$

Similarly

$$\begin{aligned} H(\psi, \phi) &= \frac{1}{2} \sum_{k=1}^{2n} \sum_{\mathbf{b}, \mathbf{d} \in \varsigma(n,k)} h(\mathbf{d}, \mathbf{b}) \\ &\quad \times \prod_{i=1}^{k} (|b_i| + |d_i| - 1)! \frac{\theta^k}{\theta_{(2n-1)}} \prod_{i=1}^{k} \left\langle \nu_0, \prod_{j \in b_i} \psi_j \prod_{j \in d_i} \phi_j \right\rangle. \end{aligned} \tag{5.35}$$

If $h(\mathbf{b}, \mathbf{d})$ is symmetric in $\mathbf{b}$ and $\mathbf{d}$, we then have $H(\phi, \psi) = H(\psi, \phi)$. The coefficient $h(\mathbf{b}, \mathbf{d})$ can calculated as follows. The contributions from the term

$$\sum_{1 \le i < j \le n} \int_{M_1(S)} \prod_{l=1}^{n} \langle \mu, \phi_l \rangle \langle \mu, \psi_i \psi_j \rangle \prod_{l \neq i,j} \langle \mu, \psi_l \rangle \Pi_{\theta, \nu_0}(d\mu)$$

are from sets d_i satisfying $|d_i| \ge 2$. For each such index i, the contribution is

$$\binom{|d_i|}{2} \frac{1}{|b_i| + |d_i| - 1} \prod_{r=1}^{k} (|b_r| + |d_r| - 1)! \frac{\theta^k}{\theta_{(2n-1)}}. \tag{5.36}$$

The contributions from the term

$$\frac{\theta}{2}\sum_{i=1}^{n}\langle \nu_0,\psi_i\rangle \int_{M_1(S)}\prod_{l=1}^{n}\langle \mu,\phi_l\rangle \prod_{l\neq i}\langle \mu,\psi_l\rangle \Pi_{\theta,\nu_0}(d\mu)$$

correspond to sets c_i with $|c_i|=1$ or equivalently to sets d_i and b_i satisfying $|d_i|=1, |b_i|=0$. The total contribution for each such i is

$$\prod_{r=1}^{k}(|b_r|+|d_r|-1)!\frac{\theta^{k-1}}{\theta_{(2n-1)}}. \tag{5.37}$$

Adding (5.36) and (5.37) together, yields

$$h(\mathbf{b},\mathbf{d})=\sum_{i:|d_i|\geq 2}\frac{|d_i|(|d_i|-1)}{|b_i|+|d_i|-1}+\sum_{i:|d_i|=1,|b_i|=0}1. \tag{5.38}$$

Set

$$\begin{aligned}
I_1&=\{1\leq i\leq k:|d_i|=0\},\ J_1=\{1\leq i\leq k:|b_i|=0\},\\
I_2&=\{1\leq i\leq k:|d_i|=1\},\ J_2=\{1\leq i\leq k:|b_i|=1\},\\
I_3&=\{1\leq i\leq k:|d_i|\geq 2\},\ J_3=\{1\leq i\leq k:|b_i|\geq 2\}.
\end{aligned}$$

It follows from (5.33) that the intersection of I_1 and J_1 is empty. The function $h(\mathbf{b},\mathbf{d})$ can be written as

$$h(\mathbf{b},\mathbf{d})=\sum_{i=1}^{k}\frac{|d_i|(|d_i|-1)}{|b_i|+|d_i|-1}\chi_{I_3}(i)+\chi_{I_2\cap J_1}(i). \tag{5.39}$$

The difference between the ith terms of $h(\mathbf{b},\mathbf{d})$ and $h(\mathbf{d},\mathbf{b})$ is

$$\begin{aligned}
&\left(\frac{|d_i|(|d_i|-1)}{|b_i|+|d_i|-1}[\chi_{I_3\cap J_1}(i)+\chi_{I_3\cap J_2}(i)+\chi_{I_3\cap J_3}(i)]+\chi_{I_2\cap J_1}(i)\right)\\
&-\left(\frac{|b_i|(|b_i|-1)}{|b_i|+|d_i|-1}[\chi_{J_3\cap I_1}(i)+\chi_{J_3\cap I_2}(i)+\chi_{J_3\cap I_3}(i)]+\chi_{J_2\cap I_1}(i)\right)\\
&=(|d_i|-|b_i|)[\chi_{J_3\cap I_3}(i)+\chi_{I_3\cap J_1}(i)+\chi_{I_3\cap J_2}(i)\\
&\quad+\chi_{I_2\cap J_1}(i)+\chi_{J_2\cap I_1}(i)+\chi_{J_3\cap I_1}(i)+\chi_{J_3\cap I_2}(i)]\\
&=(|d_i|-|b_i|),
\end{aligned}$$

which implies the symmetry of $h(\mathbf{b},\mathbf{d})$, and the reversibility of Π_{θ,ν_0}.

□

5.3 The Structure of Transition Functions

In this section, we study the structure of the transition probability function of the FV process with parent-independent mutation. An explicit representation is derived, which has an intuitive explanation. Using this result, we will show that the transition probability function of the infinitely-many-neutral-alleles process $\{\mathbf{x}(t) : t \geq 0\}$ has a density with respect to Π_θ.

For any μ in $M_1(S)$, let μ^n denote the n-fold product measure $\mu \times \cdots \times \mu$, and define

$$P(t,\mu,dv) = d_0^\theta(t)\Pi_{\theta,v_0}(dv) + \sum_{n=1}^{\infty} d_n^\theta(t) \int_{S^n} \Pi_{n+\theta,v_{n,\theta}}(dv)\mu^n(dx_1 \times \cdots dx_n), \tag{5.40}$$

where

$$v_{n,\theta} = \frac{1}{n+\theta}\sum_{i=1}^{n} \delta_{x_i} + \frac{\theta}{n+\theta} v_0, \tag{5.41}$$

and $\{d_n^\theta(t) : n = 1,\ldots;t > 0\}$ are given in Theorem 4.3.

For any $F_{\phi_1,\ldots,\phi_m}$ in $\mathscr{M}$, define

$$P_t(F_{\phi_1,\ldots,\phi_m})(\mu) = \int_{M_1(S)} F_{\phi_1,\ldots,\phi_m}(v)P(t,\mu,dv). \tag{5.42}$$

Let $(X_1,\ldots,X_n)$ have a *Dirichlet*$(1,\ldots,1)$ or uniform distribution on $\triangle_n$, and, for $s_1,\ldots,s_n$ in S, set

$$\Xi_n(s_1,\ldots,s_n) = X_1\delta_{s_1} + \cdots + X_n\delta_{s_n}.$$

Lemma 5.3. *For any $m,n \geq 1$ and $F_{\phi_1,\ldots,\phi_m}$ in $\mathscr{M}$,*

$$\int_{S^n} \mathbb{E}[F_{\phi_1,\ldots,\phi_m}(\Xi_n(s_1,\ldots,s_n))]\,\mu^n(ds_1 \times \ldots \times ds_n) = \sum_{k=1}^{m} \frac{n_{[k]}}{n_{(m)}} \sum_{\mathbf{b}\in\sigma(m,k)} |b_1|!\cdots|b_k|! \prod_{j=1}^{k}\left\langle \mu, \prod_{i\in b_j}\phi_i \right\rangle, \tag{5.43}$$

where

$$n_{[0]} = 1,\ n_{[k]} = n(n-1)\cdots(n-k+1),\ k \neq 0.$$

Proof. We prove the result by induction on n. For $n = 1$, $n_{[k]} = 0$ for $k > 1$. Thus the right-hand side of (5.43) is $\langle\mu,\prod_{i=1}^m \phi_i\rangle$. Since $\Xi_1(s_1)$ is just the Dirac measure at s_1, the left-hand side is

$$\int_S \mathbb{E}[F_{\phi_1,\ldots,\phi_m}(\Xi_1(s_1))]\,\mu(ds_1) = \left\langle \mu, \prod_{i=1}^m \phi_i \right\rangle.$$

Assume (5.43) holds for $n-1 \geq 1$; i.e., for any $m \geq 1$,

$$\begin{aligned}
&\int_{S^n} \mathbb{E}[F_{\phi_1,\ldots,\phi_m}(\Xi_{n-1}(s_1,\ldots,s_{n-1}))]\,\mu^{n-1}(ds_1\times\ldots\times ds_{n-1}) \\
&= \sum_{k=1}^m \frac{(n-1)_{[k]}}{(n-1)_{(m)}} \sum_{\mathbf{b}\in\sigma(m,k)} |b_1|!\cdots|b_k|! \prod_{j=1}^k \left\langle \mu, \prod_{i\in b_j}\phi_i \right\rangle.
\end{aligned} \tag{5.44}$$

Let $(X_1,\ldots,X_n)$ be uniformly distributed on $\triangle_n$. Set $\tilde{X}_n = X_n$ and

$$\tilde{X}_i = \frac{X_i}{1-X_n},\ i=1,\ldots,n-1.$$

By direct calculation, the joint density function of $(\tilde{X}_1,\ldots,\tilde{X}_n)$ is the product of a uniform density function on $\triangle_{n-1}$ and $(1-x_n)^{n-1}$. Therefore, $\sum_{r=1}^{n-1} X_r = 1-\tilde{X}_n$ is a $Beta(n-1,1)$ random variable, $(\tilde{X}_1,\ldots,\tilde{X}_{n-1})$ is uniform on $\triangle_{n-1}$, and $\tilde{X}_n$ is independent of $(\tilde{X}_1,\ldots,\tilde{X}_{n-1})$; so

$$\Xi_n(s_1,\ldots,s_n) \stackrel{d}{=} (1-X_n)\Xi_{n-1}(s_1,\ldots,s_{n-1}) + X_n\delta_{s_n}, \tag{5.45}$$

and

$$\begin{aligned}
&\mathbb{E}[F_{\phi_1,\ldots,\phi_m}(\Xi_n(s_1,\ldots,s_n))] = \mathbb{E}[F_{\phi_1,\ldots,\phi_m}((1-X_n)\Xi_{n-1}(s_1,\ldots,s_{n-1})+X_n\delta_{s_n})] \\
&= \sum_{B\subset\{1,\ldots,m\}} \mathbb{E}[(1-X_n)^{|B|}(X_n)^{|B^c|}] \\
&\quad\times \mathbb{E}\left[\prod_{i\in B}\langle \Xi_{n-1}(s_1,\ldots,s_{n-1}),\phi_i\rangle\right] \prod_{i\in B^c}\phi_i(s_n), \qquad (5.46) \\
&= \sum_{B\subset\{1,\ldots,m\}} \frac{(n-1)_{(|B|)}|B^c|!}{n_{(m)}} \mathbb{E}\left[\prod_{i\in B}\langle \Xi_{n-1}(s_1,\ldots,s_{n-1}),\phi_i\rangle\right] \prod_{i\in B^c}\phi_i(s_n),
\end{aligned}$$

where B^c is the complement of B in $\{1,\ldots,m\}$.

Putting together (5.44) and (5.46), we obtain that

$$\begin{aligned}
&\int_{S^n} \mathbb{E}[F_{\phi_1,\ldots,\phi_m}(\Xi_n(s_1,\ldots,s_n))]\,\mu^n(ds_1\times\ldots\times ds_n) \\
&= \sum_{B\subset\{1,\ldots,m\}} \frac{(n-1)_{(|B|)}|B^c|!}{n_{(m)}} \left\langle \mu, \prod_{i\in B^c}\phi_i \right\rangle \\
&\quad\times \int_{S^{n-1}} \mathbb{E}\left[\prod_{i\in B}\langle \Xi_{n-1}(s_1,\ldots,s_{n-1}),\phi_i\rangle\right] \mu^{n-1}(ds_1\times\cdots\times ds_{n-1}).
\end{aligned}$$

Expanding the integration leads to

$$\int_{S^n} \mathbb{E}[F_{\phi_1,\ldots,\phi_m}(\Xi_n(s_1,\ldots,s_n))]\,\mu^n(ds_1\times\ldots\times ds_n)$$
$$= \frac{m!}{n_{(m)}}\left\langle \mu, \prod_{i=1}^m \phi_i \right\rangle \tag{5.47}$$
$$+ \sum_{B\subset\{1,\ldots,m\},B\neq\emptyset} \frac{(n-1)_{(|B|)}}{n_{(m)}}\left\langle \mu, \prod_{i\in B^c}\phi_i\right\rangle$$
$$\times \sum_{k=1}^{|B|}\frac{(n-1)_{[k]}}{(n-1)_{(|B|)}} \sum_{\mathbf{c}\in\sigma(|B|,k)} |c_1|!\cdots|c_k|!|B^c|!\prod_{j=1}^k\left\langle \mu, \prod_{i\in c_j}\phi_i\right\rangle.$$

Next we show that the right-hand side of (5.43) is the same as the right-hand side of (5.47). This is achieved by comparing the coefficients of the corresponding moments. It is clear that the coefficient $\frac{m!}{n_{(m)}}$ of the moment $\langle \mu, \prod_{i=1}^m \phi_i\rangle$ is the same for both. For any $1\leq r\leq m$ and $\mathbf{b}$ in $\pi(m,r)$, the coefficient of $\prod_{j=1}^r\langle \mu\prod_{i\in b_j}\phi_i\rangle$ on the right-hand side of (5.43) is

$$\frac{n_{[r]}}{n_{(m)}}|b_1|!\cdots|b_r|!. \tag{5.48}$$

On the right-hand side of (5.47) there are $r+1$ terms, corresponding to B^c being $b_1,\ldots,b_k$, or the empty set. If $B^c=\emptyset$, the coefficient is

$$\frac{(n-1)_{[r]}}{n_{(m)}}|b_1|!\cdots|b_r|!. \tag{5.49}$$

If $B^c=b_l$ for some $l=1,\ldots,k$, the coefficient is

$$\frac{(n-1)_{[r-1]}}{n_{(m)}}|b_1|!\cdots|b_r|!. \tag{5.50}$$

The lemma now follows from the fact that

$$k(n-1)_{[r-1]}+(n-1)_{[r]}=n_{[r]}.$$

□

Lemma 5.4. *For any $m,n\geq 1$ and $F_{\phi_1,\ldots,\phi_m}$ in $\mathscr{M}$,*

$$\begin{aligned}
&\int_{S^n} \mathbb{E}^{\Pi_{n+\theta,\nu_{n,\theta}}}[F_{\phi_1,\ldots,\phi_m}(\nu)]\,\mu^n(ds_1\times\cdots\times ds_n)\\
&= \sum_{B\subset\{1,\ldots,m\}} \left\{\sum_{k=1}^{|B|} \frac{n_{[k]}}{(n+\theta)_{(m)}} \sum_{\mathbf{b}\in\sigma(|B|,k)} |b_1|!\cdots|b_k|!\prod_{r=1}^{k}\langle\mu,\prod_{i\in b_r}\phi_i\rangle\right\}\\
&\quad\times\left\{\sum_{l=1}^{|B^c|}\sum_{\mathbf{d}\in\sigma(|B^c|,l)} (|d_1|-1)!\cdots(|d_l|-1)!\theta^l\prod_{r=1}^{l}\langle\nu_0,\prod_{i\in d_r}\phi_i\rangle\right\} \qquad (5.51)\\
&= \sum_{B\subset\{1,\ldots,m\}}\sum_{k=1}^{|B|}\sum_{\mathbf{b}\in\sigma(|B|,k)}\sum_{l=1}^{|B^c|}\sum_{\mathbf{d}\in\sigma(|B^c|,l)} \frac{n_{[k]}}{(n+\theta)_{(m)}} C(B,\mathbf{b},\mathbf{d}),
\end{aligned}$$

where

$$\begin{aligned}
C(B,\mathbf{b},\mathbf{d}) &= \left\{|b_1|!\cdots|b_k|!\prod_{r=1}^{k}\left\langle\mu,\prod_{i\in b_r}\phi_i\right\rangle\right\} \qquad (5.52)\\
&\quad\times\left\{(|d_1|-1)!\cdots(|d_l|-1)!\theta^l\prod_{r=1}^{l}\left\langle\nu_0,\prod_{i\in d_r}\phi_i\right\rangle\right\}.
\end{aligned}$$

When B (B^c) is empty, the summation over k (l) is one.

Proof. Let $(X_1,\ldots,X_n)$ be uniformly distributed on $\triangle_n$, and β is a $Beta(n,\theta)$ random variable, independent of $(X_1,\ldots,X_n)$. It follows from Theorem 2.24 and Theorem 2.25 that $\Pi_{n+\theta,\nu_{n,\theta}}$ can be represented as the law of

$$\beta\Xi_n(s_1,\ldots,s_n)+(1-\beta)\Xi_{\theta,\nu_0}.$$

Hence

$$\begin{aligned}
&\mathbb{E}^{\Pi_{n+\theta,\nu_{n,\theta}}}[F_{\phi_1,\ldots,\phi_m}(\nu)]\\
&\quad= \mathbb{E}\left[\prod_{i=1}^{m}\langle\beta\Xi_n(s_1,\ldots,s_n)+(1-\beta)\Xi_{\theta,\nu_0},\phi_i\rangle\right] \qquad (5.53)\\
&\quad= \sum_{B\subset\{1,\ldots,m\}} \mathbb{E}\left[\beta^{|B|}(1-\beta)^{|B^c|}\right]\mathbb{E}^{\Pi_{\theta,\nu_0}}\left[\prod_{i\in B^c}\langle\nu,\phi_i\rangle\right]\\
&\quad\times\mathbb{E}\left[\prod_{i\in B}\langle\Xi_n(s_1,\ldots,s_n),\phi_i\rangle\right].
\end{aligned}$$

By direct calculation,

$$\mathbb{E}[\beta^{|B|}(1-\beta)^{|B^c|}] = \frac{\Gamma(n+\theta)}{\Gamma(n)\Gamma(\theta)}\frac{\Gamma(n+|B|)\Gamma(\theta+|B^c|)}{\Gamma(n+\theta+m)} = \frac{n_{(|B|)}\theta_{(|B^c|)}}{(n+\theta)_{(m)}}. \qquad (5.54)$$

It follows from Lemma 5.2 and Lemma 5.3 that

$$\mathbb{E}^{\Pi_{\theta,\nu_0}}\left[\prod_{i\in B^c}\langle \nu,\phi_i\rangle\right] \tag{5.55}$$
$$=\sum_{l=1}^{|B^c|}\sum_{\mathbf{d}\in\sigma(|B^c|,l)}(|d_1|-1)!\cdots(|d_l|-1)!\frac{\theta^l}{\theta_{(|B^c|)}}\prod_{r=1}^{l}\left\langle \nu_0,\prod_{i\in d_r}\phi_i\right\rangle,$$

and

$$\int_{S^n}\mathbb{E}\left[\prod_{i\in B}\langle \Xi_n(s_1,\ldots,s_n),\phi_i\rangle\right]\mu^n(ds_1\times\cdots\times ds_n) \tag{5.56}$$
$$=\sum_{k=1}^{|B|}\frac{n_{[k]}}{n_{(|B|)}}\sum_{\mathbf{b}\in\sigma(|B|,k)}|b_1|!\cdots|b_k|!\prod_{r=1}^{k}\left\langle \mu,\prod_{i\in b_r}\phi_i\right\rangle.$$

Putting together (5.53)–(5.56), yields (5.51). □

Lemma 5.5. *For any $F_{\phi_1,\ldots,\phi_m}$ in $\mathscr{M}$,*

$$\lim_{t\to 0}P_tF_{\phi_1,\ldots,\phi_m}(\mu)=F_{\phi_1,\ldots,\phi_m}(\mu). \tag{5.57}$$

where P_t is defined in (5.42).

Proof. Fix $m\geq 1$ and $F_{\phi_1,\ldots,\phi_m}$ in $\mathscr{M}$. Define

$$A_{\theta,\mu}^{\phi_1,\ldots,\phi_m}(n)=\int_{S^n}\mathbb{E}^{\Pi_{n+\theta,\nu_{n,\theta}}}[F_{\phi_1,\ldots,\phi_m}(\nu)]\mu^n(ds_1\times\cdots\times ds_n). \tag{5.58}$$

It follows from Lemma 5.4 that

$$\lim_{n\to\infty}A_{\theta,\mu}^{\phi_1,\ldots,\phi_m}(n)=F_{\phi_1,\ldots,\phi_m}(\mu) \tag{5.59}$$

since $n_{[k]}/(n+\theta)_{(m)}$ approaches zero for $k\neq m$. From the definition of $P(t,\mu,\cdot)$ in (5.40), it follows that

$$\int_{M_1(S)}F_{\phi_1,\ldots,\phi_m}(\nu)P(t,\mu,d\nu)=\mathbb{E}[A_{\theta,\mu}^{\phi_1,\ldots,\phi_m}(\mathbf{D}_t)]=T(t)(A_{\theta,\mu}^{\phi_1,\ldots,\phi_m})(\infty), \tag{5.60}$$

where $\mathbf{D}_t$ is the pure-death process in Section 4.2, and $T(t)$ is the corresponding semigroup. The equality (5.57) now follows from (5.59), and the fact that

$$\lim_{t\to 0}T(t)(A_{\theta,\mu}^{\phi_1,\ldots,\phi_m})(\infty)=A_{\theta,\mu}^{\phi_1,\ldots,\phi_m}(\infty)=\lim_{n\to\infty}A_{\theta,\mu}^{\phi_1,\ldots,\phi_m}(n).$$

□

Lemma 5.6. *For any $F_{\phi_1,\ldots,\phi_m}$ in $\mathscr{M}$,*

$$\frac{dP_tF_{\phi_1,\ldots,\phi_m}}{dt}=P_t(\mathscr{L}F_{\phi_1,\ldots,\phi_m}). \tag{5.61}$$

Proof. By definition, one has

$$P_t(F_{\phi_1,\ldots,\phi_m})(\mu) = \sum_{n=0}^{\infty} d_n^{\theta}(t) A_{\theta,\mu}^{\phi_1,\ldots,\phi_m}(n). \tag{5.62}$$

Taking the derivative with respect to t, one obtains

$$\begin{aligned}\frac{dP_t F_{\phi_1,\ldots,\phi_m}}{dt} &= \sum_{n=0}^{\infty} \frac{d_n^{\theta}(t)}{dt} A_{\theta,\mu}^{\phi_1,\ldots,\phi_m}(n) \\ &= \sum_{n=0}^{\infty} (-\lambda_n d_n^{\theta} + \lambda_{n+1} d_{n+1}^{\theta}(t)) A_{\theta,\mu}^{\phi_1,\ldots,\phi_m}(n),\end{aligned} \tag{5.63}$$

where $\lambda_n = n(n-1+\theta)/2$, the interchange of summation and differentiation in the first equality follows from (4.22) in Corollary 4.4, and the second equality is due to the Kolmogorov forward equation for the process $\mathbf{D}(t)$.

On the other hand,

$$\begin{aligned}P_t(\mathscr{L} F_{\phi_1,\ldots,\phi_m}) = &\sum_{1 \le i < j \le m} P_t F_{\phi_1,\ldots,\phi_m}^{(i,j)} \\ &+ \frac{\theta}{2} \sum_{i=1}^{m} \langle \nu_0, \phi_i \rangle P_t F_{\phi_1,\ldots,\phi_m}^{i} \\ &- \lambda_m P_t F_{\phi_1,\ldots,\phi_m}.\end{aligned} \tag{5.64}$$

The coefficient of the term $C(B, \mathbf{b}, \mathbf{d})$ on the right-hand side of (5.63) is

$$\begin{aligned}&\sum_{n=0}^{\infty} (-\lambda_n d_n^{\theta} + \lambda_{n+1} d_{n+1}^{\theta}(t)) \frac{n_{[k]}}{(n+\theta_{(m)})} \\ &= \sum_{n=0}^{\infty} \lambda_{n+1} d_{n+1}^{\theta}(t) \left[\frac{n_{[k]}}{(n+\theta)_{(m)}} - \frac{(n+1)_{[k]}}{(n+1+\theta)_{(m)}}\right] \\ &= \sum_{n=1}^{\infty} d_n^{\theta}(t) \frac{n_{[k]}}{(n+\theta)_{(m)}} \left[\frac{(n-k)(n+m+\theta-1) - n(n+\theta-1)}{2} + \lambda_m - \lambda_m\right] \\ &= \sum_{n=0}^{\infty} d_n^{\theta}(t) \frac{n_{[k]}}{(n+\theta)_{(m)}} \left[\frac{(m-k)(n+m+\theta-1)}{2} - \lambda_m\right],\end{aligned} \tag{5.65}$$

where, in the last equality, the term corresponding to $n = 0$ is always zero since $0_{[k]} = 0$ for $k \ge 1$ and $\frac{(m-k)(n+m+\theta-1)}{2}$ equals λ_m for $k = 0$.

The coefficient of the term $C(B, \mathbf{b}, \mathbf{d})$ on the right-hand side of (5.64) can be calculated as follows. First we calculate the contributions from

$$\sum_{1 \le i < j \le m} P_t F_{\phi_1,\ldots,\phi_m}^{(i,j)}.$$

If both i and j belong to b_r for some $r = 1,\dots,k$, then the contribution from term $P_t F^{(i,j)}_{\phi_1,\dots,\phi_m}$ is $\frac{1}{|b_r|}\sum_{n=0}^{\infty} d_n^{\theta}(t)\frac{n_{[k]}}{(n+\theta)_{(m-1)}}$. The total of such contributions is

$$I_1 = \sum_{n=0}^{\infty} d_n^{\theta}(t)\frac{n_{[k]}}{(n+\theta)_{(m-1)}} \sum_{\substack{1\le r\le k\\ |b_r|\ge 2}} \binom{|b_r|}{2}\frac{1}{|b_r|} \tag{5.66}$$

$$= \sum_{n=0}^{\infty} d_n^{\theta}(t)\frac{n_{[k]}}{(n+\theta)_{(m)}}(n+\theta+m-1) \sum_{\substack{1\le r\le k\\ |b_r|\ge 1}} \frac{|b_r|-1}{2}.$$

If both i and j belong to d_r for some $r = 1,\dots,l$, then the contribution is

$$\frac{1}{|d_r|-1}\sum_{n=0}^{\infty} d_n^{\theta}(t)\frac{n_{[k]}}{(n+\theta)_{(m-1)}}.$$

The total of such contributions is

$$I_2 = \sum_{n=0}^{\infty} d_n^{\theta}(t)\frac{n_{[k]}}{(n+\theta)_{(m-1)}} \sum_{\substack{1\le r\le l\\ |d_r|\ge 2}} \binom{|d_r|}{2}\frac{1}{|d_r|-1} \tag{5.67}$$

$$= \sum_{n=0}^{\infty} d_n^{\theta}(t)\frac{n_{[k]}}{(n+\theta)_{(m)}}(n+\theta+m-1) \sum_{\substack{1\le r\le l\\ |d_r|\ge 2}} \frac{|d_r|}{2}.$$

Next we calculate the contributions from the term $\frac{\theta}{2}\sum_{i=1}^{m}\langle \nu_0,\phi_i\rangle P_t F^{i}_{\phi_1,\dots,\phi_m}$. In order to have contributions in this case, there must exist an d_r containing only one element for some $1\le r\le l$. The total contribution is

$$I_3 = \sum_{n=0}^{\infty} d_n^{\theta}(t)\frac{n_{[k]}}{(n+\theta)_{(m-1)}} \sum_{\substack{1\le r\le l\\ |d_r|=1}} \frac{\theta}{2}\frac{1}{\theta}. \tag{5.68}$$

The contribution from the last term in (5.64) is clearly

$$I_4 = -\lambda_m \sum_{n=0}^{\infty} d_n^{\theta}(t)\frac{n_{[k]}}{(n+\theta)_{(m)}}. \tag{5.69}$$

The coefficient can now be calculated by adding up all these terms as

$$\begin{aligned} I &= I_1+I_2+I_3+I_4 = \sum_{n=0}^{\infty} d_n^{\theta}(t)\frac{n_{[k]}}{(n+\theta)_{(m)}} \\ &\times \left\{\frac{n+\theta+m-1}{2}\left[\sum_{r:1\le r\le k}(|b_r|-1) + \sum_{r:1\le r\le l}|d_r|\right] - \lambda_m\right\} \\ &= \sum_{n=0}^{\infty} d_n^{\theta}(t)\frac{n_{[k]}}{(n+\theta)_{(m)}}\left[\frac{(m-k)(n+m+\theta-1)}{2} - \lambda_m\right], \end{aligned} \tag{5.70}$$

which is the same as (5.65).

□

On the basis of these preliminary results, we are now ready to prove the main result of this section.

Theorem 5.5. *The function $P(t,\mu,d\nu)$ is the probability transition function of the FV process with parent-independent mutation.*

Proof. For every $t>0$ and μ in $M_1(S)$, let $Q(t,\mu,d\nu)$ be the probability transition function of the FV process with parent-independent mutation. For any $m\geq 1$, and any $F_{\phi_1,\ldots,\phi_m}$ in $\mathscr{M}$, define

$$Q_t F_{\phi_1,\ldots,\phi_m}(\mu)=\int_{M_1(S)} F_{\phi_1,\ldots,\phi_m}(\nu)Q(t,\mu,d\nu).$$

Then

$$Q_t F_{\phi_1,\ldots,\phi_m}(\mu)=F_{\phi_1,\ldots,\phi_m}(\mu)+\int_0^t Q_s\mathscr{L}F_{\phi_1,\ldots,\phi_m}(\mu)ds. \tag{5.71}$$

On the other hand, integrating both sides of (5.61) from 0 to t and taking Lemma 5.5 into account, one gets

$$P_t F_{\phi_1,\ldots,\phi_m}(\mu)=F_{\phi_1,\ldots,\phi_m}(\mu)+\int_0^t P_s\mathscr{L}F_{\phi_1,\ldots,\phi_m}(\mu)ds. \tag{5.72}$$

Define

$$H_m(t)=\sup_{\phi_i\in C(S),\|\phi_i\|\leq 1,1\leq i\leq m}|(P_t-Q_t)F_{\phi_1,\ldots,\phi_m}(\mu)|. \tag{5.73}$$

Then

$$\begin{aligned}
&|(P_t-Q_t)\mathscr{L}F_{\phi_1,\ldots,\phi_m}(\mu)|\leq \lambda_m|(P_t-Q_t)F_{\phi_1,\ldots,\phi_m}(\mu)|\\
&+\sum_{1\leq i<j\leq m}|(P_t-Q_t)F^{(i,j)}_{\phi_1,\ldots,\phi_m}(\mu)|+\frac{\theta}{2}\sum_{i=1}^m|(P_t-Q_t)F^i_{\phi_1,\ldots,\phi_m}(\mu)|\\
&\leq 2\lambda_m H_m(t),
\end{aligned} \tag{5.74}$$

where the monomials $F^i_{\phi_1,\ldots,\phi_m}$ and $F^{(i,j)}_{\phi_1,\ldots,\phi_m}$ can be considered as being of order m, by inserting the constant function 1.

Thus

$$H_m(t)\leq 2\lambda_m\int_0^t H_m(s)ds, \tag{5.75}$$

and the theorem follows from Gronwall's lemma.

□

Remark: It is clear from the theorem that at every positive time, the values of the process are pure atomic probability measures, and the transition probability function does not have a density with respect to Π_{θ,ν_0}. The structure of the transition function provides a good picture of the genealogy of the population. With probability one, the population at time $t>0$ has only finite number of distinct ancestors at time zero.

Given n distinct ancestors, each one, independently, will choose a type according to μ. After the n ancestors and their types are fixed, the distribution of the allele frequency in the population is a $Dirichlet(\theta\nu_0 + \sum_{i=1}^n \delta_{x_i})$ process.

It is expected that a similar representation can be obtained for the probability transition function of the infinitely-many-neutral-alleles model. What is more surprising is that the transition function in this case has a density with respect to the Poisson–Dirichlet distribution.

Define

$$\Phi : M_1(S) \to \nabla, \mu \mapsto (x_1, x_2, \ldots), \tag{5.76}$$

where $x_1, x_2, \ldots,$ are the masses of the atoms of μ arranged in descending order.

If ν_0 is diffuse, then it is clear from the definition of Dirichlet process that

$$\Pi_\theta = \Pi_{\theta,\nu_0} \circ \Phi^{-1}. \tag{5.77}$$

In the remaining part of this section, we assume that ν_0 is diffuse.

Theorem 5.6. *Let μ_t be the FV process with parent-independent mutation starting at μ. Then the process $\mathbf{x}(t) = \Phi(\mu_t)$ is the infinitely-many-neutral-alleles process starting at $(x_1, x_2, \ldots) = \Phi(\mu)$.*

Proof. For each $r \geq 1$, and $\mathbf{n} = (n_1, \ldots, n_r)$ with $n = \sum_{i=1}^r n_i$, let

$$g_{\mathbf{n}}(\mathbf{x}) = \varphi_{n_1}(\mathbf{x}) \cdots \varphi_{n_r}(\mathbf{x}),$$

where φ_i is defined in (5.5). Let f denote the indicator function of the set

$$\{(s_1, \ldots, s_n) : s_1 = \cdots = s_{n_1}, \ldots, s_{n-n_r+1} = \cdots = s_{n_r}\}.$$

Then for μ in $M_1(S)$ with atomic part $\sum_{i=1}^\infty x_i \delta_{u_i}$,

$$F(\mu) = \langle \mu^n, f \rangle \equiv \left\langle \left(\sum_{i=1}^\infty x_i \delta_{u_i} \right)^n, f \right\rangle = g_{\mathbf{n}}(\Phi(\mu)). \tag{5.78}$$

It follows from direct calculation that

$$\begin{aligned} \mathscr{L}F(\mu) &= \sum_{i=1}^r \binom{n_i}{2} \varphi_{n_i-1}(\Phi(\mu)) \prod_{j \neq i} \varphi_{n_j}(\Phi(\mu)) \\ &\quad + \sum_{1 \leq i < j \leq r} n_i n_j \varphi_{n_i+n_j-1} \prod_{l \neq i,j} \varphi_{n_l}(\Phi(\mu)) - \lambda_n g_{\mathbf{n}}(\Phi(\mu)) \\ &= L g_{\mathbf{n}}(\Phi(\mu)). \end{aligned} \tag{5.79}$$

The result follows from the existence and uniqueness of Markov process associated with each of $\mathscr{L}$ and L, and fact that the domain of L is the span of $\{\varphi_i : i = 1, 2, \ldots\}$.

□

Lemma 5.7. *For each $n \geq 1$ and any μ, ν in $M_1(S)$, if $\Phi(\mu) = \Phi(\nu)$, then*

$$\int_{S^n} \mu^n(ds_1 \times \cdots \times ds_n) \Pi_{n+\theta, \nu_{n,\theta}}(\Phi^{-1}(\cdot)) \\ = \int_{S^n} \nu^n(ds_1 \times \cdots \times ds_n) \Pi_{n+\theta, \nu_{n,\theta}}(\Phi^{-1}(\cdot)), \tag{5.80}$$

where

$$\nu_{n,\theta} = (n+\theta)^{-1} \left[\sum_{i=1}^{n} \delta_{s_i} + \theta \nu_0 \right].$$

Proof. For each $n \geq 1$, as in the proof of Lemma 5.4, $\Pi_{n+\theta, \nu_{n,\theta}}$ can be represented as the law of

$$\beta \Xi_n(s_1, \ldots, s_n) + (1-\beta) \Xi_{\theta, \nu_0} = \beta \sum_{i=1}^{n} X_i \delta_{s_i} + (1-\beta) \sum_{i=1}^{\infty} Y_i \delta_{\xi_i}.$$

Thus

$$\int_{S^n} \mu^n(ds_1 \times \cdots \times ds_n) \Pi_{n+\theta, \nu_{n,\theta}}(\Phi^{-1}(\cdot)) \\ = \int_{S^n} \mu^n(ds_1 \times \cdots \times ds_n) \mathbb{P}\{\Phi(\beta \Xi_n(s_1, \ldots, s_n) + (1-\beta) \Xi_{\theta, \nu_0}) \in \cdot\}. \tag{5.81}$$

If μ does not have any atoms, then the left-hand side of (5.81) becomes

$$\mathbb{P}\{(P_1, P_2, \ldots) \in \cdot\},$$

where $(P_1, P_2, \ldots)$ is $\beta X_1, \ldots, \beta X_n, (1-\beta) Y_1, (1-\beta) Y_2, \ldots$ arranged in descending order, and the integral does not depend on μ.

If μ has an atomic part, then the atoms in $\Xi_n(s_1, \ldots, s_n)$ will depend on the partition of $\{1, \ldots, n\}$ generated by $s_1, \ldots, s_n$ such that i, j are in the same subset if $s_i = s_j$. The distribution of these partitions under $\mu^n(ds_1 \times \cdots \times ds_n)$ clearly depends only on the masses of atoms of μ. Thus the left-hand side of (5.81) depends on μ only through $\Phi(\mu)$, which implies (5.80).

□

It follows from (5.80) that for any $\mathbf{x}$ in ∇ and $A \subset \nabla$, the transition probability function of the process $\mathbf{x}_t$ can be written as

$$Q(t, \mathbf{x}, A) = P(t, \mu, \Phi^{-1}(A)), \tag{5.82}$$

where μ is any measure in $M_1(S)$ such that $\Phi(\mu) = \mathbf{x}$. Our next task is to show that $Q(t, \mathbf{x}, A)$ has a density with respect to the Poisson–Dirichlet distribution Π_θ.

Lemma 5.8. *For any $t > 0, \mathbf{x} \in \nabla$, $A \subset \nabla$, and any $\mu \in M_1(S)$ satisfying $\Phi(\mu) = \mathbf{x}$,*

$$Q(t,\mathbf{x},A) = \Pi_\theta(A) + \sum_{m=2}^{\infty} \frac{2m-1+\theta}{m!} e^{-\lambda_m t} \tag{5.83}$$
$$\times \sum_{n=0}^{m} (-1)^{m-n} \binom{m}{n} (n+\theta)_{m-1} \int_{S^n} \mu^n(ds_1 \times \cdots \times ds_n) \Pi_{n+\theta,\nu_{n,\theta}}(\Phi^{-1}(A)).$$

Proof. Let β be a $Beta(1,\theta)$ random variable and, independently of β, let $(Y_1,\ldots)$ have the Poisson–Dirichlet distribution with parameter θ. Let $V_1, V_2, \ldots$ be the size-biased permutation of $Y_1, Y_2, \ldots$ in Theorem 2.7. Then for each Borel subset A of ∇,

$$\int_S \mu(ds) \Pi_{1+\theta,\nu_{1,\theta}}(\Phi^{-1}(A)) = \mathbb{P}\{(Z_1, Z_2, \ldots) \in A\}, \tag{5.84}$$

where $(Z_1,\ldots)$ is $\beta, (1-\beta)V_1, (1-\beta)V_2, \ldots$ in descending order. Clearly $(Z_1,\ldots)$ has the Poisson–Dirichlet distribution with parameter θ and

$$\int_S \mu(ds) \Pi_{1+\theta,\nu_{1,\theta}}(\Phi^{-1}(A)) = \Pi_\theta(A). \tag{5.85}$$

It follows from (5.40) and (5.82) that

$$Q(t,\mathbf{x},A) = d_0^\theta(t) \Pi_{\theta,\nu_0}(\Phi^{-1}(A) \tag{5.86}$$
$$+ \sum_{n=1}^{\infty} d_n^\theta(t) \int_{S^n} \Pi_{n+\theta,\nu_{n,\theta}}(\Phi^{-1}(A)) \mu^n(dx_1 \times \cdots dx_n).$$

Applying Theorem 4.3, we obtain

$$Q(t,\mathbf{x},A)$$
$$= \Pi_\theta(A) - \sum_{m=1}^{\infty} \frac{(2m-1+\theta)}{m!} (-1)^{m-1} \theta_{(m-1)} e^{-\lambda_m t} \Pi_\theta(A)$$
$$+ \sum_{m=1}^{\infty} \frac{(2m-1+\theta)}{m!} (-1)^{m-1} \binom{m}{1} (1+\theta)_{(m-1)} e^{-\lambda_m t} \int_S \mu(ds) \Pi_{1+\theta,\nu_{1,\theta}}(A)$$
$$+ \sum_{n=2}^{\infty} \sum_{m=n}^{\infty} \frac{(2m-1+\theta)}{m!} (-1)^{m-1} \binom{m}{n} (n+\theta)_{(m-1)} e^{-\lambda_m t}$$
$$\times \int_{S^n} \mu^n(ds_1 \times \cdots \times ds_n) \Pi_{n+\theta,\nu_{n,\theta}}(\Phi^{-1}(A)) \tag{5.87}$$
$$= \Pi_\theta(A) - \sum_{m=2}^{\infty} \frac{(2m-1+\theta)}{m!} e^{-\lambda_m t} (-1)^{m-1} \theta_{(m-1)} \Pi_\theta(A)$$
$$+ \sum_{m=2}^{\infty} \frac{(2m-1+\theta)}{m!} e^{-\lambda_m t} (-1)^{m-1} \binom{m}{1} (1+\theta)_{(m-1)} \int_S \mu(ds) \Pi_{1+\theta,\nu_{1,\theta}}(A)$$
$$+ \sum_{m=2}^{\infty} \frac{(2m-1+\theta)}{m!} e^{-\lambda_m t}$$
$$\times \sum_{n=2}^{\infty} (-1)^{m-1} \binom{m}{n} (n+\theta)_{(m-1)} \int_{S^n} \mu^n(ds_1 \times \cdots \times ds_n) \Pi_{n+\theta,\nu_{n,\theta}}(\Phi^{-1}(A)),$$

where in the last equality, the terms corresponding to $m = 1$ cancel each other, and the interchange of summations is justified by their absolute convergence. It is now clear that (5.83) follows by collecting the last three terms together.

□

Note that by Lemma 5.7,

$$\int_{S^n} \mu^n(ds_1 \times \cdots \times ds_n) \Pi_{n+\theta,\nu_{n,\theta}}(\Phi^{-1}(\cdot))$$

depends on μ only through the distribution of partitions of $\{1,\ldots,n\}$ induced by $s_1,\ldots,s_n$. Hence we can write

$$\int_{S^n} \mu^n(ds_1 \times \cdots \times ds_n) \Pi_{n+\theta,\nu_{n,\theta}}(\Phi^{-1}(\cdot)) = \sum_{\mathbf{n} \in I_n} C(\mathbf{n}) \psi_{\mathbf{n}}(\mathbf{x}) P_{\mathbf{n}}(\cdot), \tag{5.88}$$

where

$$I_n = \left\{ \mathbf{n} = (n_1,\ldots,n_l) : n_1 \geq n_2 \geq \cdots \geq n_l \geq 1, \sum_{k=1}^{l} n_k = n,\ l = 1,\ldots,n \right\},$$

and for $\mathbf{n} = (n_1,\ldots,n_l)$,

$$C(\mathbf{n}) = \binom{n}{n_1,\ldots,n_l} \frac{1}{\prod_{i=1}^{n} a_i!},$$

with $a_i = \#\{k : n_k = i, 1 \leq k \leq l\}$.

For a given $\mathbf{n} = (n_1,\ldots,n_l)$ in I_n, let $(X_1,\ldots,X_l)$ have the $Dirichlet(n_1,\ldots,n_l)$ distribution. Independent of $X_1,\ldots,X_l$, let $(Y_1,\ldots)$ have the Poisson–Dirichlet distribution. The random variable β has a $Beta(n,\theta)$ distribution, and is independent of both $(X_1,\ldots,X_l)$ and $(Y_1,\ldots)$. Then $P_{\mathbf{n}}$ is the law of the descending order statistics of $\beta X_1,\ldots,\beta X_l,(1-\beta)Y_1,\ldots$.

Lemma 5.9. *Let* $\mathbf{n}$ *be a partition of* $\{1,\ldots,n\}$ *with* $\mathbf{n} = (n_1,\ldots,n_l)$. *For each partition* $\mathbf{m} = (m_1,\ldots,m_k)$ *of* $\{1,\ldots,m\}$, *set*

$$\psi_{\mathbf{m}}(\mathbf{z}) = \sum_{i_1,\ldots,i_k,\ \text{distinct}} z_{i_1}^{m_1} \cdots z_{i_k}^{m_k}, \tag{5.89}$$

and for notational convenience, we set $\psi_{\mathbf{0}} \equiv 1$.

Then

$$\int_{\nabla} \psi_{\mathbf{m}}(\mathbf{z}) dP_{\mathbf{n}} = \sum_{i=0}^{l \wedge k} \sum_{\substack{B \subset \{1,\ldots,k\} \\ |B|=i}} \sum_{\pi \in \mathscr{C}(B)} \prod_{j \in B} (n_{\pi(j)})_{(m_j)} \prod_{j \notin B} (m_j - 1)! \frac{\theta^{k-i}}{(n+\theta)_{(m)}} \tag{5.90}$$

where $\mathscr{C}(B)$ *is the collection of all one-to-one maps from* B *into* $\{1,\ldots,l\}$.

Proof. Since $P_{\mathbf{n}}$ is the law of the descending order statistics of $\beta X_1,\ldots,\beta X_l,(1-\beta)Y_1,\ldots$, each term $z_{i_1}^{m_1} \cdots z_{i_k}^{m_k}$ in $\psi_{\mathbf{m}}$ under $P_{\mathbf{n}}$ can be written as

$$\beta^{\sum_{j\in B} m_j}(1-\beta)^{m-\sum_{j\in B} m_j} \prod_{j\in B} X_{\pi(j)}^{m_j} \prod_{j\notin B} Y_{r_j}^{m_j} \tag{5.91}$$

for certain $B \subset \{1,\ldots,k\}$ with $|B| \le l$, π in $\mathscr{C}(B)$, and $\{r_j : j \notin B\} \subset \{1,2,\ldots\}$.

Thus it follows from the independence that

$$\begin{aligned} \int_\nabla \psi_{\mathbf{m}}(\mathbf{z})dP_{\mathbf{n}} = \sum_{i=0}^{l\wedge k} \sum_{\substack{B\subset\{1,\ldots,k\}\\ |B|=i}} \sum_{\pi\in\mathscr{C}(B)} \mathbb{E}[\beta^{\sum_{j\in B} m_j}(1-\beta)^{m-\sum_{j\in B} m_j}] \\ \times \mathbb{E}\left[\prod_{j\in B} X_{\pi(j)}^{m_j}\right] \mathbb{E}\left[\sum_{j\notin B, r_j \text{ distinct}} \prod_{\iota\notin B} Y_{r_\iota}^{m_\iota}\right]. \end{aligned} \tag{5.92}$$

By direct calculation,

$$\mathbb{E}[\beta^{\sum_{j\in B} m_j}(1-\beta)^{m-\sum_{j\in B} m_j}] = \frac{\Gamma(n+\sum_{j\in B} m_j)}{\Gamma(n)} \theta_{(m-\sum_{j\in B} m_j)} \frac{1}{(n+\theta)_{(m)}}, \tag{5.93}$$

and

$$\begin{aligned} \mathbb{E}\left[\prod_{j\in B} X_{\pi(j)}^{m_j}\right] &= \frac{\Gamma(n)}{(\prod_{j\in B}\Gamma(n_{\pi(j)}))\Gamma(n-\sum_{j\in B} n_{\pi(j)})} \\ &\quad \times \frac{(\prod_{j\in B}\Gamma(m_j+n_{\pi(j)}))\Gamma(n-\sum_{j\in B} n_{\pi(j)})}{\Gamma(n+\sum_{j\in B} m_j)} \\ &= \frac{\Gamma(n)}{\Gamma(n+\sum_{j\in B} m_j)} \prod_{j\in B} (n_{\pi(j)})_{(m_j)}. \end{aligned} \tag{5.94}$$

Let $\mathbf{m}_B$ denote the partition of $\{1,\ldots,m-\sum_{j\in B} m_j\}$ with $\mathbf{m}_B = \{m_j : j \notin B\}$. Then

$$\begin{aligned} \mathbb{E}\left[\sum_{j\notin B, r_j \text{ distinct}} \prod_{\iota\notin B} Y_{r_\iota}^{m_\iota}\right] &= \mathbb{E}^{\Pi_\theta}[\psi_{\mathbf{m}_B}(\mathbf{z})] \\ &= \prod_{j\notin B} (m_j-1)! \frac{\theta^{k-i}}{\theta_{(m-\sum_{j\in B} m_j)}}, \end{aligned} \tag{5.95}$$

where (5.20) is used in deriving the last equality. It is now clear that (5.90) follows by multiplying (5.93), (5.94), and (5.95) together.

□

Theorem 5.7. *The transition function $Q(t,\mathbf{x},A)$ has a density $q(t,\mathbf{x},\mathbf{y})$ with respect to Π_θ, and*

$$q(t,\mathbf{x},\mathbf{y}) = 1 + \sum_{m=2}^{\infty} \frac{2m-1+\theta}{m!} e^{-\lambda_m t} \times \sum_{n=0}^{m} (-1)^{m-n} \binom{m}{n} (n+\theta)_{(m-1)} p_n(\mathbf{x},\mathbf{y}), \tag{5.96}$$

where

$$p_n(\mathbf{x},\mathbf{y}) = \sum_{\mathbf{n}\in I_n} C(\mathbf{n}) \frac{\psi_{\mathbf{n}}(\mathbf{x})\psi_{\mathbf{n}}(\mathbf{y})}{\mathbb{E}^{\Pi_\theta}[\psi_{\mathbf{n}}(\mathbf{z})]}. \tag{5.97}$$

Proof. For each partition $\mathbf{m} = (m_1,\ldots,m_k)$ of $\{1,\ldots,m\}$, and each partition $\mathbf{n} = (n_1,\ldots,n_l)$ of $\{1,\ldots,n\}$,

$$\begin{aligned}
\psi_{\mathbf{m}}(\mathbf{z})\psi_{\mathbf{n}}(\mathbf{z}) &= \sum_{i_1,\ldots,i_k,\ \text{distinct}} z_{i_1}^{m_1}\cdots z_{i_k}^{m_k} \sum_{j_1,\ldots,j_l,\ \text{distinct}} z_{j_1}^{n_1}\cdots z_{j_l}^{n_l} \\
&= \sum_{r=0}^{k\wedge l} \sum_{\substack{B\subset\{1,\ldots,k\}\\ |B|=r}} \sum_{\pi\in\mathscr{C}(B)} \{ \sum_{\text{distinctindex}} \prod_{\tau\notin B} z_{i_\tau}^{m_\tau} \prod_{\tau\notin\pi(B)} z_{j_\tau}^{n_\tau} \prod_{\tau\in B} z_{i_\tau}^{m_\tau+n_{\pi(\tau)}} \} \\
&= \sum_{r=0}^{k\wedge l} \sum_{\substack{B\subset\{1,\ldots,k\}\\ |B|=r}} \sum_{\pi\in\mathscr{C}(B)} \psi_{\mathbf{mBn}}(\mathbf{z}),
\end{aligned} \tag{5.98}$$

where

$$\mathbf{mBn} = \{m_\tau, n_\iota, m_\omega + n_{\pi(\omega)} : \tau\in\{1,\ldots,k\}\setminus B, \iota\in\{1,\ldots,l\}\setminus\pi(B), \omega\in B\}$$

is a partition of $\{1,\ldots,n+m\}$.

It follows from (5.20) that

$$\mathbb{E}^{\Pi_\theta}[\psi_{\mathbf{n}}(\mathbf{z})] = (n_1-1)!\cdots(n_l-1)!\frac{\theta^l}{\theta_{(n)}}, \tag{5.99}$$

and

$$\mathbb{E}^{\Pi_\theta}[\psi_{\mathbf{mBn}}(\mathbf{z})] = \prod_{\tau\notin B}(m_\tau-1)! \prod_{\iota\notin\pi(B)}(n_\iota-1)! \times \prod_{\omega\in B}(m_\omega+n_{\pi(\omega)}-1)!\frac{\theta^{k+l-r}}{\theta_{(n+m)}}, \tag{5.100}$$

which implies that

$$\begin{aligned}
&\frac{\mathbb{E}^{\Pi_\theta}[\psi_{\mathbf{mBn}}(\mathbf{z})]}{\mathbb{E}^{\Pi_\theta}[\psi_{\mathbf{n}}(\mathbf{z})]} \\
&= \sum_{r=0}^{k\wedge l} \sum_{\substack{B\subset\{1,\ldots,k\}\\ |B|=r}} \sum_{\pi\in\mathscr{C}(B)} \prod_{\omega\in B} (n_{\pi(\omega)})_{(m_\omega)} \prod_{\tau\notin B}(m_\tau-1)!\frac{\theta^{k-r}}{(n+\theta)_{(m)}}.
\end{aligned} \tag{5.101}$$

Taking account of (5.88) and (5.90), one obtains for $n \geq 1$ and any partition $\mathbf{m}$ of $\{1,\ldots,m\}$,

$$\int_{\nabla} \psi_{\mathbf{m}}(\mathbf{z}) d\left[\int_{S^n} \mu^n(ds_1 \times \cdots \times ds_n) \Pi_{n+\theta, \nu_{n,\theta}}(\Phi^{-1}(\cdot))\right] = \int_{\nabla} \psi_{\mathbf{m}}(\mathbf{z}) p_n(\mathbf{x},\mathbf{z}) d\Pi_{\theta}. \tag{5.102}$$

It follows from (5.83) that

$$\int_{\nabla} \psi_{\mathbf{m}}(\mathbf{z}) Q(t,\mathbf{x},d\mathbf{z}) = \int_{\nabla} \psi_{\mathbf{m}}(\mathbf{z}) q(t,\mathbf{x},\mathbf{z}) d\Pi_{\theta}. \tag{5.103}$$

For any $n_1,\ldots,n_k$, the equality

$$\varphi_{n_1}(\mathbf{z}) \cdots \varphi_{n_k}(\mathbf{z}) = \sum_{i=1}^{k} \sum_{\mathbf{b} \in \sigma(k,i)} \psi_{n(b_1),\ldots,n(b_i)}(\mathbf{z}) \tag{5.104}$$

holds on the space ∇_∞, where $n(b_r) = \sum_{j \in b_r} n_j, r = 1,\ldots,i$. Hence the family $\{\psi_{\mathbf{m}} : \mathbf{m} \in I_m, m = 0,\ldots\}$ is a determining class for probability measures on ∇_∞.

The theorem now follows from the fact that both $\Pi_\theta(\cdot)$ and $Q(t,\mathbf{x},\cdot)$ are concentrated on the space ∇_∞.

□

5.4 A Measure-valued Branching Diffusion with Immigration

It was indicated in Chapter 1 that the finite-dimensional Wright–Fisher diffusion can be derived from a family of finite-dimensional independent diffusions, through normalization and conditioning. This is a dynamical analog of the relation between the gamma and Dirichlet distributions. In this section, we will consider an infinite dimensional generalization of this structure. In particular, we will describe a derivation of the Fleming–Viot process with parent-independent mutation through an infinite-dimensional diffusion process, called the measure-valued branching diffusion with immigration, by conditioning the total mass to be 1.

Let $M(S)$ denote the space of finite Borel measures on S equipped with weak topology, and define

$$\begin{aligned}\tilde{\mathscr{D}} = \{F : F(\mu) = f(\langle \mu, \phi_1 \rangle, \ldots, \langle \mu, \phi_k \rangle), f \in C_b^2(\mathbb{R}^k), \phi_i \in B(S),\\ k = 1,2,\ldots; i = 1,\ldots,k, \mu \in M(S)\}.\end{aligned} \tag{5.105}$$

Then the generator of the measure-valued branching diffusion with immigration has the form

$$\mathscr{G}F(\mu) = \frac{\theta}{2}\int_S \frac{\delta F(\mu)}{\delta\mu(x)}(\nu_0(dx) - \mu(dx)) + \frac{1}{2}\int_S\int_S \Big(\frac{\delta^2 F(\mu)}{\delta\mu(x)\delta\mu(x)}\Big)\mu(dx),\ F \in \tilde{\mathscr{D}}, \tag{5.106}$$

where the immigration is described by ν_0, a diffuse probability measure in $M_1(S)$. For a non-zero measure μ in $M(S)$, set $|\mu| = \langle \mu, 1\rangle$ and $\hat{\mu} = \frac{\mu}{|\mu|}$. For f in $C_b^2(\mathbb{R})$ and ϕ in $B(S)$, let $F(\mu) = f(\langle \hat{\mu}, \phi\rangle) = H(\hat{\mu})$. Then

$$\begin{aligned}
\frac{\delta F(\mu)}{\delta\mu(x)} &= \frac{f'(\langle\hat{\mu},\phi\rangle)}{|\mu|}(\phi(x) - \langle\hat{\mu},\phi\rangle),\\
\frac{\delta^2 F(\mu)}{\delta\mu(x)\delta\mu(x)} &= f''(\langle\hat{\mu},\phi\rangle)\left(\frac{\phi(x)-\langle\hat{\mu},\phi\rangle}{|\mu|}\right)^2\\
&\quad -2\frac{f'(\langle\hat{\mu},\phi\rangle)}{|\mu|^2}(\phi(x) - \langle\hat{\mu},\phi\rangle).
\end{aligned}$$

Substituting this into (5.106), yields

$$\begin{aligned}
\mathscr{G}F(\mu) &= \frac{f'(\langle\hat{\mu},\phi\rangle)}{|\mu|}\frac{\theta}{2}\int_S \phi(x)(\nu_0(dx) - \hat{\mu}(dx))\\
&\quad + \frac{f''(\langle\hat{\mu},\phi\rangle)}{2|\mu|}\int_S (\phi^2(x) - \langle\hat{\mu},\phi\rangle^2)\hat{\mu}(dx) \qquad (5.107)\\
&= \frac{1}{|\mu|}\mathscr{L}H(\hat{\mu}),
\end{aligned}$$

where $\mathscr{L}$ is the generator of the FV process. Formally, conditioning on $|\mu| = 1$, $\mathscr{G}$ becomes $\mathscr{L}$. Let μ_t be the measure-valued branching diffusion with immigration. Then the total mass process $x_t = \langle \mu_t, 1\rangle$ is a diffusion on $[0,\infty)$ with generator

$$\frac{x}{2}\frac{d^2 f}{dx^2} + \frac{\theta}{2}(1-x)\frac{df}{dx}.$$

Since x_t may take on the value zero, a rigorous proof of the conditional result is technically involved.

The measure-valued branching diffusion with immigration is reversible with the gamma random measure as the reversible measure. The log-Laplace functional of the gamma random measure is given by

$$\theta\int_S \log(1+\theta\phi(x))\nu_0(dx).$$

In comparison with the construction of the Poisson–Dirichlet distribution from the gamma process in Chapter 2, it is expected that the conditioning result would hold.

For any $a > 0, \nu \in M(S)$, let $\Gamma_{a,\nu}$ denote the gamma random measure with log-Laplace functional

$$\int_S \log(1+a^{-1}\phi(x))\nu(dx).$$

Then the measure-valued branching diffusion with immigration has the following transition function representation that is similar to the representation of the FV process

$$\begin{aligned} &P_{\theta,\nu_0}(t,\mu,d\nu) \\ &\quad = \exp(-c_t|\mu|)\Gamma_{e^{\theta t/2}c_t,\theta\nu_0}(d\nu) \qquad (5.108) \\ &\quad + \sum_{n=1}^{\infty} \exp(-c_t|\mu|)\frac{c_t^n}{n!}\int_{S^n} \mu^n(dx_1,\ldots,dx_n)\Gamma_{e^{\theta t/2}c_t,\theta\nu_0+\sum_{k=1}^n \delta_{x_k}}(d\nu), \end{aligned}$$

where

$$c_t = \frac{\theta e^{-\theta t/2}}{1-e^{-\theta t/2}}.$$

5.5 Two-parameter Generalizations

As shown in Chapter 3, the two parameter Poisson–Dirichlet distribution possesses many structures that are similar to the Poisson–Dirichlet distribution, including the urn construction, GEM representation, and sampling formula. It is thus natural to investigate the two-parameter analog of the infinitely-many-neutral-alleles model. Two such models will be described in this section.

Model I: GEM Process for $\theta > 1,\ 0 \le \alpha < 1$

For each $i \ge 1$, let $a_i = \frac{1-\alpha}{2}$, $b_i = \frac{\theta+i\alpha}{2}$. The process $u_i(t)$ is the unique strong solution of the stochastic differential equation

$$du_i(t) = (a_i - (a_i+b_i)u_i(t))dt + \sqrt{u_i(t)(1-u_i(t))}dB_i(t), u_i(0) \in [0,1],$$

where $\{B_i(t) : i = 1,2,\ldots\}$ are independent one-dimensional Brownian motions. It is known that the process $u_i(t)$ is reversible, and $Beta(2a_i, 2b_i)$ distribution is the reversible measure. By direct calculation, the scale function of $u_i(\cdot)$ is given by

$$s_i(x) = \left(\frac{1}{4}\right)^{a_i+b_i} \int_{1/2}^{x} \frac{dy}{y^{2a_i}(1-y)^{2b_i}}.$$

Since $\theta > 1, \alpha \ge 0$, it follows that $\lim_{x\to 1} s_i(x) = +\infty$ for all i. Therefore, starting from any point in $[0,1)$, with probability one the process $u_i(t)$ will not hit the boundary 1. Let $E = [0,1)^{\mathbb{N}}$. The process

$$\mathbf{u}(t) = (u_1(t), u_2(t), \ldots)$$

is then an E-valued Markov process. Since, defined on E, the map

$$\mathbb{H}(u_1, u_2, \ldots) = (u_1, (1-u_1)u_2, \ldots)$$

is one-to-one, the process $\mathbf{x}(t) = \mathbb{H}(\mathbf{u}(t))$ is again a Markov process. We call it the *GEM process*. It is clear from the construction that the GEM process is reversible, with reversible measure $GEM(\alpha, \theta)$, the two-parameter GEM distribution.

Model II: Two-Parameter Infinitely-Many-Neutral-Alleles Model

Recall the definition of $\mathscr{D}(L)$ in (5.4). For any $0 \le \alpha < 1, \theta > -\alpha$, and f in $\mathscr{D}(L)$, let

$$L_{\alpha,\theta} f(\mathbf{x}) = \frac{1}{2}\left[\sum_{i,j=1}^{\infty} x_i(\delta_{ij} - x_j)\frac{\partial^2 f}{\partial x_i \partial x_j} - \sum_{i=1}^{\infty}(\theta x_i + \alpha)\frac{\partial f}{\partial x_i}\right]. \tag{5.109}$$

For $m_1, \ldots, m_k \in \{2, 3, \ldots\}$ and $k \ge 1$, it follows from direct calculation that

$$\begin{aligned}
&L_{\alpha,\theta}(\varphi_{m_1} \cdots \varphi_{m_k}) \\
&= \sum_{i=1}^{k}\left[\binom{m_i}{2} - \frac{m_i\alpha}{2}\right]\varphi_{m_i-1}\prod_{j\neq i}\varphi_{m_j} + \sum_{i<j} m_i m_j \varphi_{m_i+m_j-1}\prod_{l\neq i,j}\varphi_{m_l} \\
&\quad - \left\{\sum_{i=1}^{k}\left[\binom{m_i}{2} + \frac{m_i\theta}{2}\right] + \sum_{i<j} m_i m_j\right\}\prod_{i=1}^{k}\varphi_{m_i} \qquad (5.110)\\
&= \sum_{i=1}^{k}\left[\binom{m_i}{2} - \frac{m_i\alpha}{2}\right]\varphi_{m_i-1}\prod_{j\neq i}\varphi_{m_j} + \sum_{i<j} m_i m_j \varphi_{m_i+m_j-1}\prod_{l\neq i,j}\varphi_{m_l} \\
&\quad - \frac{1}{2}m(m-1+\theta)\prod_{i=1}^{k}\varphi_{m_i}.
\end{aligned}$$

For any $1 \le l \le k$, denote by $\{n_1, n_2, \ldots, n_l\}$ an arbitrary partition of the set $\{m_1, m_2, \ldots, m_k\}$. By the Pitman sampling formula, we get

$$\begin{aligned}
&\mathbb{E}_{\alpha,\theta}\left[L_{\alpha,\theta}(\varphi_{m_1} \cdots \varphi_{m_k})\right] \\
&= \sum_{n_1, n_2, \ldots, n_l}\left\{\sum_{i=1}^{l}\frac{n_i}{2}(n_i - 1 - \alpha)\frac{(-\frac{\theta}{\alpha})(-\frac{\theta}{\alpha}-1)\cdots(-\frac{\theta}{\alpha}-l+1)}{\theta(\theta+1)\cdots(\theta+m-2)}\right. \\
&\quad \cdot\left(\prod_{j\neq i}(-\alpha)\cdots(-\alpha+n_j-1)\right)(-\alpha)\cdots(-\alpha+n_i-2) \\
&\quad - \frac{1}{2}m(m-1+\theta)\frac{(-\frac{\theta}{\alpha})(-\frac{\theta}{\alpha}-1)\cdots(-\frac{\theta}{\alpha}-l+1)}{\theta(\theta+1)\cdots(\theta+m-1)} \\
&\quad \left.\cdot\prod_{j}(-\alpha)\cdots(-\alpha+n_j-1)\right\} = 0,
\end{aligned}$$

where the value of the right-hand side is obtained by continuity when $\alpha = 0$ or $\theta = 0$. Thus

$$\mathbb{E}_{\alpha,\theta}[L_{\alpha,\theta} f] = 0, \ \forall f \in \mathscr{D}(L). \tag{5.111}$$

Similarly, we can further check that

$$\mathbb{E}_{\alpha,\theta}[g \cdot L_{\alpha,\theta}(f)] = \mathbb{E}_{\alpha,\theta}[f \cdot L_{\alpha,\theta}(g)], \ \forall f,g \in \mathscr{D}(L). \tag{5.112}$$

These two identities lead to the next result.

Theorem 5.8. (Petrov [147])
(1) *The operator $L_{\alpha,\theta}$ is closable in $C(\nabla)$, and its closure generates a ∇-valued diffusion process.*
(2) *The Markov process associated with $L_{\alpha,\theta}$, called the two-parameter infinitely-many-neutral-alleles model, is reversible with respect to $PD(\alpha,\theta)$, the two-parameter Poisson–Dirichlet distribution.*
(3) *The complete set of eigenvalues of operator $L_{\alpha,\theta}$ is $\{0, -\lambda_2, -\lambda_3, \ldots\}$, which is the same as that of operator L.*

5.6 Notes

The infinitely-many-neutral-alleles model is due to Kimura and Crow [124]. Watterson [181] formulated the model as the limit of a sequence of finite-dimensional diffusions. The formulation and the approach taken in Section 5.1 follow Ethier and Kurtz [61]. The stationarity of Π_K in Theorem 5.1 is shown in Wright [186]. The reversibility is due to Griffiths [92].

The first FV process appears in Fleming and Viot [82] where a class of a probability-valued diffusion processes was introduced. The mutation process in [82] is the Brownian motion. Ethier and Kurtz [62] reformulated the infinitely-many-neutral-alleles model as an FV process with parent-independent mutation. The proofs of Lemma 5.2, Theorem 5.3, and Theorem 5.4 are from [56] with minor modifications. In [134] reversibility has been shown to be a unique feature of parent-independent mutation. Without selection, parent-independent mutation is the only motion that makes the FV process reversible.

The transition density expansion is obtained in Griffiths [92] for the finitely-many-alleles diffusions. The results in Section 5.3 follow Ethier and Griffiths [59] for the most part. Theorem 5.7 was originally proved in Ethier [57]. The treatment given here, is to derive the result directly through the expansion in Theorem 5.5. Theorem 5.6 is from Ethier and Kurtz [64]. The equality (5.104) is from Kingman [126].

Another stochastic dynamic that has the Poisson–Dirichlet distribution as an invariant measure is the coagulation–fragmentation process. The basic model is a Markov chain with partitions of the unit interval as the state space. The transition mechanism involves splitting, merging and reordering. For more details and generalizations, see [172], [154], [136], [32], and the references therein.

One can find introductions to measure-valued processes in [49] and [52]. For a more detailed study of measure-valued processes and related topics, [20], [133], and [144] are good sources. The relation between the measure-valued branching diffusion with immigration and the FV process is part of a general structure that connects the measure-valued branching process or superprocess with the FV process. Konno and Shiga [131] discovered that the FV process can be derived from the superprocess by normalization and a random time change. By viewing a measure as the product of a radial part and an angular part, Etheridge and March [53] showed that the original FV process can be derived from the Dawson–Watanabe process or super-Brownian motion, by conditioning the total mass to be one. A more general result, the Perkins disintegration theorem, was established in Perkins [143] where the law of a superprocess is represented as the integration, with respect to the law of the total mass process, of the law of a family of FV processes with sampling rate inversely proportional to the total mass. The representation of the transition function for the measure-valued branching process with immigration is obtained in [60]. Further discussions and generalizations can be found in [166] and [167]. Results on functional inequalities, not included here, can be found in [165], [78], [45], and [77]. General references on functional inequalities include [96], [97], [179], and [18].

Several papers have emerged recently investigating stochastic dynamics associated with the two-parameter Poisson–Dirichlet distribution and the two-parameter Dirichlet process. The GEM process is from [78]. The two-parameter infinitely-many-neutral-alleles model first appeared in [147]. In [76], symmetric diffusion processes, including the two-parameter infinitely-many-neutral-alleles model, are constructed and studied for both the two-parameter Poisson–Dirichlet distribution and the two-parameter Dirichlet process using techniques from the theory of Dirichlet forms. It is still an open problem to generalize the FV process with parent-independent mutation to the two-parameter setting. In [13], $PD(\alpha,\theta)$ is shown to be the unique reversible measure of a continuous-time Markov process constructed through an exchangeable fragmentation–coagulation process.

Chapter 6
Particle Representation

The Fleming–Viot process studied in Chapter 5 describes the macroscopic evolution of genotype distributions of a large population under the influence of parent-independent mutation and random genetic drift or random sampling. In this chapter, we focus on a microscopic system, a special case of the Donnelly–Kurtz particle model, with macroscopic average following the FV process.

The Donnelly–Kurtz particle model is motivated by an infinite exchangeable particle system studied by Dawson and Hochberg in [24]. Due to exchangeability, particles can be labeled in a way that the genealogy of the population is explicitly represented. The empirical process of n particles of the system is shown to converge to the FV process as n tends to infinity. Our objective here is to explore the close relation, provided by this model, between the backward-looking coalescent of Chapter 4 and the forward-looking stochastic dynamics of Chapter 5. We would like to emphasize that the particle representation developed by Donnelly and Kurtz is much more general and powerful than what has been presented here.

6.1 Exchangeability and Random Probability Measures

Let E be a Polish space, and $\mathscr{E}$ the Borel σ-field on E. The space $M_1(E)$ is the collection of all probability measures on E equipped with the weak topology. Let $(\Omega,\mathscr{F})$ be a measurable space. A random probability measure on E is a probability kernel $\xi(\omega,B)$, defined on $\Omega\times\mathscr{E}$ such that $\xi(\omega,\cdot)$ is in $M_1(E)$ for each ω in Ω, and $\xi(\omega,B)$ is measurable in ω for each B in $\mathscr{E}$. It can be viewed as an $M_1(E)$-valued random variable defined on $(\Omega,\mathscr{F})$. The law of a random probability measure belongs to $M_1(M_1(E))$, the space of probability measures on $M_1(E)$. Let $\mathbb{N}_+$ denote the set of strictly positive integers. A finite permutation π on $\mathbb{N}_+$ is a one-to-one map between $\mathbb{N}_+$ and $\mathbb{N}_+$ such that $\pi(k)=k$ for all but a finite number of integers.

Definition 6.1. A finite or infinite sequence $Z=(Z_1,Z_2,\ldots)$ of E-valued random variables defined on $(\Omega,\mathscr{F})$ is said to be exchangeable if for any finite permutation π on $\mathbb{N}_+$

S. Feng, *The Poisson–Dirichlet Distribution and Related Topics*,
Probability and its Applications, DOI 10.1007/978-3-642-11194-5_6,

$$(Z_1, Z_2, \ldots) \stackrel{d}{=} (Z_{\pi(1)}, Z_{\pi(2)}, \ldots),$$

where $\stackrel{d}{=}$ denotes equality in distribution.

For any $n \geq 1$, and any random probability measure ξ, the nth moment measure M_n, of ξ is a probability measure on E^n given by

$$M_n(dx_1, \ldots, dx_n) = \mathbb{E}[\xi(dx_1) \cdots \xi(dx_n)].$$

Clearly if the random vector $(Z_1, , \ldots, Z_n)$ has probability law M_n, then $(Z_1, \ldots, Z_n)$ is exchangeable. Since the family $\{M_n(dx_1, \ldots, dx_n) : n \geq 1\}$ is consistent, the following theorem follows from Kolmogorov's extension theorem.

Theorem 6.1. *For each* $\mathbf{P}$ *in* $M_1(M_1(E))$, *there exists an infinite sequence* $\{Z_n : n \geq 1\}$ *of* E*-valued exchangeable random variables such that for each* $n \geq 1$, *the joint distribution of* $(Z_1, \ldots, Z_n)$ *is given by*

$$\mathbb{E}^{\mathbf{P}}[\mu(dx_1) \cdots \mu(dx_n)] \equiv \int \mu(dx_1) \cdots \mu(dx_n) \mathbf{P}(d\mu).$$

For an infinite exchangeable sequence $\{Z_n : n \geq 1\}$, and any $m \geq 1$, let $\mathscr{C}_m$ be the σ-field generated by events in $\sigma\{Z_1, Z_2, \ldots\}$ that are invariant under permutations of $(Z_1, \ldots, Z_m)$. The σ-field $\mathscr{C} = \bigcap_{m=1}^{\infty} \mathscr{C}_m$ is called the *exchangeable σ-field*. The following theorem provides a converse to Theorem 6.1.

Theorem 6.2. (de Finetti's theorem) *Conditional on* $\mathscr{C}$, *the infinite exchangeable sequence* $\{Z_n : n \geq 1\}$ *is i.i.d.*

Proof. For any $m \geq 1, 1 \leq i \leq m$, it follows from exchangeability that (Z_1, Y) equals (Z_i, Y) in distribution, where $Y = (h(Z_1, \ldots, Z_m), Z_{m+1}, \ldots)$ for a symmetric measurable function h. Therefore, for any bounded measurable function g on E,

$$\begin{aligned}
\mathbb{E}[g(Z_1) \mid \mathscr{C}_m] &= \mathbb{E}[g(Z_i) \mid \mathscr{C}_m] \\
&= \mathbb{E}\left[\frac{1}{m} \sum_{j=1}^{m} g(Z_j) \mid \mathscr{C}_m\right] \\
&= \frac{1}{m} \sum_{j=1}^{m} g(Z_j).
\end{aligned}$$

By the reversed martingale convergence theorem, one gets that

$$\frac{1}{m} \sum_{j=1}^{m} g(Z_j) \to \mathbb{E}[g(Z_1) \mid \mathscr{C}] \text{ almost surely.} \tag{6.1}$$

For fixed $k \geq 1$ and $m \geq k$, set

$$I_{k,m} = \{(j_1, \ldots, j_k) : 1 \leq j_r \leq m, j_r \neq j_l \text{ for } r \neq l\}.$$

Let $g_1,\ldots,g_k$ be bounded measurable functions on E. Then by exchangeability,

$$\mathbb{E}\left[\prod_{r=1}^{k} g_r(Z_r) \mid \mathscr{C}_m\right] = \frac{1}{m(m-1)\cdots(m-k+1)} \sum_{(j_1,\ldots,j_k)\in I_{k,m}} \prod_{r=1}^{k} g_{j_r}(Z_{j_r}). \quad (6.2)$$

Noting that as m goes to infinity

$$\frac{1}{m(m-1)\cdots(m-k+1)} \sum_{1\leq j_i\leq m,(j_1,\ldots,j_k)\notin I_{k,m}} \prod_{r=1}^{k} g_{j_r}(Z_{j_r}) \to 0,$$

and

$$\frac{1}{m^k} \sum_{1\leq j_i\leq m,(j_1,\ldots,j_k)\notin I_{k,m}} \prod_{r=1}^{k} g_{j_r}(Z_{j_r}) \to 0,$$

it follows from the reversed martingale theorem and (6.1) that, with probability one,

$$\begin{aligned}\lim_{m\to\infty} \mathbb{E}\left[\prod_{r=1}^{k} g_r(Z_r) \mid \mathscr{C}_m\right] &= \mathbb{E}\left[\prod_{r=1}^{k} g_r(Z_r) \mid \mathscr{C}\right] \\ &= \prod_{r=1}^{k} \mathbb{E}[g_r(Z_1) \mid \mathscr{C}].\end{aligned} \quad (6.3)$$

Thus, conditional on $\mathscr{C}$, $\{Z_n : n\geq 1\}$ are i.i.d. □

Since $\{Z_n : n\geq 1\}$ are conditionally i.i.d., it is natural to get:

Theorem 6.3. *For each $m\geq 1$, set $\xi_m = \frac{1}{m}\sum_{i=1}^{m} \delta_{Z_i}$. As m goes to infinity, ξ_m converges almost surely in $M_1(E)$ to a random probability measure ξ, a version of the regular conditional probability $\mathbb{P}[Z_1 \in \cdot \mid \mathscr{C}]$.*

Proof. Since E is Polish, there exists a countable set $\{f_l : l\geq 1\}$ of bounded continuous functions that is convergence determining in $M_1(E)$. The topology on $M_1(E)$ is then the same as the topology generated by the metric

$$\rho(\mu,\nu) = \sum_{l=1}^{\infty} 2^{-l}(1\wedge|\langle\mu-\nu,f_l\rangle|), \mu,\nu\in M_1(E).$$

Noting that $\langle\xi_m,f_l\rangle = \frac{1}{m}\sum_{i=1}^{m} f_l(Z_i)$, it follows from (6.1) that

$$\lim_{m\to\infty} \rho(\xi_m,\xi) = 0, \text{ almost surely.}$$

□

6.2 The Moran Process and the Fleming–Viot Process

Let S be the compact metric space in Section 5.2, and A be the parent-independent mutation generator defined in (5.16). For each $n \geq 1$, the Moran particle system with n particles is an S^n-valued Markov process $(X_1^n(t),\ldots,X_n^n(t))$ that evolves as follows:

1. **Motion process.** Each particle, independently of all other particles, moves in S according to A for a period of time exponentially distributed with parameter $\frac{n-1}{2}$.
2. **Sampling replacement.** At the end of its waiting period, the particle jumps at random to the location of one of the other $(n-1)$ particles.
3. **Renewal.** After the jump, the particle resumes the A-motion, independent of all other particles. This process is repeated for all particles.

Thus the generator of the Moran particle system is given by

$$Kf(x_1,\ldots,x_n) = \sum_{i=1}^{n} A_i f(x_1,\ldots,x_n) + \frac{1}{2}\sum_{i\neq j}[\Phi_{ij}f(x_1,\ldots,x_n) - f(x_1,\ldots,x_n)], \quad (6.4)$$

where f is in $C(S^n)$, A_i is A acted on the ith coordinate, and $\Phi_{ij}f$ is defined by

$$\Phi_{ij}f(x_1,\ldots,x_n) = f(\ldots,x_{j-1},x_i,x_{j+1},\ldots).$$

By relabeling, $\Phi_{ij}f$ can be viewed as a function in $C(S^{n-1})$.

Theorem 6.4. *For any $n \geq 1$, the operator K generates a unique Markov process with semigroup V_t^n satisfying*

$$\begin{aligned} V_t^n f(x_1,\ldots,x_n) &= e^{-\frac{n(n-1)t}{2}} T_t^{\otimes n} f(x_1,\ldots,x_n) \\ &\quad + \frac{1}{2}\sum_{i\neq j}\int_0^t e^{-\frac{n(n-1)u}{2}} T_u^{\otimes n}(\Phi_{ij}(V_{t-u}^n f))(x_1,\ldots,x_n)du, \end{aligned} \quad (6.5)$$

where f belongs to $C(S^n)$, T_t is the semigroup generated by A and $T_t^{\otimes n}$ is the semigroup of n independent A-motions.

Proof. Let $\{\tau_k : k \geq 1\}$ be a family of i.i.d. exponential random variables with parameter $\frac{n(n-1)}{2}$. Starting with $(X_0^n(0),\ldots,X_n^n(0))$, we run n independent A-motions until time τ_1. Here, we assume that the A-motions and the τ_k's are independent of each other. At time τ_1, a jump occurs involving a particular pair (i,j), and the $X_j^n(\tau_1)$ takes the value of $X_i^n(\tau_1)$. After the jump, the pattern is repeated until $\tau_1+\tau_2$. This continues for each time interval $[\sum_{i=1}^k \tau_i, \sum_{i=1}^{k+1}\tau_i)$ for all $k \geq 1$. Since

$$\lim_{k\to\infty}\sum_{i=1}^{k}\tau_i = \infty,$$

we have constructed explicitly a Markov process with generator K. Let

$$(X_1^n(t),\ldots,X_n^n(t))$$

be any Markov process generated by K and, for any f in $C(S^n)$, define

$$V_t^n f(x_1,\ldots,x_n) = \mathbb{E}[f(X_1^n(t),\ldots,X_n^n(t)) \mid X_1^n(0)=x_1,\ldots,X_n^n(0)=x_n]. \qquad (6.6)$$

Let $\{\tau_{ij} : i,j=1,\ldots,n; i\neq j\}$ be a family of i.i.d. exponential random variables with parameter $1/2$. Then $\tau = \inf\{\tau_{ij} : i,j=1,\ldots,n; i\neq j\}$ is the first time that a sampling replacement occurs. Clearly τ is an exponential random variable with parameter $\frac{n(n-1)}{2}$ and $\mathbb{P}\{\tau=\tau_{ij}\} = \frac{1}{n(n-1)}$. Then, conditioning on τ, it follows that

$$\begin{aligned}
&V_t^n f(x_1,\ldots,x_n)\\
&= \mathbb{E}[f(X_1^n(t),\ldots,X_n^n(t))(\chi_{\{\tau>t\}}+\chi_{\{\tau\le t\}}) \mid X_1^n(0)=x_1,\ldots,X_n^n(0)=x_n]\\
&= \mathbb{E}[f(X_1^n(t),\ldots,X_n^n(t))\chi_{\{\tau>t\}} \mid X_1^n(0)=x_1,\ldots,X_n^n(0)=x_n] \qquad (6.7)\\
&+\sum_{i\neq j}\mathbb{E}[f(X_1^n(t),\ldots,X_n^n(t))\chi_{\{\tau=\tau_{ij}\le t\}} \mid X_1^n(0)=x_1,\ldots,X_n^n(0)=x_n]\\
&= e^{-\frac{n(n-1)t}{2}} T_t^{\otimes n} f(x_1,\ldots,x_n)\\
&+\frac{1}{2}\sum_{i\neq j}\int_0^t e^{-\frac{n(n-1)u}{2}} T_u^{\otimes n}(\Phi_{ij}(V_{t-u}^n f))(x_1,\ldots,x_n)du.
\end{aligned}$$

Considering functions of the forms

$$\sum_{i=1}^n g_i(x_i),\ \sum_{i,j=1}^n g_{ij}(x_i,x_j),$$

and so on, one obtains the uniqueness of solutions to equation (6.5).

□

Remarks:

1. Let $\mathscr{F}_t$ denote the σ-algebra generated by $(X_1^n(t),\ldots,X_n^n(t))$ up to time t and thus $\{\mathscr{F}_t : t\ge 0\}$ be the natural filtration of $(X_1^n(t),\ldots,X_n^n(t))$. Then the process can be formulated as the unique process such that for any f in $C(S^n)$ and any $t>0$, $f(X_1^n(t),\ldots,X_n^n(t)) - \int_0^t Kf(X_1^n(u),\ldots,X_n^n(u))du$ is a martingale with respect to its natural filtration. In this framework, we say that $(X_1^n(t),\ldots,X_n^n(t))$ is the unique solution to the *martingale problem* associated with K.
2. Note that both the motion and the sampling replacement are symmetric. A permutation in the initial state will result in the same permutation at a later time. Thus if $(X_1^n(0),\ldots,X_n^n(0))$ is exchangeable, then the martingale problem associated with K is determined by symmetric functions and for each $t>0$, $(X_1^n(t),\ldots,X_n^n(t))$ is exchangeable.

Let

$$\eta_n(t) = \frac{1}{n}\sum_{i=1}^n \delta_{X_i^n(t)} \qquad (6.8)$$

be the empirical process of the Moran particle system. If $(X_1^n(0),\ldots,X_n^n(0))$ is exchangeable, then $\eta_n(\cdot)$ is a measure-valued Markov process. For any

$$f \in C_b^3(\mathbb{R}^k), \phi_1,\ldots,\phi_k \in C(S), 1 \leq i,j \leq k,$$

let f_i denote the first order partial derivative f with respect to the ith coordinate, and f_{ij} be the corresponding second order partial derivative. Set

$$\begin{aligned} F(\mu) &= f(\langle \mu,\phi_1\rangle,\ldots,\langle \mu,\phi_k\rangle),\\ F_i(\mu) &= f_i(\langle \mu,\phi_1\rangle,\ldots,\langle \mu,\phi_k\rangle),\\ F_{ij}(\mu) &= f_{ij}(\langle \mu,\phi_1\rangle,\ldots,\langle \mu,\phi_k\rangle). \end{aligned}$$

Then the generator of the process $\eta_n(\cdot)$ is given by

$$\begin{aligned} \mathscr{L}_n F(\mu) = &\sum_{i=1}^k F_i(\mu)\langle \mu, A\phi_i\rangle \\ &+\frac{1}{2}\sum_{i,j=1}^k [\langle \mu,\phi_i\phi_j\rangle - \langle \mu,\phi_i\rangle\langle \mu,\phi_j\rangle]F_{ij}(\mu) + o(n^{-1}). \end{aligned} \tag{6.9}$$

Clearly when n goes to infinity, $\mathscr{L}_n F(\mu)$ converges to $\mathscr{L}F(\mu)$, with $\mathscr{L}$ being the generator of the FV process with parent-independent mutation. Thus we have:

Theorem 6.5. *Let $D([0,+\infty),M_1(S))$ be the space of càdlàg functions equipped with the Skorodhod topology. Assume that $\eta_n(0)$ converges to $\eta(0)$ in $M_1(S)$ as n tends to infinity. Then the process $\eta_n(t)$ converges in distribution to the FV process η_t in $D([0,+\infty),M_1(S))$ as n goes to infinity; that is, the FV process can be derived through the empirical processes of a sequence of Moran particle systems.*

Proof. See Theorem 2.7.1 in [20]. □

For any $f(x_1,\ldots,x_n)$ in $C(S^n)$, let $\mu^n = \mu\times\cdots\times\mu$ and

$$\langle \mu^n, f\rangle = \int_{S^n} f(x_1,\ldots,x_n)\mu(dx_1)\cdots\mu(dx_n).$$

It follows from direct calculation that

$$\begin{aligned} \frac{\delta\langle \mu^n,f\rangle}{\delta\mu(x)} &= \sum_{i=1}^n \langle \mu^{n-1}, f^{i\to x}\rangle \\ \frac{\delta^2\langle \mu^n,f\rangle}{\delta\mu(x)\delta\mu(y)} &= \sum_{i\neq j}\langle \mu^{n-2}, f^{i\to x,j\to y}\rangle, \end{aligned}$$

where $f^{i\to x}$ and $f^{i\to x,j\to y}$ are obtained from f by setting $x_i = x$, and $x_i = x, x_j = y$, respectively. Thus

$$\begin{aligned}\mathscr{L}\langle\mu^n, f\rangle &= \left\langle \mu^n, \sum_{i=1}^{n} A_i f \right\rangle + \frac{1}{2}\sum_{i\neq j}[\langle\mu^{n-1}, \Phi_{ij} f\rangle - \langle\mu^n, f\rangle] \\ &= \langle\mu^n, Kf\rangle.\end{aligned} \tag{6.10}$$

Let $\eta(t)$ denote the FV process started at μ in $M_1(S)$. For any $n \geq 1$, let $M_n(t,\mu;dx_1,\ldots,dx_n)$ denote the nth moment measure of $\eta(t)$. It then follows from (6.10) that for any $f(x_1,\ldots,x_n)$ in $C(S^n)$,

$$\begin{aligned}&\int\cdots\int f(x_1,\ldots,x_n)M_n(t,\mu;dx_1,\ldots,dx_n) \\ &= \int\cdots\int V_t^n f(x_1,\ldots,x_n)\mu(dx_1)\cdots\mu(dx_n),\end{aligned} \tag{6.11}$$

where V_t^n is the semigroup of the Moran particle system with n particles, and the system has initial distribution μ^n.

The equality (6.11) indicates a *duality* relation between the Moran particle systems and the FV process. To be precise, let S_1 and S_2 be two topological spaces, and $B(S_1)$ and $B(S_2)$ the respective set of bounded measurable functions on S_1 and S_2. Similarly, we define $B(S_1 \times S_2)$ to be the set of bounded measurable functions on $S_1 \times S_2$.

Definition 6.2. Let X_t and Y_t be two Markov processes with respective state spaces S_1 and S_2. The process X_t and the process Y_t are said to be dual to each other with respect to an F in $B(S_1 \times S_2)$ if for any $t > 0$, x in S_1 and y in S_2,

$$\mathbb{E}[F(X_t, y) \mid X_0 = x] = \mathbb{E}[F(x, Y_t) \mid Y_0 = y]. \tag{6.12}$$

Remark: The use of dual processes is an effective tool in the study of stochastic processes. The basic idea of duality is to associate a dual process to the process of interest. The properties of the original process become more tractable if the dual process is simpler or easier to handle than the original process. The mere existence of a dual process usually guarantees the uniqueness of the martingale problem associated with the original process. Even when the dual process is not much simpler, it may still provide new perspectives on the original process. In the context of population genetics, dual processes relate naturally to the genealogy of the population.

Note that $\{M_n(t,\mu;\cdot) \in M_1(S^n) : n \geq 1\}$ forms a consistent family of probability measures, and for each n, $M_n(t,\mu;dx_1,\ldots,dx_n)$ is exchangeable; that is,

$$M_n(t,\mu;B_1 \times \cdots \times B_n) = M_n(t,\mu;B_{\pi(1)} \times \cdots \times B_{\pi(n)})$$

for any permutation π of the set $\{1,\ldots,n\}$. By Kolmogorov's extension theorem, this family of measures defines a sequence of exchangeable S-valued random variables $(X_1(t), X_2(t),\ldots)$ with a probability law P on $(S^\infty, \mathscr{S}^\infty)$, where $\mathscr{S}^\infty$ is the P-completion of the product σ-algebra. On the other hand, by de Finetti's theorem, the particle system determines a random measure $\tilde{\eta}(t)$ with moment measures given by

$$\{M_n(t,\mu;\cdot) \in M_1(S^n) : n \geq 1\}.$$

Therefore one could construct the FV process from an infinite exchangeable particle system. Due to the exchangeable structure, one could go one-step further to construct a particle system that incorporates the genealogical structure of the population. This brings us to the main topic of this chapter: particle representation.

6.3 The Donnelly–Kurtz Look-down Process

For each $n \geq 1$ and f in $C(S^n)$, define

$$\tilde{K}f(x_1,\ldots,x_n) = \sum_{i=1}^{n} A_i f(x_1,\ldots,x_n) + \sum_{i<j}[\Phi_{ij} f(x_1,\ldots,x_n) - f(x_1,\ldots,x_n)]. \quad (6.13)$$

Clearly $\tilde{K}$ and K differ only in the aspect of sampling replacement. This seemingly minor change in the sampling term leads to a new and more informative particle system: the *Donnelly–Kurtz look-down process.*

The existence and uniqueness of a Feller process with generator $\tilde{K}$ can be established by an argument similar to that used in the proof of Theorem 6.4.

The n-particle look-down process, denoted by $(Y_1(t),\ldots,Y_n(t))$ with each index representing a "level", is a Markov process that evolves as follows:

1. **Motion process.** The particle at level k, independently of all other particles, moves in S according to A for a period of time exponentially distributed with parameter $k-1$.
2. **Sampling replacement.** At the end of the waiting period, the particle at level k "looks down" at a particle chosen at random from the first $k-1$ levels and assumes its type.
3. **Renewal.** After the jump, the particle resumes the A-motion, independent of all other particles. This process is repeated for all particles.

Even though the transition mechanism is asymmetric, the look-down process still preserves exchangeability.

Theorem 6.6. *If* $(Y_1(0),\ldots,Y_n(0))$ *is exchangeable with a law of the form*

$$Q_n(dy_1\cdots dy_n) = \int \mu(dy_1)\cdots\mu(dy_n)\Theta(d\mu)$$

for some Θ *in* $M_1(M_1(S))$, *then* $(Y_1(t),\ldots,Y_n(t))$ *is exchangeable for each* $t > 0$.

Proof. First note that for any μ in $M_1(S)$, f in $C(S^n)$, and $1 \leq i < j \leq n$,

$$\langle \mu^{n-1}, \Phi_{ij} f\rangle = \langle \mu^{n-1}, \Phi_{ji} f\rangle.$$

Thus

$$\langle \mu^n, Kf \rangle = \langle \mu^n, \tilde{K} f \rangle,$$

and, by (6.10),

$$\mathscr{L}\langle \mu^n, f \rangle = \langle \mu^n, \tilde{K} f \rangle.$$

This relation of duality ensures that

$$\mathbb{E}^{\Theta}[\langle \eta(t)^n, f \rangle] = \mathbb{E}^{Q_n}[f(Y_1(t), \ldots, Y_n(t))], \tag{6.14}$$

where $\mathbb{E}^{\Theta}$ denotes the expectation with respect to the FV process with initial distribution Θ, and $\mathbb{E}^{Q_n}$ denotes the expectation with respect to the look-down process $(Y_1(t), \ldots, Y_n(t))$ with initial distribution Q_n. By the dominated convergence theorem, f can be chosen as any bounded measurable function on S^n. Therefore, the equality (6.14) implies that for any measurable sets $B_1, \ldots, B_n$ in S,

$$\mathbb{P}\{Y_i(t) \in B_i, i = 1, \ldots, n\} = \mathbb{E}^{\Theta}[\eta(t, B_1) \cdots \eta(t, B_n)] \tag{6.15}$$

from which the result follows.

□

For any $n \geq 1$, a measurable function $f(x_1, \ldots, x_n)$ on S^n is *symmetric* if

$$f(x_1, \ldots, x_n) = f(x_{\pi(1)}, \ldots, x_{\pi(n)})$$

for every permutation π of $\{1, \ldots, n\}$. Let $C_{sym}(S^n)$ be the set of all continuous, symmetric functions on S^n. By definition, each f in $C_{sym}(S^n)$ can be written as

$$f(x_1, \ldots, x_n) = \frac{1}{n!} \sum_{\pi} f(x_{\pi(1)}, \ldots, x_{\pi(n)}). \tag{6.16}$$

Substituting this into (6.4) and (6.13), yields

$$\begin{aligned}
&Kf(x_1, \ldots, x_n) \\
&= \frac{1}{n!} \sum_{\pi} Kf(x_{\pi(1)}, \ldots, x_{\pi(n)}) \\
&= \sum_{i=1}^{n} A_i f(x_1, \ldots, x_n) + \frac{1}{2} \sum_{i \neq j} (\Phi_{ij}(x_1, \ldots, x_n) - f(x_1, \ldots, x_n)) \\
&= \tilde{K} f(x_1, \ldots, x_n) \in C_{sym}(S^n).
\end{aligned}$$

Therefore, the generators K and $\tilde{K}$ coincide on $C_{sym}(S^n)$. Let $\mathscr{B}_1$ denote the σ-algebra on S generated by functions in $C_{sym}(S^n)$, and $\mathscr{B}_2$ the σ-algebra generated by the map

$$\eta : S^n \to M_1(S), \; (x_1, \ldots, x_n) \mapsto \frac{1}{n} \sum_{i=1}^{n} \delta_{x_i}.$$

Then it is known (cf. [20]) that $\mathscr{B}_1 = \mathscr{B}_2$. It is thus natural to expect:

Theorem 6.7. *For any $n \geq 1$, let $(X_1^n(t),\ldots,X_n^n(t))$ be the n-particle Moran process with generator K, and $(Y_1^n(t),\ldots,Y_n^n(t))$ the n-particle look-down process generated by $\tilde{K}$. Let $\eta_n(t)$ be defined as in (6.8) and*

$$\tilde{\eta}_n(t) = \frac{1}{n}\sum_{i=1}^{n} \delta_{Y_i(t)}. \tag{6.17}$$

Then the $M_1(S)$-valued processes $\eta_n(t)$ and $\tilde{\eta}_n(t)$ have the same distribution, provided that

$$(X_1^n(0),\ldots,X_n^n(0)) \stackrel{d}{=} (Y_1^n(0),\ldots,Y_n^n(0))$$

is exchangeable.

Proof. See Lemma 2.1 in [38]. A different proof can be found in Theorem 11.3.1 of [20].

□

For the Moran particle systems, a change in n will change the whole system. But for the look-down process, the n-particle process is naturally embedded into the $(n+1)$-particle process. This makes it possible to keep track of the genealogical structure of the population. By taking the projective limit, we end up with an infinite particle system $(Y_1(t),Y_2(t),\ldots)$ such that the first n coordinates form the n-particle look-down process. We call the infinite particle system $(Y_1(t),Y_2(t)\ldots)$ the *Donnelly–Kurtz infinite look-down process.*

Corollary 6.8 *If $(Y_1(0),Y_2(0),\ldots)$ is exchangeable, then $(Y_1(t),Y_2(t),\ldots)$ is exchangeable for all $t > 0$.*

Proof. By de Finetti's theorem, the exchangeability of $(Y_1(0),Y_2(0),\ldots)$ ensures that the law of $(Y_1(0),\ldots,Y_n(0))$ is of the form in the hypothesis of Theorem 6.6 for each $n \geq 1$. Thus the result follows from the fact that $(Y_1(t),\ldots,Y_n(t))$ is exchangeable for each $n \geq 1$.

□

Now we are ready to present the main result of this section.

Theorem 6.9. *Let $(Y_1(t),Y_2(t)\ldots)$ be the Donnelly–Kurtz infinite look-down process with an exchangeable initial law. Then*

$$\tilde{\eta}(t) = \lim_{n\to\infty} \tilde{\eta}_n(t) \;\; a.s. \tag{6.18}$$

and is a version of the FV process $\eta(t)$.

Proof. By Corollary 6.8, $(Y_1(t),Y_2(t)\ldots)$ is exchangeable for any $t > 0$. It follows from de Finetti's theorem and Theorem 6.3 that for each $t > 0$

$$\tilde{\eta}(t) = \lim_{n\to\infty} \tilde{\eta}_n(t) \;\; a.s.$$

The fact that $\tilde{\eta}(t)$ is a version of the FV processes, follows from Theorem 6.5 and Theorem 6.7.

□

6.4 Embedded Coalescent

The FV process in Chapter 5 is concerned with the forward evolution of a population under the influence of mutation and sampling replacement. At the infinity horizon, the population stabilizes at an equilibrium state. The model of the coalescent in Chapter 4 took a backward view. The same equilibrium state can be reached by tracing the most recent common ancestors. The close relation between the two is demonstrated in the representation of the transition functions in Chapter 5. The urn models studied in Chapter 2 are just different views of the coalescent.

We have seen how the FV process is embedded in the particle representation. It will be shown now that the coalescent is also naturally embedded in the look-down process. Thus the backward and forward viewpoints are unified through the particle representation.

Let $\{N_{ij}(\cdot) : i, j \geq 1\}$ be a family of independent Poisson processes on the whole real line with the Lebesgue measure as the common mean measure. Let each point in $N_{ij}(\cdot)$ correspond to a look-down event. Noting that there is a stationary process with generator A and stationary distribution ν_0, then by associating each point in $N_{ij}(\cdot)$ with an independent version of this stationary A-motion and following the look-down mechanism, we will have an explicit construction of the look-down process on the whole real line. Running this process backwards in time, for any $s < t \leq 0$, let

$$\sigma_j^1(t) = \sup\left\{u : \sum_{i<j} N_{ij}((u,t]) > 0\right\}$$

be the time of the most recent look-down from level j, and the level that is looked down at time $\sigma_j^1(t)$ is denoted by $l_j^1(t)$. Define $\sigma_j^2(t) = \sigma^1_{l_j^1(t)}(\sigma_j^1(t))$ and the level reached after the second most recent look-down from level j is denoted by $l_j^2(t)$. Similarly one can define $\sigma_j^k(t), l_j^k(t)$ for $k \geq 3$. The ancestor at time s of the level j particle at time t is then given by

$$\gamma_j(s,t) = \begin{cases} j, & \sigma_j^1(t) \leq s \\ l_j^k(t), & \sigma_j^{k+1}(t) \leq s < \sigma_j^k(t). \end{cases}$$

Consider the present time as zero. Take a sample of size n from the population at time zero. For any $t \geq 0$, let

$$\Upsilon(t) = \{\gamma_j(-t,0) : j = 1,\ldots,n\}$$

be the collection of ancestors at time t in the past of the sample at time zero. Introduce an equivalence relation $R_n(t)$ on the set $\{1,2,\ldots,n\}$ so that (i,j) belongs to $R_n(t)$ if and only if $\gamma_i(-t,0) = \gamma_j(-t,0)$. Clearly there is a one-to-one correspondence between the indices in $\Upsilon(t)$ and the equivalence classes derived from $R_n(t)$. Set

$$D_n(t) = |\Upsilon(t)| \equiv \text{the cardinality of } \Upsilon(t).$$

Theorem 6.10. *The process $\{R_n(u) : u \geq 0\}$ is Kingman's n-coalescent and thus $D_n(t)$, the number of equivalence classes in $R_n(t)$, is a pure-death process starting at n with transition rates*

$$q_{k,k-1} = \frac{k(k-1)}{2}, \; k = 2, \ldots, n.$$

Proof. By definition, $R_n(0) = \{(i,i) : i = 1, \ldots, n\}$ and $\{R_n(t) : t \geq 0\}$ is a pure jump Markov chain. Since each jump corresponds to a point in the Poisson processes $\{N_{ij}(\cdot) : i, j \geq 1\}$, the transition rate is thus one. At the time of a jump, the ancestor of one equivalence class looks down to the level associated another equivalence class. As a result, the two equivalence classes coalesce. The theorem now follows from Definition 4.1.

□

6.5 Notes

The material in Section 6.1 comes from [24], [2], and [20]. The derivation, in Section 6.2, of the FV process from the Moran particle system was obtained in [25]. The look-down process in Section 6.3 was introduced and studied in [38]. In [40], the particle representations and the corresponding genealogical processes were obtained for the FV process with mutation, selection and recombination. The particle representation was generalized in [41], to the measure-valued branching process. Further generalizations can be found in [132], where particle representations were studied for the measure-valued branching process with spatially varying birth rates.

Part II
Asymptotic Behaviors

Chapter 7
Fluctuation Theorems

Consider a family of random variables $\{Y_\lambda : \lambda > 0\}$. Assume that a weak law of large numbers holds; i.e., Y_λ converges in probability to a constant as λ tends to infinity. A *fluctuation theorem* such as the central limit theorem, refers to the existence of constants $a(\lambda)$ and $b(\lambda)$ such that the "normalized" family $a(\lambda)Y_\lambda - b(\lambda)$ converges in distribution to a nontrivial random variable as λ tends to infinity. When a fluctuation theorem is established, we will say the family $\{Y_\lambda : \lambda > 0\}$ or the family of laws of Y_λ satisfies the fluctuation theorem. In the first two sections of this chapter, fluctuation theorems are established for the Poisson–Dirichlet distribution, the Dirichlet process, and their two-parameter counterparts, when the scaled mutation rate θ tends to infinity. The last section includes several Gaussian limits associated with both the one- and two-parameter Poisson–Dirichlet distributions.

7.1 The Poisson–Dirichlet Distribution

In this section, we discuss the fluctuation theorems of the Poisson–Dirichlet distribution and its two-parameter generalization. The main idea is to exploit the subordinator representation and marginal distributions obtained in Chapter 2 and Chapter 3.

Let $Z_1 \geq Z_2 \geq \cdots$ be the ranked jump sizes up to time one of the gamma process $\{\gamma_t : t \geq 0\}$ with Lévy measure $\Lambda(dx) = \theta x^{-1} e^{-x} dx$. It follows from Theorem 2.2 that the law of

$$\mathbf{P}(\theta) = (P_1(\theta), P_2(\theta), \ldots) = \left(\frac{Z_1}{\gamma_1}, \frac{Z_2}{\gamma_1}, \ldots\right)$$

is the Poisson–Dirichlet distribution Π_θ. Let $N(\cdot)$ be the Poisson random measure associated with the jump sizes of γ_t over the interval $[0,1]$. Then for any $z \geq 0$,

$$\begin{aligned} \mathbb{P}\{Z_1 \leq z\} &= \mathbb{P}(N((z,+\infty)) = 0) \\ &= e^{-\Lambda((z,+\infty))} \\ &= \exp\{-\theta E_1(z)\}, \end{aligned} \tag{7.1}$$

S. Feng, *The Poisson–Dirichlet Distribution and Related Topics*, Probability and its Applications, DOI 10.1007/978-3-642-11194-5_7,

where $E_1(z) = \int_z^\infty u^{-1}e^{-u}du$. The density function of Z_1 is thus given by

$$\frac{\theta}{z}e^{-z}\exp\{-\theta E_1(z)\},\ z > 0.$$

Lemma 7.1. *For any $i \geq 1$, $P_i(\theta)$ converges in probability to zero as θ tends to infinity.*

Proof. It suffices to verify the case of $i = 1$. By direct calculation, we have

$$\mathbb{E}[\gamma_1] = \theta, \quad \mathbb{E}[\gamma_1^2] = (\theta+1)\theta,$$

and

$$\lim_{\theta\to\infty} \mathbb{E}\left[\left(\frac{\gamma_1}{\theta} - 1\right)^2\right] = 0.$$

Thus,

$$\frac{\gamma_1}{\theta} \to 1 \text{ in probability as } \theta \to \infty. \tag{7.2}$$

Since γ_1 is independent of $\frac{Z_1}{\gamma_1}$, it follows that

$$\mathbb{E}[P_1(\theta)] = \frac{\mathbb{E}[Z_1]}{\theta} = \int_0^\infty e^{-\theta E_1(z)}e^{-z}dz \longrightarrow 0 \quad \text{as} \quad \theta \to \infty.$$

□

Lemma 7.2. *As θ tends to infinity, the family of random variables $\sqrt{\theta}(\frac{\gamma_1}{\theta} - 1)$ converges in distribution to a standard normal random variable.*

Proof. Let $\varphi_\theta(t)$ denote the characteristic function of $\sqrt{\theta}(\frac{\gamma_1}{\theta} - 1)$. By direct calculation,

$$\begin{aligned}\log\varphi_\theta(t) &= \theta\log\left(1 - \frac{it}{\sqrt{\theta}}\right) - \frac{it}{\sqrt{\theta}}\\ &\longrightarrow -\frac{t^2}{2} \text{ as } \theta \to \infty,\end{aligned}$$

which yields the lemma.

□

Consider a Poisson random measure $N(\cdot)$ on $\mathbb{R}$ with mean measure

$$e^{-u}du,\ u \in \mathbb{R}.$$

Let $\zeta_1 \geq \zeta_2 \geq \cdots$ be the sequence of points of the Poisson random measure, in descending order. Then for each $r \geq 1$, the joint density of $(\zeta_1, \ldots, \zeta_r)$ is

$$e^{-\sum_{k=1}^r u_k}e^{-e^{-u_r}},\ -\infty < u_r < \cdots < u_1 < \infty, \tag{7.3}$$

and for each $k = 1, 2, \ldots$, the density function of ζ_k is

$$\frac{1}{\Gamma(k)} e^{-ku - e^{-u}}, \ u \in \mathbb{R}. \tag{7.4}$$

For

$$\beta(\theta) = \log\theta - \log\log\theta, \tag{7.5}$$

we have:

Lemma 7.3. *The family of the scaled random variables $\theta P_1(\theta) - \beta(\theta)$ converges in distribution to ζ_1 as θ tends to infinity.*

Proof. By L'Hospital's rule,

$$\lim_{x \to \infty} x e^x E_1(x) = 1. \tag{7.6}$$

Thus for any real number u,

$$\begin{aligned} &\lim_{\theta\to\infty} \exp\{-\theta E_1((u+\beta(\theta)))\} \\ &\quad = \lim_{\theta\to\infty} \exp\left\{-\theta \frac{(u+\beta(\theta))e^{(u+\beta(\theta))}E_1(u+\beta(\theta))}{(u+\beta(\theta))e^{(u+\beta(\theta))}}\right\} \\ &\quad = \lim_{\theta\to\infty} \exp\left\{-\theta \frac{(1+o(\frac{1}{\theta}))}{(u+\beta(\theta))e^u \cdot e^{\log\theta - \log\log\theta}}\right\} \\ &\quad = \lim_{\theta\to\infty} \exp\left\{-\frac{\log\theta}{(u+\log\theta-\log\log\theta)e^u}\left(1+o\left(\frac{1}{\theta}\right)\right)\right\} \\ &\quad = e^{-e^{-u}}. \end{aligned} \tag{7.7}$$

This, combined with (7.1), implies that $Z_1(\theta) - \beta(\theta)$ converges in distribution to ζ_1. Since $\frac{\theta}{\gamma_1}$ converges to 1 in probability, and $\sqrt{\theta}(\frac{\gamma_1}{\theta} - 1)$ converges in distribution to the standard normal random variable, it follows that

$$\theta P_1(\theta) - \beta(\theta) = \frac{\theta}{\gamma_1}(Z_1 - \beta(\theta)) - \frac{\theta}{\gamma_1}\sqrt{\theta}\left(\frac{\gamma_1}{\theta} - 1\right)\frac{\beta(\theta)}{\sqrt{\theta}} \tag{7.8}$$

and $Z_1 - \beta(\theta)$ both converge in distribution to ζ_1 as θ tends to infinity.

□

To establish the fluctuation theorem for the Poisson–Dirichlet distribution, we need a classical result in probability theory, Scheffé's theorem.

Theorem 7.1. (Scheffé) *Let $\{X_n : n \geq 1\}$ be a sequence of random variables taking values in $\mathbb{R}^d$ with corresponding density functions $\{f_n : n \geq 1\}$. If f_n converges point-wise to the density function f of an $\mathbb{R}^d$-valued random variable X, then X_n converges in distribution to X.*

Proof. For any Borel measurable set B in $\mathbb{R}^d$, it follows from the dominated convergence theorem that

$$
\begin{aligned}
|\int_B f_n(\mathbf{x})d\mathbf{x} - \int_B f(\mathbf{x})d\mathbf{x}| &\le \int_B |f_n(\mathbf{x}) - f(\mathbf{x})|d\mathbf{x} \\
&\le 2\int_B (f(\mathbf{x}) - f_n(\mathbf{x}))^+ d\mathbf{x} \\
&\le 2\int_B f(\mathbf{x})d\mathbf{x} \to 0,\ n \to \infty,
\end{aligned}
$$

which implies the result.

□

Now we are ready to prove the fluctuation theorem for the Poisson–Dirichlet distribution.

Theorem 7.2. *As θ tends to infinity,*

$$(\theta P_1(\theta) - \beta(\theta), \theta P_2(\theta) - \beta(\theta), \ldots)$$

converges in distribution to $(\zeta_1, \zeta_2, \ldots)$.

Proof. What we need to show is that the law of

$$(\theta P_1(\theta) - \beta(\theta), \theta P_2(\theta) - \beta(\theta), \ldots)$$

converges weakly to the law of $(\zeta_1, \zeta_2, \ldots)$ in the space $M_1(\nabla)$ as θ tends to infinity. By the Stone–Weierstrass theorem, it suffices to verify that for each $m \ge 1$, $(\theta P_1(\theta) - \beta(\theta), \ldots, \theta P_m(\theta) - \beta(\theta))$ converges in distribution to $(\zeta_1, \ldots, \zeta_m)$ as θ tends to infinity.

Following (7.8), we have that for any $t_1, \ldots, t_m$ in $\mathbb{R}$

$$
\begin{aligned}
\sum_{i=1}^{m} t_i(\theta P_1(\theta) - \beta(\theta)) &= \frac{\theta}{\gamma_1}\sum_{i=1}^{m} t_i(Z_i - \beta(\theta)) \\
&\quad - \frac{\theta}{\gamma_1}\sqrt{\theta}\left(\frac{\gamma_1}{\theta} - 1\right)\frac{\beta(\theta)}{\sqrt{\theta}}\sum_{i=1}^{m} t_i.
\end{aligned}
$$

Thus $\sum_{i=1}^{m} t_i(\theta P_1(\theta) - \beta(\theta))$ and $\sum_{i=1}^{m} t_i(Z_i - \beta(\theta))$ have the same limiting distribution, or equivalently,

$$(\theta P_1(\theta) - \beta(\theta), \ldots, \theta P_m(\theta) - \beta(\theta)) \text{ and } (Z_1 - \beta(\theta), \ldots, Z_m - \beta(\theta))$$

have the same limiting distribution.

By direct calculation, the joint density function of $(Z_1, \ldots, Z_m)$ is

$$\frac{\theta^m}{z_1 \cdots z_m} \exp\left\{-\sum_{i=1}^{m} z_i\right\} \exp\{-\theta E_1(z_m)\},\ z_1 \ge \cdots \ge z_m > 0,$$

and for any $u_1 \ge u_2 \ge \cdots \ge u_m$ and large enough θ, the joint density function of

$$(Z_1 - \beta(\theta), \ldots, Z_m - \beta(\theta))$$

is

$$\frac{\theta^m}{(u_1+\beta(\theta))\cdots(u_m+\beta(\theta))}\exp\left\{-\sum_{i=1}^m u_i - m\beta(\theta) - \theta E_1(u_m+\beta(\theta))\right\}$$
$$=\frac{(\log\theta)^m}{(u_1+\beta(\theta))\cdots(u_m+\beta(\theta))}\exp\left\{-\sum_{i=1}^m u_i\right\}\exp\{-\theta E_1(u_m+\beta(\theta))\},$$

which, by (7.6), converges to

$$\exp\left\{-\sum_{i-1}^m u_i\right\}\exp\{-e^{-u_m}\}$$

as θ tends to infinity. It follows by Theorem 7.1 that $(Z_1-\beta(\theta),\ldots,Z_m-\beta(\theta))$ converges in distribution to $(\zeta_1,\ldots,\zeta_m)$ as θ tends to infinity.

□

By the continuity of the projection map, we get:

Corollary 7.3 *For each $n\geq 2$, the family $\{\theta P_n(\theta)-\beta(\theta):\theta>0\}$ converges in distribution to ζ_n as θ tends to infinity.*

Remarks:

1. Let $V_1,V_2,\ldots$ be the GEM representation (2.13) of $\mathbf{P}(\theta)$. By direct calculation, we have

$$E[V_i]=\left(\frac{\theta}{\theta+1}\right)^{i-1}\frac{1}{\theta+1},\ i=1,2,\ldots,$$
$$E[P_1(\theta)]=\int_0^\infty e^{-u}e^{-\theta E_1(u)}du$$
$$\geq e^{-\beta(\theta)}e^{-\theta E_1(\beta(\theta))}\approx\frac{\log\theta}{\theta}.$$

This, combined with the fluctuation theorem, shows that the value of $P_1(\theta)$ obtained by the ordering of $V_1,V_2,\ldots$ increases its value by a scale factor of $\log\theta$ for large θ.

2. In the classical extreme value theory (cf. [19]), the limiting distributions in the fluctuation theorems for the maximum of n independent and identically distributed random variables consist of three families of distributions: the Gumbel, Fréchet, and Weibull. The distribution of ζ_1 belongs to the family of Gumbel distributions.

Next we turn to the fluctuation theorem for the two-parameter Poisson–Dirichlet distribution $PD(\alpha,\theta)$ with $0<\alpha<1,\theta>-\alpha$. Since θ will eventually tend to infinity, we may assume $\theta>0$. Let $\mathbf{P}(\alpha,\theta)=(P_1(\alpha,\theta),P_2(\alpha,\theta),\ldots)$ have the law $PD(\alpha,\theta)$. Then by Proposition 3.7, $\mathbf{P}(\alpha,\theta)$ can be represented as

$$\left(\frac{J_1(\alpha,\theta)}{\sigma_{\alpha,\theta}},\frac{J_2(\alpha,\theta)}{\sigma_{\alpha,\theta}},\ldots\right),$$

which is independent of the $Gamma(\theta, 1)$ random variable $\sigma_{\alpha,\theta}$.

Let

$$\beta(\alpha,\theta) = \log\theta - (\alpha+1)\log\log\theta - \log\Gamma(1-\alpha). \tag{7.9}$$

Then the following holds.

Lemma 7.4. *The family* $\{\theta P_1(\alpha,\theta) - \beta(\alpha,\theta) : \theta > 0\}$ *converges in distribution to* ζ_1 *as* θ *tends to infinity.*

Proof. First note that $\sigma_{\alpha,\theta}$ and γ_1 have the same distribution. Since

$$\begin{aligned} \theta P_1(\alpha,\theta) - \beta(\alpha,\theta) &= \frac{\theta}{\sigma_{\alpha,\theta}}(J_1(\alpha,\theta) - \beta(\alpha,\theta)) \\ &\quad - \frac{\theta}{\sigma_{\alpha,\theta}}\sqrt{\theta}\left(\frac{\sigma_{\alpha,\theta}}{\theta} - 1\right)\frac{\beta(\alpha,\theta)}{\sqrt{\theta}}, \end{aligned} \tag{7.10}$$

it follows from an argument similar to that used in the proof of Lemma 7.3, that $\theta P_1(\alpha,\theta) - \beta(\alpha,\theta)$ and $J_1(\alpha,\theta) - \beta(\alpha,\theta)$ have the same limiting distribution.

For any u in $\mathbb{R}$, choose θ large enough so that

$$\frac{\alpha}{\Gamma(1-\alpha)}\int_{u+\beta(\alpha,\theta)}^{\infty} x^{-(1+\alpha)} e^{-x} dx < 1.$$

Then

$$\begin{aligned} &\mathbb{P}\{J_1(\alpha,\theta) - \beta(\alpha,\theta) \le u\} \\ &\quad = \mathbb{E}[\mathbb{E}[J_1(\alpha,\theta) \le u + \beta(\alpha,\theta) | \gamma_{1/\alpha}]] \\ &\quad = \mathbb{E}\left[\exp\left\{\frac{\alpha}{\Gamma(1-\alpha)}\gamma_{1/\alpha}\int_{u+\beta(\alpha,\theta)}^{\infty} x^{-(1+\alpha)} e^{-x} dx\right\}\right] \\ &\quad = \left(1 - \frac{\alpha}{\Gamma(1-\alpha)}\int_{u+\beta(\alpha,\theta)}^{\infty} x^{-(1+\alpha)} e^{-x} dx\right)^{-\theta/\alpha}. \end{aligned} \tag{7.11}$$

The lemma now follows from the fact that

$$\begin{aligned} &\lim_{\theta\to\infty}\left(1 - \frac{\alpha}{\Gamma(1-\alpha)}\int_{u+\beta(\alpha,\theta)}^{\infty} x^{-(1+\alpha)} e^{-x} dx\right)^{-\theta/\alpha} \\ &\quad = \lim_{\theta\to\infty}\left(1 - \frac{\alpha}{\Gamma(1-\alpha)}(u+\beta(\alpha,\theta))^{-(1+\alpha)} e^{-(u+\beta(\alpha,\theta))}\right)^{-\theta/\alpha} \\ &\quad = \lim_{\theta\to\infty}\left(1 - \frac{\alpha e^{-u}}{\theta}\left(\frac{\log\theta}{u+\beta(\alpha,\theta)}\right)^{(1+\alpha)}\right)^{-\theta/\alpha} \\ &\quad = e^{-e^{-u}}. \end{aligned}$$

□

Theorem 7.4. *As θ tends to infinity,*

$$(\theta P_1(\alpha,\theta)-\beta(\alpha,\theta),\theta P_2(\alpha,\theta)-\beta(\alpha,\theta),\ldots)$$

converges in distribution to $(\zeta_1,\zeta_2,\ldots)$.

Proof. Following an argument similar to that used in the proof of Theorem 7.2, it suffices to verify that for each $m>1$, $(\theta P_1(\alpha,\theta)-\beta(\alpha,\theta),\ldots,\theta P_m(\alpha,\theta)-\beta(\alpha,\theta))$ converges in distribution to $(\zeta_1,\ldots,\zeta_m)$ as θ tends to infinity. For any $u_1\geq u_2\cdots\geq u_m>-\infty$, choose θ large enough such that

$$\frac{u_i+\beta(\alpha,\theta)}{\theta}>0,\ i=1,\ldots,m,$$

and

$$\sum_{i=1}^{m}\frac{u_i+\beta(\alpha,\theta)}{\theta}<1.$$

By Theorem 3.6, the joint density function of

$$(\theta P_1(\alpha,\theta)-\beta(\alpha,\theta),\ldots,\theta P_m(\alpha,\theta)-\beta(\alpha,\theta))$$

is

$$\frac{C_{\alpha,\theta}c_{\alpha}^{m}}{C_{\alpha,\theta+m\alpha}}\theta^m\frac{(1-\frac{\sum_{i=1}^{m}(u_i+\beta(\alpha,\theta))}{\theta})^{\theta+m\alpha-1}}{[(u_1+\beta(\alpha,\theta))\cdots(u_m+\beta(\alpha,\theta))]^{(1+\alpha)}}$$
$$\times\mathbb{P}\left\{\theta P_1(\alpha,\theta+m\alpha)\leq\frac{u_m+\beta(\alpha,\theta)}{1-\frac{\sum_{i=1}^{m}(u_i+\beta(\alpha,\theta))}{\theta}}\right\}.$$

Since

$$\lim_{\theta\to\infty}\frac{C_{\alpha,\theta}c_{\alpha}^{m}}{C_{\alpha,\theta+m\alpha}}=\frac{1}{\Gamma(1-\alpha)^m},$$

$$e^{-\sum_{i=1}^{m}(u_i+\beta(\alpha,\theta))}=e^{-\sum_{i=1}^{m}u_i}\theta^{-m}\Gamma(1-\alpha)^m(\log\theta)^{m(1+\alpha)},$$

and

$$\lim_{\theta\to\infty}e^{\sum_{i=1}^{m}(u_i+\beta(\alpha,\theta))}\left(1-\frac{\sum_{i=1}^{m}(u_i+\beta(\alpha,\theta))}{\theta}\right)^{\theta+m\alpha-1}=1,$$

it follows that

$$\lim_{\theta\to\infty}\frac{C_{\alpha,\theta}c_{\alpha}^{m}}{C_{\alpha,\theta+m\alpha}}\theta^m\frac{(1-\frac{\sum_{i=1}^{m}(u_i+\beta(\alpha,\theta))}{\theta})^{\theta+m\alpha-1}}{[(u_1+\beta(\alpha,\theta))\cdots(u_m+\beta(\alpha,\theta))]^{(1+\alpha)}}=e^{-\sum_{i=1}^{m}u_i}. \tag{7.12}$$

On the other hand,

$$\mathbb{P}\left\{\theta P_1(\alpha,\theta+m\alpha) \le \frac{u_m+\beta(\alpha,\theta)}{1-\frac{\sum_{i=1}^m(u_i+\beta(\alpha,\theta)}{\theta}}\right\}$$
$$= \mathbb{P}\{(\theta+m\alpha)P_1(\alpha,\theta+m\alpha)-\beta(\alpha,\theta+m\alpha)+\delta(\alpha,\theta)\le u_m\}.$$

where

$$\delta(\alpha,\theta)=\beta(\alpha,\theta)-\beta(\alpha,\theta+m\alpha)-\frac{m\alpha+\sum_{i=1}^m(u_i+\beta(\alpha,\theta))}{\theta}\,\frac{u_m+\beta(\alpha,\theta)}{1-\frac{\sum_{i=1}^m(u_i+\beta(\alpha,\theta))}{\theta}}$$

converges to zero as θ tends to infinity. By Lemma 7.4,

$$\lim_{\theta\to\infty}\mathbb{P}\left\{\theta P_1(\alpha,\theta+m\alpha)\le\frac{u_m+\beta(\alpha,\theta)}{1-\frac{\sum_{i=1}^m(u_i+\beta(\alpha,\theta)}{\theta}}\right\}=e^{-e^{-u_m}}, \tag{7.13}$$

which, combined with (7.12) and Theorem 7.1, implies the theorem.

□

7.2 The Dirichlet Process

Let ν_0 be a diffuse probability measure on $[0,1]$, and set $h(t)=\nu_0([0,t])$. Then $h(t)$ is a nondecreasing, continuous function on $[0,1]$ with $h(0)=0$, $h(1)=1$. For the sequence ξ_k, $k=1,2,\ldots$, of independent random variables with common distribution ν_0, let

$$\Xi_{\theta,\nu_0}=\sum_{k=1}^{\infty}P_k(\theta)\delta_{\xi_k}$$

and

$$\Xi_{\theta,\alpha,\nu_0}=\sum_{k=1}^{\infty}P_k(\alpha,\theta)\delta_{\xi_k}$$

be the respective one-parameter and two-parameter Dirichlet processes. For every t in $[0,1]$, $\Xi_{\theta,\nu_0}([0,t])$ and $\Xi_{\theta,\alpha,\nu_0}([0,t])$ have the same distributions as $\frac{\gamma(h(t))}{\gamma_1}$ and $\frac{\sigma(h(t)\gamma_{1/\alpha}(C\Gamma(1-\alpha))^{-1})}{\sigma_{\alpha,\theta}}$, respectively.

Let B_t denote the standard Brownian motion. For $0\le t\le 1$, the process

$$\hat{B}(t)=B_t-tB_1$$

is called the *Brownian bridge*. Both Brownian motion and the Brownian bridge appear as the limits in fluctuation theorems associated with subordinators. Due to the subordinator representations, it is natural to expect functional fluctuation theorems for the Dirichlet process and its two-parameter counterpart.

Let $\mathbb{D} = D([0,1])$ be the space of real-valued functions on $[0,1]$ that are right continuous and have left-hand limits (and left continuity at 1), and $\mathbb{H}$ denote the set of all strictly increasing and continuous functions from $[0,1]$ onto itself, keeping the end points fixed. For any x,y in $\mathbb{D}$, set

$$d(x,y) = \inf\{\sup_{0\le t\le 1} |r(t)-t| \sup_{0\le t\le 1} |x(r(t)) - y(t)| : r \in \mathbb{H}\}.$$

Then $(\mathbb{D},d)$ is a complete, separable metric space. The set $\mathbb{C}$ of all continuous functions on $[0,1]$ is a closed subset of $\mathbb{D}$ and the subspace topology of $\mathbb{C}$ is the topology of uniform convergence. For any $m \ge 1$, and $0 \le t_1 < \cdots < t_m \le 1$, the projection map

$$\pi_{t_1,\ldots,t_m} : \mathbb{D} \to \mathbb{R}^m,\ x(\cdot) \to (x(t_1),\ldots,x(t_m))$$

is measurable but not continuous on $\mathbb{D}$.

The keys to the proof of our fluctuation theorems are the following criterion for convergence in distribution and the comparison theorem.

Theorem 7.5. *Let $Y(t)$ belong to $\mathbb{C}$. Suppose that for any $m \ge 1$, and any $0 \le t_1 < \cdots < t_m \le 1$,*

$$\pi_{t_1,\ldots,t_m} Y_\lambda \Rightarrow \pi_{t_1,\ldots,t_m} Y,\ \lambda \to \infty; \tag{7.14}$$

and there exist $a > 0, b > \frac{1}{2}$, and a nondecreasing continuous function g on $[0,1]$ such that for any $0 \le t_1 \le t \le t_2 \le 1$,

$$\mathbb{E}[(|Y_\lambda(t) - Y_\lambda(t_1)| \cdot |Y_\lambda(t_2) - Y_\lambda(t)|)^a] \le (g(t_2) - g(t_1))^{2b}. \tag{7.15}$$

Then $Y_\lambda \Rightarrow Y$ as λ tends to infinity.

Proof. This follows from Theorem 15.6 in [14]. □

Theorem 7.6. (Comparison theorem) *Let $\{Y_\lambda(t) : \lambda > 0\}$ and $\{\tilde{Y}_\lambda(t) : \lambda > 0\}$ be two families of processes in $\mathbb{D}$. Assume that for any $\varepsilon > 0$,*

$$\lim_{\lambda\to\infty} \mathbb{P}\{\sup_{0\le t\le 1} |Y_\lambda(t) - \tilde{Y}_\lambda(t)| > \varepsilon\} = 0. \tag{7.16}$$

Then $Y_\lambda \Rightarrow Y$ if and only if $\tilde{Y}_\lambda \Rightarrow Y$ as λ tends to infinity.

Proof. Due to symmetry, it suffices to verify the "if" part. Assume that $\tilde{Y}_\lambda \Rightarrow Y$. Then the following hold:

(1) For any $b > 0$, there exists an $c > 0$ such that for all $\lambda > 0$,

$$\mathbb{P}\{\sup_{0\le t\le 1} |\tilde{Y}_\lambda(t)| \ge c\} \le b. \tag{7.17}$$

(2) For any $\varepsilon > 0$ and $b > 0$, there exist a $0 < \delta < 1$, and an $\lambda_0 > 0$ such that for $\lambda \ge \lambda_0$

$$\mathbb{P}\{w'_{\tilde{Y}_\lambda}(\delta) \geq \varepsilon\} \leq b, \tag{7.18}$$

where

$$w'_{\tilde{Y}_\lambda}(\delta) = \inf\{ \sup_{0<i\leq r; s,t\in[t_{i-1},t_i)} |\tilde{Y}_\lambda(t) - \tilde{Y}_\lambda(s)| : r \geq 1, 0 = t_0 < \cdots < t_r = 1,$$
$$t_k - t_{k-1} > \delta, k = 1,\ldots,r\}.$$

It follows from (7.16), that $Y_\lambda(t)$ also satisfies conditions (7.17) and (7.18). Therefore, the family of laws of $Y_\lambda(\cdot)$ is tight. Applying (7.16) again we obtain that $Y_\lambda \Rightarrow Y$.

□

Theorem 7.7. *Set*

$$X_\theta(t) = \sqrt{\theta}\left(\frac{\gamma(h(t))}{\theta} - h(t)\right),\ X(t) = B(h(t)).$$

Then $X_\theta(\cdot)$ converges in distribution to $X(\cdot)$ in $\mathbb{D}$ as θ tends to infinity.

Proof. First note that $X(\cdot)$ is in $\mathbb{C}$. For any $m \geq 1$, $0 = t_0 \leq t_1 < \cdots < t_m \leq 1$, let $s_k = h(t_k), k = 1,\ldots,m$. For any $\lambda_1,\ldots,\lambda_m$ in $\mathbb{R}$ and large θ, it follows by direct calculation that

$$\begin{aligned}
&\mathbb{E}\left\{\exp\left[\sum_{k=1}^m i\lambda_k(X_\theta(t_k) - X_\theta(t_{k-1}))\right]\right\}\\
&\quad = \exp\left\{-\sum_{k=1}^m \left[\theta(s_k - s_{k-1})\log\left(1 - i\frac{\lambda_k}{\sqrt{\theta}}\right) + i\lambda_k\sqrt{\theta}(s_k - s_{k-1})\right]\right\}\\
&\quad = \exp\left\{-\sum_{k=1}^m \frac{(s_k - s_{k-1})\lambda_k^2}{2} + O\left(\frac{1}{\sqrt{\theta}}\right)\right\}\\
&\quad \to \exp\left\{-\sum_{k=1}^m \frac{(s_k - s_{k-1})\lambda_k^2}{2}\right\},\ \theta \to \infty,
\end{aligned}$$

which implies that

$$(X_\theta(t_1) - X_\theta(t_0),\ldots,X_\theta(t_m) - X_\theta(t_{m-1})) \Rightarrow (X(t_1) - X(t_0),\ldots,X(t_m) - X(t_{m-1})),$$

as θ converges to infinity. Since the map

$$f(x_1,\ldots,x_m) = (x_1, x_1 + x_2, \ldots, x_1 + \cdots + x_m)$$

is continuous, it follows from the continuous-mapping theorem ([14]) that

$$(X_\theta(t_1),\ldots,X_\theta(t_m)) \Rightarrow (X(t_1),\ldots,X(t_m)),\ \theta \to \infty. \tag{7.19}$$

On the other hand, for any $0 \leq t_1 < t < t_2 \leq 1$,

$$
\begin{aligned}
&\mathbb{E}[(X_\theta(t)-X_\theta(t_1))^2(X_\theta(t_2)-X_\theta(t))^2]\\
&\quad=\theta\mathbb{E}\left[\left(\frac{\gamma(h(t))-\gamma(h(t_1))}{\theta}-(h(t)-h(t_1))\right)^2\right]\\
&\quad\quad\times\mathbb{E}\left[\left(\frac{\gamma(h(t_2))-\gamma(h(t))}{\theta}-(h(t_2)-h(t))\right)^2\right] \qquad (7.20)\\
&\quad=(h(t)-h(t_1))(h(t_2)-h(t))\\
&\quad\le(h(t_2)-h(t_1))^2,
\end{aligned}
$$

which is (7.15) with $a=2, b=1, g=h$. Therefore the result follows from Theorem 7.5.

□

Next we turn to the fluctuation theorem for $\Xi_{\theta,\nu_0}([0,t])$ or equivalently its subordinator representation through $\frac{\gamma(h(t))}{\gamma_1}$.

Theorem 7.8. *As θ tends to infinity, the process $\sqrt{\theta}(\frac{\gamma(h(\cdot))}{\gamma_1}-h(\cdot))$ converges in distribution to $\hat{B}(h(\cdot))$ in $\mathbb{D}$.*

Proof. By direct calculation,

$$
\begin{aligned}
\sqrt{\theta}\left(\frac{\gamma(h(t))}{\gamma_1}-h(t)\right)&=X_\theta(t)+\sqrt{\theta}\frac{\gamma(h(t))}{\theta}\left(\frac{\theta}{\gamma_1}-1\right) \qquad (7.21)\\
&=X_\theta(t)-h(t)X_\theta(1)-\left(\frac{\gamma(h(t))}{\gamma_1}-h(t)\right)X_\theta(1).
\end{aligned}
$$

It follows from Theorem 7.7 that $X_\theta(\cdot)\Rightarrow B(h(\cdot))$ and $X_\theta(1)\Rightarrow B(1)$. Since the map $x(\cdot)\mapsto x(\cdot)-h(\cdot)x(1)$ is continuous on $\mathbb{D}$, it follows from the continuous-mapping theorem that

$$X_\theta(\cdot)-h(\cdot)X_\theta(1)\Rightarrow\hat{B}(h(\cdot)),\ \theta\to\infty.$$

For any $\varepsilon>0$,

$$
\begin{aligned}
&\mathbb{P}\left\{\sup_{0\le t\le 1}\left|\left(\frac{\gamma(h(t))}{\gamma_1}-h(t)\right)X_\theta(1)\right|\ge\varepsilon\right\}\\
&\le\mathbb{P}\left\{\sup_{0\le t\le 1}\left|\left[\left(\frac{\gamma(h(t))}{\theta}-h(t)\right)+\left(\frac{\theta}{\gamma_1}-1\right)\frac{\gamma(h(t))}{\theta}\right]X_\theta(1)\right|\ge\varepsilon\right\}\\
&\le\mathbb{P}\left\{\sup_{0\le t\le 1}\left|\left(\frac{\gamma(h(t))}{\theta}-h(t)\right)X_\theta(1)\right|\ge\varepsilon/2\right\}\\
&\quad+\mathbb{P}\left\{\left|\left(\frac{\theta}{\gamma_1}-1\right)\frac{\gamma_1}{\theta}X_\theta(1)\right|\ge\varepsilon/2\right\}.
\end{aligned}
$$

Since $\frac{\gamma(h(t))}{\theta} - h(t)$ is a martingale, it follows from the Cauchy–Schwarz inequality and Doob's inequality that

$$\begin{aligned}
&\mathbb{P}\left\{\sup_{0\le t\le 1}\left|\left(\frac{\gamma(h(t))}{\gamma_1} - h(t)\right) X_\theta(1)\right| \ge \varepsilon\right\} \\
&\le \frac{2}{\varepsilon}\left(\mathbb{E}\left[\sup_{0\le t\le 1}\left(\frac{\gamma(h(t))}{\theta} - h(t)\right)^2\right]\right)^{1/2} \left(\mathbb{E}[X_\theta^2(1)]\right)^{1/2} \\
&\qquad\qquad + \mathbb{P}\{X_\theta^2(1) \ge \sqrt{\theta}\varepsilon/2\} \qquad\qquad (7.22)\\
&\le \frac{4}{\varepsilon}\left(\mathbb{E}\left[\left(\frac{\gamma_1}{\theta} - 1\right)^2\right]\right)^{1/2} \left(\mathbb{E}[X_\theta^2(1)]\right)^{1/2} + \mathbb{P}\{X_\theta^2(1) \ge \sqrt{\theta}\varepsilon/2\} \\
&\le \frac{6}{\sqrt{\theta}\varepsilon}\mathbb{E}[X_\theta^2(1)] \\
&= \frac{6}{\sqrt{\theta}\varepsilon},
\end{aligned}$$

where the last equality is due to $\mathbb{E}[X_\theta^2(1)] = 1$. The theorem now follows from Theorem 7.6 and Theorem 7.7.

□

The fact that the gamma process has independent increments plays a key role in the proof of Theorem 7.7. In the subordinator representation of the two-parameter Dirichlet process, the increments of the process $\tilde{\sigma}(t) = \sigma(\gamma_{1/\alpha}(C\Gamma(1-\alpha))^{-1})t)$ are exchangeable instead of independent. Thus conditional arguments are needed to establish the corresponding fluctuation theorem.

Theorem 7.9. *Let*

$$Y_\theta(t) = \sqrt{\theta}\left(\frac{\tilde{\sigma}(h(t))}{\sigma_{\alpha,\theta}} - h(t)\right).$$

Then $Y_\theta(\cdot)$ converges in distribution to $\sqrt{1-\alpha}\hat{B}(h(\cdot))$ in $\mathbb{D}$ as θ tends to infinity.

Proof. First note that $\tilde{\sigma}(1) = \sigma_{\alpha,\theta}$. For every s in $[0,1]$, we have

$$\begin{aligned}
\mathbb{E}[\tilde{\sigma}(s)] &= s\theta, \\
\mathbb{E}[\tilde{\sigma}^2(s)] &= \mathbb{E}[\mathbb{E}[\tilde{\sigma}^2(s) \mid \gamma_{1/\alpha}]] \\
&= \mathbb{E}[Var[\tilde{\sigma}(s) \mid \gamma_{1/\alpha}] + (\mathbb{E}[\tilde{\sigma}(s) \mid \gamma_{1/\alpha}])^2] \\
&= (1-\alpha)\theta s + (\theta+\alpha)\theta s^2,
\end{aligned}$$

and

$$Var\left[\frac{\tilde{\sigma}(s)}{\theta}\right] = \frac{(1-\alpha)s + \alpha s^2}{\theta}.$$

Thus for every $0 \le t \le 1$,

$$\frac{\tilde{\sigma}(h(t))}{\theta} \to h(t), \text{ in probability.} \tag{7.23}$$

Set $Z_\theta(t) = \frac{1}{\sqrt{\theta}}(\tilde{\sigma}(h(t)) - h(t)\tilde{\sigma}(1))$. It is not hard to check that

$$Y_\theta(t) = Z_\theta(t) - \tilde{X}_\theta(1)\left(\frac{\tilde{\sigma}(h(t))}{\tilde{\sigma}(1)} - h(t)\right), \tag{7.24}$$

where

$$\tilde{X}_\theta(1) = \sqrt{\theta}\left(\frac{\tilde{\sigma}(1)}{\theta} - 1\right)$$

has the same distribution as $X_\theta(1)$ in (7.22).

By an argument similar to that used in (7.22), it follows that for any $\varepsilon > 0$,

$$\begin{aligned}
&\mathbb{P}\left\{\sup_{0\le t\le 1}\left|\tilde{X}_\theta(1)\left(\frac{\tilde{\sigma}(h(t))}{\tilde{\sigma}(1)} - h(t)\right)\right| \ge \varepsilon\right\}\\
&\quad\le \mathbb{P}\left\{\sup_{0\le t\le 1}\left|\tilde{X}_\theta(1)\left[\left(\frac{\tilde{\sigma}(h(t))}{\theta} - h(t)\right) + \left(\frac{\theta}{\tilde{\sigma}(1)} - 1\right)\left(\frac{\tilde{\sigma}(h(t))}{\theta}\right)\right]\right| \ge \varepsilon\right\}\\
&\quad\le \mathbb{P}\left\{\sup_{0\le t\le 1}\left|\left(\frac{\tilde{\sigma}(h(t))}{\theta} - h(t)\right)\tilde{X}_\theta(1)\right| \ge \varepsilon/2\right\}\\
&\qquad\qquad + \mathbb{P}\left\{\left|\left(\frac{\tilde{\sigma}(1)}{\theta} - 1\right)\tilde{X}_\theta(1)\right| \ge \varepsilon/2\right\} \qquad (7.25)\\
&\quad\le \frac{2}{\varepsilon}\left(\mathbb{E}\left[\sup_{0\le t\le 1}\left(\frac{\tilde{\sigma}(h(t))}{\theta} - h(t)\right)^2\right]\right)^{1/2} + \mathbb{P}\{\tilde{X}_\theta^2(1) \ge \sqrt{\theta}\varepsilon/2\}\\
&\quad\le \frac{2}{\varepsilon}\left(\mathbb{E}\left[\mathbb{E}\left[\sup_{0\le t\le 1}\left(\frac{\tilde{\sigma}(h(t))}{\theta} - h(t)\right)^2\Big|\gamma_{1/\alpha}\right]\right]\right)^{1/2} + \frac{2}{\sqrt{\theta}\varepsilon}\\
&\quad\le \frac{4}{\varepsilon}\left(\mathbb{E}\left[\mathbb{E}\left[\left(\frac{\tilde{\sigma}(1)}{\theta} - 1\right)^2\Big|\gamma_{1/\alpha}\right]\right]\right)^{1/2} + \frac{2}{\sqrt{\theta}\varepsilon}\\
&\quad= \frac{6}{\sqrt{\theta}\varepsilon},
\end{aligned}$$

where in the last inequality, Doob's inequality and the martingale property of $\frac{\tilde{\sigma}(h(t))}{\theta} - h(t)$ are used under the conditional law given $\gamma_{1/\alpha}$. By Theorem 7.6, $Y_\theta(\cdot)$ and $Z_\theta(\cdot)$ have the same limit in distribution.

To prove the theorem it suffices to verify conditions (7.14) and (7.15) in Theorem 7.5 for the process $Z_\theta(\cdot)$.

Let $m \ge 1$, $0 = t_0 \le t_1 < \cdots < t_m \le 1$, $s_k = h(t_k)$, $k = 1, \ldots, m$, and $\lambda_1, \ldots, \lambda_m$ in $\mathbb{R}$. By choosing λ_m to be zero if necessary, we can take $t_m = 1$.

Set

$$\hat{\lambda}_k = \lambda_k - \sum_{i=1}^m \lambda_i(s_i - s_{i-1}).$$

Then we have

$$\sum_{k=1}^{m} \hat{\lambda}_k (s_k - s_{k-1}) = 0,$$

$$\sum_{k=1}^{m} \hat{\lambda}_k^2 (s_k - s_{k-1}) = \sum_{k=1}^{m} (s_k - s_{k-1}) \lambda_k^2 - \left(\sum_{k=1}^{m} (s_k - s_{k-1}) \lambda_k \right)^2.$$

By direct calculation,

$$\sum_{i=1}^{m} \lambda_k (\hat{B}(s_k) - \hat{B}(s_{k-1})) = \sum_{k=1}^{m} \hat{\lambda}_k (B(s_k) - B(s_{k-1})).$$

Thus,

$$\begin{aligned}
&\mathbb{E}\left[\exp\left\{ i\sqrt{1-\alpha} \sum_{k=1}^{m} \lambda_k (\hat{B}(s_k) - \hat{B}(s_{k-1})) \right\}\right] \\
&\quad = \mathbb{E}\left[\exp\left\{ i\sqrt{1-\alpha} \sum_{k=1}^{m} \hat{\lambda}_k (B(s_k) - B(s_{k-1})) \right\}\right] \\
&\quad = \exp\left\{ -\frac{1-\alpha}{2} \sum_{k=1}^{m} \hat{\lambda}_k (s_k - s_{k-1}) \right\} \\
&\quad = \exp\left\{ -\frac{1-\alpha}{2} \left[\sum_{k=1}^{m} (s_k - s_{k-1}) \lambda_k^2 - \left(\sum_{k=1}^{m} (s_k - s_{k-1}) \lambda_k \right)^2 \right] \right\}.
\end{aligned}$$

On the other hand, by writing

$$\sum_{i=1}^{m} \lambda_k (Z_\theta(t_k) - Z_\theta(t_{k-1})) = \sum_{k=1}^{m} \frac{\hat{\lambda}_k}{\sqrt{\theta}} (\tilde{\sigma}(s_k) - \tilde{\sigma}(s_{k-1})),$$

it follows that for large θ,

$$\begin{aligned}
&\mathbb{E}\left[\exp\left\{ \sum_{k=1}^{m} i\lambda_k (Z_\theta(t_k) - Z_\theta(t_{k-1})) \right\}\right] \\
&= \mathbb{E}\left[\exp\left\{ \frac{\gamma_{1/\alpha}}{C\Gamma(1-\alpha)} \sum_{k=1}^{m} (s_k - s_{k-1}) \int_0^\infty (e^{\frac{\hat{\lambda}_k}{\sqrt{\theta}}x} - 1)\alpha C x^{-(1+\alpha)} e^{-x} dx \right\}\right] \\
&= \exp\left\{ -\frac{\theta}{\alpha} \log\left(1 - \sum_{k=1}^{m} (s_k - s_{k-1}) \frac{\alpha \int_0^\infty (e^{\frac{\hat{\lambda}_k}{\sqrt{\theta}}x} - 1) x^{-(1+\alpha)} e^{-x} dx}{\Gamma(1-\alpha)} \right) \right\} \\
&= \exp\left\{ -\sum_{k=1}^{m} \frac{(1-\alpha)(s_k - s_{k-1})\hat{\lambda}_k^2}{2} + O\left(\frac{1}{\sqrt{\theta}}\right) \right\}.
\end{aligned}$$

As θ tends to infinity, we get

$$\mathbb{E}\left[\exp\left\{\sum_{k=1}^{m} i\lambda_k(Z_\theta(t_k)-Z_\theta(t_{k-1}))\right\}\right]$$
$$\to \exp\left\{-\frac{1-\alpha}{2}\left[\sum_{k=1}^{m}(s_k-s_{k-1})\lambda_k^2-\left(\sum_{k=1}^{m}(s_k-s_{k-1})\lambda_k\right)^2\right]\right\},$$

which, combined with the continuous-mapping theorem, implies that

$$(Z_\theta(t_1),\ldots,Z_\theta(t_m)) \Rightarrow \sqrt{1-\alpha}(\hat{B}(h(t_1)),\ldots,\hat{B}(h(t_m))),\ \theta\to\infty. \tag{7.26}$$

For $t_0=0,\ t_4=1,\ \lambda_1=\lambda_4=0$, and any $0\le t_1<t_2<t_3\le 1$, expanding

$$\mathbb{E}[\exp\{i\sqrt{\theta}[(\lambda_2(Z_\theta(t_2)-Z_\theta(t_1))+\lambda_3(Z_\theta(t_3)-Z_\theta(t_2))]\}]$$
$$=\mathbb{E}\left[\exp\left\{i\sqrt{\theta}\sum_{k=1}^{4}\lambda_k(Z_\theta(t_k)-Z_\theta(t_{k-1})\right\}\right]$$
$$=\mathbb{E}\left[\exp\left\{\frac{\alpha\gamma_{1/\alpha}}{\Gamma(1-\alpha)}\int_0^\infty\left[\sum_{k=1}^{4}(s_k-s_{k-1})(e^{i\hat{\lambda}_k x}-1)\right]x^{-(1+\alpha)}e^{-x}dx\right\}\right]$$

as power series in λ_2 and λ_3, and equating the coefficients of $\lambda_2^2\lambda_3^2$, we obtain

$$\mathbb{E}[(Z_\theta(t)-Z_\theta(t_1))^2(Z_\theta(t_2)-Z_\theta(t))^2]=\frac{I_1+I_2}{\theta^2},$$

where

$$\begin{aligned}
I_1 &= \frac{6\theta\Gamma(4-\alpha)}{\Gamma(1-\alpha)}[(1-(s_3-s_1))(s_2-s_1)^2(s_3-s_2)^2\\
&\quad + (s_2-s_1)(1-(s_2-s_1))^2(s_3-s_2)^2+(s_3-s_2)(1-(s_3-s_2))^2(s_2-s_1)^2]\\
&\le C_1(\alpha)\theta(h(t_3)-h(t_2))(h(t_2)-h(t_1)),\\
I_2 &= 6(1-\alpha)^2(\theta+\alpha)\theta(s_2-s_1)(s_3-s_2)[(1-(s_2-s_1))(1-(s_3-s_2))\\
&\quad + 2(s_2-s_1)(s_3-s_2)]\\
&\le C_2(\alpha)\theta^2(h(t_2)-h(t_1))(h(t_3)-h(t_2)).
\end{aligned}$$

It follows by choosing $C_3(\alpha)=C_1(\alpha)+C_2(\alpha)$, that

$$\begin{aligned}
&\mathbb{E}[(Z_\theta(t)-Z_\theta(t_1))^2(Z_\theta(t_2)-Z_\theta(t))^2]\\
&\qquad\le C_3(\alpha)(h(t_2)-h(t_1))(h(t_3)-h(t_2))
\end{aligned} \tag{7.27}$$

which, combined with (7.26), implies the result.

□

Remark: In comparison with the fluctuation theorem of the one-parameter Dirichlet process, the impact of the parameter α is reflected from the factor $\sqrt{1-\alpha}$.

7.3 Gaussian Limits

Consider a population of individuals whose types are labeled by $\{1,2,\ldots\}$. Let $\mathbf{p} = (p_1, p_2, \ldots)$ denote the relative frequencies of all types with p_i denoting the relative frequency of type i for $i \geq 1$. For any $n \geq 2$, and a random sample of size n from the population, the probability that all individuals in the sample are of the same type is given by

$$\sum_{i=1}^{\infty} p_i^n,$$

which is the function $\varphi_n(\mathbf{p})$ defined in (5.5).

If the relative frequencies of all types in the population follow the one-parameter Poisson–Dirichlet distribution, then $\varphi_2(\mathbf{P}(\theta))$ is known as *homozygosity* of the population in population genetics. For general $n \geq 2$, we call $\varphi_n(\mathbf{P}(\theta))$ the *homozygosity of order n*. It follows from the Ewens sampling formula that

$$\mathbb{E}[\varphi_n(\mathbf{P}(\theta))] = \frac{(n-1)!}{(\theta+1)_{(n-1)}} \to 0,\ \theta \to \infty.$$

This implies that $\varphi_n(\mathbf{P}(\theta))$ converges to zero in probability. It is thus natural to consider fluctuation theorems for $\{\varphi_n(\mathbf{P}(\theta)) : n \geq 2\}$ as θ tends to infinity.

Theorem 7.10. *Let*

$$H_j(\theta) = \sqrt{\theta}\left(\frac{\theta^{j-1}}{\Gamma(j)}\varphi_j(\mathbf{P}(\theta)) - 1\right),\ j = 2,3,\ldots$$

and

$$\mathbf{H}_\theta = (H_2(\theta), H_3(\theta), \ldots).$$

Then

$$\mathbf{H}_\theta \Rightarrow \mathbf{H},\ \theta \to \infty, \tag{7.28}$$

where $\mathbf{H} = (H_2, H_3, \ldots)$ *is a* $\mathbb{R}^\infty$*-valued random element and for each* $r \geq 2$, $(H_2, \ldots, H_r)$ *has a multivariate normal distribution with zero means, and covariance matrix*

$$Cov(H_j, H_l) = \frac{\Gamma(j+l) - \Gamma(j+1)\Gamma(l+1)}{\Gamma(j)\Gamma(l)},\ j,l = 2,\ldots,r. \tag{7.29}$$

Proof. First note that

$$\mathbf{P}(\theta) = \left(\frac{Z_1}{\gamma_1}, \frac{Z_2}{\gamma_1}, \ldots\right).$$

For each $j \geq 1$, set

$$M_j(\theta) = \sqrt{\theta}\left(\frac{1}{\Gamma(j)\theta}\sum_{i=1}^{\infty} Z_i^j - 1\right),$$
$$\mathbf{M}_\theta = (M_1(\theta), \ldots).$$

For each fixed $r \geq 1$, and any $(\lambda_1, \ldots, \lambda_r)$ in $\mathbb{R}^r$, set

$$f_\theta(x) = \sum_{j=1}^{r} \frac{1}{\Gamma(j)\sqrt{\theta}} \lambda_j x^j.$$

It follows from Theorem A.6, that

$$\begin{aligned}
&\mathbb{E}\left(e^{it\sum_{j=1}^r \lambda_j M_j(\theta)}\right) \\
&= e^{-it\sum_{j=1}^r \lambda_j \sqrt{\theta}}\, \mathbb{E}\left(e^{it\sum_{l=1}^\infty f_\theta(Z_l)}\right) \qquad (7.30)\\
&= e^{-it\sum_{j=1}^r \lambda_j \sqrt{\theta}} \exp\left\{\theta \int_0^\infty (e^{itf_\theta(y)} - 1) y^{-1} e^{-y} dy\right\} \\
&\to \exp\left\{-\frac{t^2}{2} \sum_{j,l=1}^{r} \lambda_j \lambda_l \frac{\Gamma(j+l)}{\Gamma(j)\Gamma(l)}\right\}.
\end{aligned}$$

Let $\mathbf{M} = (M_1, \ldots,)$ be such that for each $r \geq 1$, $(M_1, \ldots, M_r)$ is a multivariate normal random vector with zero mean and covariance matrix

$$\frac{\Gamma(j+l)}{\Gamma(j)\Gamma(l)}, \; j,l = 1, \ldots, r. \qquad (7.31)$$

Then (7.30) implies that $\mathbf{M}_\theta$ converges in distribution to $\mathbf{M}$. It follows from direct calculation, that for any $j \geq 2$

$$H_j(\theta) = M_j(\theta) + \sqrt{\theta}\left(\left(\frac{\theta}{\gamma_1}\right)^j - 1\right)\left(\sum_{l=1}^{\infty} \frac{Z_l^j}{\Gamma(j)\theta}\right). \qquad (7.32)$$

Since $\mathbf{M}_\theta \Rightarrow \mathbf{M}$, it follows that

$$\sum_{l=1}^{\infty} \frac{Z_l^j}{\Gamma(j)\theta} \to 1 \text{ in distribution.} \qquad (7.33)$$

By Theorem 7.7 and basic algebra, one gets

$$\sqrt{\theta}\left(\left(\frac{\theta}{\gamma_1}\right)^j - 1\right) \Rightarrow -jM_1, \qquad (7.34)$$

which, combined (7.32) and (7.33), yields

$$\sum_{j=2}^{r} \alpha_j H_j(\theta) \Rightarrow \sum_{j=2}^{r} \alpha_j H_j, \tag{7.35}$$

where $H_j = M_j - jM_1$.

The theorem now follows from the fact that the covariance of H_j and H_l is

$$\frac{\Gamma(j+l) - \Gamma(j+1)\Gamma(l+1)}{\Gamma(j)\Gamma(l)}.$$

□

Next consider the case that the relative frequencies of all types in the population follow the two-parameter Poisson–Dirichlet distribution with parameters $0 < \alpha < 1, \theta > -\alpha$. By the Pitman sampling formula, we have that for any $n \geq 2$,

$$\mathbb{E}[\varphi_n(\mathbf{P}(\alpha,\theta))] = \frac{(1-\alpha)_{(n-1)}}{(\theta+1)_{(n-1)}} \to 0, \ \theta \to \infty.$$

Therefore, $\varphi_n(\mathbf{P}(\alpha,\theta))$ also converges to zero in probability when θ tends to infinity. The corresponding fluctuation result for $\{\varphi_n(\mathbf{P}(\alpha,\theta)) : n \geq 2\}$ is established in the next theorem.

Theorem 7.11. *Let*

$$H_j(\alpha,\theta) = \sqrt{\theta}\left(\frac{\Gamma(1-\alpha)}{\Gamma(j-\alpha)}\theta^{j-1}\varphi_j(\mathbf{P}(\alpha,\theta)) - 1\right), \ j = 2,3,\ldots$$

and

$$\mathbf{H}_\theta^\alpha = (H_2(\alpha,\theta), H_3(\alpha,\theta), \ldots).$$

Then

$$\mathbf{H}_\theta^\alpha \Rightarrow \mathbf{H}^\alpha, \ \theta \to \infty, \tag{7.36}$$

where $\mathbf{H}^\alpha = (H_2^\alpha, H_3^\alpha, \ldots)$ *is a* $\mathbb{R}^\infty$*-valued random element and for each* $r \geq 2$, $(H_2^\alpha, \ldots, H_r^\alpha)$ *has a multivariate normal distribution with zero mean and covariance matrix*

$$Cov(H_j^\alpha, H_l^\alpha) = \frac{\Gamma(1-\alpha)\Gamma(j+l-\alpha)}{\Gamma(j-\alpha)\Gamma(l-\alpha)} + \alpha - jl, \ j,l = 2,\ldots,r. \tag{7.37}$$

Proof. Similar to the proof of Theorem 7.10, we appeal to the subordinator representation

$$\left(\frac{J_1(\alpha,\theta)}{\sigma_{\alpha,\theta}}, \frac{J_2(\alpha,\theta)}{\sigma_{\alpha,\theta}}, \ldots\right)$$

of $\mathbf{P}(\alpha,\theta)$ and define

$$M_j(\alpha,\theta) = \sqrt{\theta}\left(\frac{\Gamma(1-\alpha)}{\Gamma(j-\alpha)}\sum_{l=1}^{\infty}\frac{J_l^j(\alpha,\theta)}{\theta} - 1\right), \ j \geq 1,$$

$$\mathbf{M}_\theta^\alpha = (M_1(\alpha,\theta), M_2(\alpha,\theta),\ldots).$$

For each fixed $r \geq 1$, any $(\lambda_1,\ldots,\lambda_r)$ in $\mathbb{R}^r$, set

$$g_\theta(x) = \sum_{j=1}^r \frac{\Gamma(1-\alpha)}{\Gamma(j-\alpha)\sqrt{\theta}}\lambda_j x^j.$$

By direct calculation,

$$\begin{aligned}
&\int_0^\infty (e^{itg_\theta(x)}-1)x^{-(1+\alpha)}e^{-x}dx \\
&\quad = \frac{it\Gamma(1-\alpha)\sum_{j=1}^r \lambda_j}{\sqrt{\theta}} \\
&\quad - \sum_{j,l=1}^r \frac{\Gamma^2(1-\alpha)\Gamma(j+l-\alpha)t^2}{2\theta\Gamma(j-\alpha)\Gamma(l-\alpha)}\lambda_j\lambda_l + o\left(\frac{1}{\theta}\right).
\end{aligned} \tag{7.38}$$

This, combined with Theorem A.6 applied to the conditional law given $\gamma_{1/\alpha}$, implies that for θ large enough

$$\begin{aligned}
&\mathbb{E}\left(e^{it\sum_{j=1}^r \lambda_j M_j(\alpha,\theta)}\right) \\
&= e^{-it\sum_{j=1}^r \lambda_j\sqrt{\theta}}\mathbb{E}\left(e^{it\sum_{l=1}^\infty g_\theta(J_l(\alpha,\theta))}\right) \\
&= e^{-it\sum_{j=1}^r \lambda_j\sqrt{\theta}}\mathbb{E}\left(\mathbb{E}[e^{it\sum_{l=1}^\infty g_\theta(J_l(\alpha,\theta))} \mid \gamma_{1/\alpha}]\right) \\
&= e^{-it\sum_{j=1}^r \lambda_j\sqrt{\theta}} \\
&\quad \times\mathbb{E}\left(\exp\left\{\frac{\alpha\gamma_{1/\alpha}}{\Gamma(1-\alpha)}\int_0^\infty (e^{itg_\theta(x)}-1)x^{-(1+\alpha)}e^{-x}dx\right\}\right) \qquad (7.39) \\
&= \exp\left\{-it\sum_{j=1}^r \lambda_j\sqrt{\theta}\right\} \\
&\quad \times\exp\left\{-\frac{\theta}{\alpha}\log\left[1-\frac{\alpha}{\Gamma(1-\alpha)}\int_0^\infty (e^{itg_\theta(x)}-1)x^{-(1+\alpha)}e^{-x}dx\right]\right\} \\
&= \exp\left\{-\frac{t^2}{2}\left[\sum_{j,l=1}^r \lambda_j\lambda_l\frac{\Gamma(1-\alpha)\Gamma(j+l-\alpha)}{\Gamma(j-\alpha)\Gamma(l-\alpha)}+\alpha\left(\sum_{j=1}^r \lambda_j\right)^2\right]+o\left(\frac{1}{\theta}\right)\right\} \\
&\to \exp\left\{-\frac{t^2}{2}\sum_{j,l=1}^r \lambda_j\lambda_l\left(\frac{\Gamma(1-\alpha)\Gamma(j+j-\alpha)}{\Gamma(j-\alpha)\Gamma(l-\alpha)}+\alpha\right)\right\}.
\end{aligned}$$

For each $r \geq 1$, let $(M_1^\alpha,\ldots,M_r^\alpha)$ be a multivariate normal random vector with zero mean and covariance matrix

$$\frac{\Gamma(1-\alpha)\Gamma(j+l-\alpha)}{\Gamma(j-\alpha)\Gamma(l-\alpha)},\ j,l=1,\ldots,r, \tag{7.40}$$

and set $\mathbf{M}^\alpha = (M_1^\alpha, M_2^\alpha, \ldots)$. Then it follows from (7.39) that $\mathbf{M}_\theta^\alpha$ converges in distribution to $\mathbf{M}^\alpha$.

Noting that for each $j \geq 2$

$$H_j(\alpha,\theta) = M_j(\alpha,\theta) + \sqrt{\theta}\left(\left(\sum_{i=1}^{\infty}\frac{J_i(\alpha,\theta)}{\theta}\right)^{-j} - 1\right)\left(\frac{M_j(\alpha,\theta)}{\sqrt{\theta}} + 1\right),$$

$$\sum_{i=1}^{\infty}\frac{J_i(\alpha,\theta)}{\theta} \to 1, \text{ in probability,}$$

and

$$\sqrt{\theta}\left(\sum_{i=1}^{\infty}\frac{J_i(\alpha,\theta)}{\theta} - 1\right) \Rightarrow M_1^\alpha,$$

it follows that for any $r \geq 2$

$$\sum_{j=2}^{r}\lambda_j H_j(\alpha,\theta) \Rightarrow \sum_{j=2}^{r}\lambda_j(M_j^\alpha - jM_1^\alpha). \tag{7.41}$$

Choosing $H_j^\alpha = M_j^\alpha - jM_1^\alpha$ for each $j \geq 2$, we get the result.

□

For any $n \geq 1$, let $\mathscr{A}_n$ be the collection of allelic partitions of n defined in (2.14). Consider a random sample of size n from a population with relative frequencies of different alleles given by $\mathbf{P} = (P_1, P_2, \ldots)$. Given that $\mathbf{P} = \mathbf{p} = (p_1, p_2, \ldots)$, the conditional probability of the random partition $\mathbf{A}_n$ is given by

$$\mathbb{P}\{\mathbf{A}_n = \mathbf{a} \mid \mathbf{P} = \mathbf{p}\} = \phi_{\mathbf{a}}(\mathbf{p}) = C(\mathbf{a})\sum\prod_{i=1}^{n}\prod_{j=1}^{a_i} p_{l_{ij}}^{i}, \tag{7.42}$$

where the summation is over distinct $l_{ij}, i = 1, \ldots, n; j = 1, \ldots, a_i$; and

$$C(\mathbf{a}) = \frac{n!}{\prod_{i=1}^{n}(i!)^{a_i}a_i!}. \tag{7.43}$$

Let $k = \sum_{i=1}^{n} a_i$ be the total number of different alleles in the sample, and

$$\{i : a_i > 0, i = 1, \ldots, n\}$$

the corresponding allelic frequencies. Denote the allelic frequencies in descending order by $n_1 \geq \cdots \geq n_k \geq 1$. Then $n_1, \ldots, n_k$ is a partition of n and

$$\phi_{\mathbf{a}}(\mathbf{p}) = C(\mathbf{a}) \sum_{distinct\ i_1,\ldots,i_k} p_{i_1}^{n_1}\cdots p_{i_k}^{n_k}. \tag{7.44}$$

It follows from (5.104) and (7.44) that

$$\prod_{i=1}^{n}\varphi_i^{a_i}(\mathbf{p}) = \prod_{i=1}^{k}\varphi_{n_i}(\mathbf{p}) \tag{7.45}$$
$$= C^{-1}(\mathbf{a})\phi_{\mathbf{a}}(\mathbf{p}) + \sum_{j=1}^{k-1}\sum_{\mathbf{b}\in\sigma(k,j)} C^{-1}(\mathbf{a}(j,\mathbf{b}))\phi_{\mathbf{a}(j,\mathbf{b})}(\mathbf{p}),$$

where $\sigma(k,j)$ is defined in (5.17), and $\mathbf{a}(j,\mathbf{b})$ is obtained from $\mathbf{a}$ by coalescing the k different alleles into j different alleles according to $\mathbf{b}$.

Let $\mathbb{E}_\theta$ and $\mathbb{E}_{\alpha,\theta}$ denote the expectations with respect to the one-parameter and the two-parameter Poisson–Dirichlet distributions, respectively. Then by Theorem 2.8 and Theorem 3.8, we have

$$ESF(\theta,\mathbf{a}) = \mathbb{E}_\theta[\phi_{\mathbf{a}}(\mathbf{P})], PSF(\alpha,\theta,\mathbf{a}) = \mathbb{E}_{\alpha,\theta}[\phi_{\mathbf{a}}(\mathbf{P})]. \tag{7.46}$$

Since increasing θ increases the mean number of alleles in the population, the number of different alleles in a random sample of size n converges to n under both the one-parameter Poisson–Dirichlet distribution and the two-parameter Poisson–Dirichlet distribution as θ tends to infinity. In other words, in the large-θ limit, the random partition $\mathbf{A}_n$ converges in probability to $\mathbf{a}^1 = (n,0,\ldots,0)$, and

$$\lim_{\theta\to\infty} ESF(\theta,\mathbf{a}^1) = \lim_{\theta\to\infty} PSF(\alpha,\theta,\mathbf{a}^1) = 1, \tag{7.47}$$
$$\lim_{\theta\to\infty} ESF(\theta,\mathbf{a}) = \lim_{\theta\to\infty} PSF(\alpha,\theta,\mathbf{a}) = 0 \text{ for } \mathbf{a}\neq\mathbf{a}^1. \tag{7.48}$$

The last theorem in this section gives the asymptotic normality of $\phi_{\mathbf{a}}(\mathbf{P})$ under both the one-parameter Poisson–Dirichlet distribution and the two-parameter Poisson–Dirichlet distribution. The scaling factors depend naturally on whether $\mathbf{a}$ equals $\mathbf{a}^1$ or not.

Theorem 7.12. *Fix $n \geq 2$. Let $\mathbf{H}$ and $\mathbf{H}^\alpha$ be defined as in Theorem 7.10 and Theorem 7.11.*

(1) *For any allelic partition $\mathbf{a}\neq\mathbf{a}^1$, we have*

$$\sqrt{\theta}\left(\frac{\phi_{\mathbf{a}}(\mathbf{P}(\theta)) - ESF(\theta,\mathbf{a})}{ESF(\theta,\mathbf{a})}\right) \Rightarrow \sum_{i=1}^{n} a_i H_i, \tag{7.49}$$
$$\sqrt{\theta}\left(\frac{\phi_{\mathbf{a}}(\mathbf{P}(\alpha,\theta)) - PSF(\alpha,\theta,\mathbf{a})}{PSF(\alpha,\theta,\mathbf{a})}\right) \Rightarrow \sum_{i=1}^{n} a_i H_i^\alpha. \tag{7.50}$$

(2) *For $\mathbf{a}=\mathbf{a}^1$ and $\mathbf{a}^2 = (n-2,1,0,\ldots,0)$, we have*

$$\sqrt{\theta}\left(\frac{\phi_{\mathbf{a}^1}(\mathbf{P}(\theta)) - ESF(\theta,\mathbf{a}^1)}{ESF(\theta,\mathbf{a}^2)}\right) \Rightarrow H_2, \tag{7.51}$$
$$\sqrt{\theta}\left(\frac{\phi_{\mathbf{a}^1}(\mathbf{P}(\alpha,\theta)) - PSF(\alpha,\theta,\mathbf{a}^1)}{PSF(\alpha,\theta,\mathbf{a}^2)}\right) \Rightarrow H_2^\alpha. \tag{7.52}$$

Proof. We will focus on the proof of the two-parameter results. The one-parameter results will follow by taking $\alpha = 0$. Replace $\mathbf{p}$ with $\mathbf{P}(\alpha,\theta)$ in (7.45), and multiply both sides by $\prod_{i=1}^n (\theta^{i-1}\frac{\Gamma(1-\alpha)}{\Gamma(i-\alpha)})^{a_i}$. Since

$$\sum_{i=1}^n a_i(i-1) = n-k,$$

$$(PSF(\alpha,\theta,\mathbf{a}))^{-1} = \frac{\theta_{(n)}}{\prod_{l=0}^{k-1}(\theta+l\alpha)} C^{-1}(\mathbf{a}) \prod_{i=1}^n \left(\frac{\Gamma(1-\alpha)}{\Gamma(i-\alpha)}\right)^{a_i},$$

it follows that

$$\prod_{i=2}^n \left(\frac{H_i(\alpha,\theta)}{\sqrt{\theta}}+1\right)^{a_i} = \frac{\theta^n \prod_{l=0}^{k-1}(\theta+l\alpha)}{\theta_{(n)}\theta^k} \frac{\phi_{\mathbf{a}}(\mathbf{P}(\alpha,\theta))}{PSF(\alpha,\theta,\mathbf{a})} + R(\theta), \tag{7.53}$$

where

$$R(\theta) = \theta^{n-k} \prod_{i=2}^n \left(\frac{\Gamma(1-\alpha)}{\Gamma(i-\alpha)}\right)^{a_i} \left(\sum_{j=1}^{k-1} \sum_{\mathbf{b}\in\sigma(k,j)} C^{-1}(\mathbf{a}(j,\mathbf{b}))\phi_{\mathbf{a}(j,\mathbf{b})}(\mathbf{P}(\alpha,\theta))\right).$$

Noting that the number of different alleles j in $\mathbf{a}(j,\mathbf{b})$ is less than k, and the order of

$$PSF(\alpha,\theta,\mathbf{a}(j,\mathbf{b}))$$

is θ^{j-n}, it follows that, as θ tends to infinity,

$$\sqrt{\theta}R(\theta) \to 0, \text{ in probability}. \tag{7.54}$$

Since

$$\frac{\theta^n \prod_{l=0}^{k-1}(\theta+l\alpha)}{\theta_{(n)}\theta^k} = 1 + O\left(\frac{1}{\theta}\right), \tag{7.55}$$

it follows from Theorem 7.11 that $\frac{\phi_{\mathbf{a}}(\mathbf{P}(\alpha,\theta))}{PSF(\alpha,\theta,\mathbf{a})}$ converges to 1 in probability.

Rewrite (7.53) as

$$\sqrt{\theta}\left(\prod_{i=2}^n \left(\frac{H_i(\alpha,\theta)}{\sqrt{\theta}}+1\right)^{a_i} - 1\right) = \sqrt{\theta}\left(\frac{\phi_{\mathbf{a}}(\mathbf{P}(\alpha,\theta))}{PSF(\alpha,\theta,\mathbf{a})} - 1\right) + R_1(\theta) + R_2(\theta), \tag{7.56}$$

where

$$R_1(\theta) = \sqrt{\theta}\left(\frac{\theta^n \prod_{l=0}^{k-1}(\theta+l\alpha)}{\theta_{(n)}\theta^k} - 1\right)\frac{\phi_{\mathbf{a}}(\mathbf{P}(\alpha,\theta))}{PSF(\alpha,\theta,\mathbf{a})}, \quad R_2(\theta) = \sqrt{\theta}R(\theta).$$

By (7.54) and (7.55), we have $R_1(\theta)+R_2(\theta)$ converges to zero in probability. Thus

$$\sqrt{\theta}\left(\prod_{i=2}^{n}\left(\frac{H_i(\alpha,\theta)}{\sqrt{\theta}}+1\right)^{a_i}-1\right) \text{ and } \sqrt{\theta}\left(\frac{\phi_{\mathbf{a}}(\mathbf{P}(\alpha,\theta))}{PSF(\alpha,\theta,\mathbf{a})}-1\right)$$

have the same limit in distribution. Expanding the product

$$\prod_{i=2}^{n}\left(\frac{H_i(\alpha,\theta)}{\sqrt{\theta}}+1\right)^{a_i},$$

and applying the continuous-mapping theorem, it follows that

$$\sqrt{\theta}\left(\prod_{i=2}^{n}\left(\frac{H_i(\alpha,\theta)}{\sqrt{\theta}}+1\right)^{a_i}-1\right) \Rightarrow \sum_{i=2}^{n} a_i H_i^{\alpha} \tag{7.57}$$

which leads to (7.50).

It remains to verify (7.52). Note that

$$\sum_{\mathbf{a}} \phi_{\mathbf{a}}(\mathbf{P}(\alpha,\theta)) = 1.$$

Therefore

$$\begin{aligned}
&\frac{\phi_{\mathbf{a}^1}(\mathbf{P}(\alpha,\theta)) - PSF(\alpha,\theta,\mathbf{a}^1)}{PSF(\alpha,\theta,\mathbf{a}^2)}\\
&\quad= \frac{PSF(\alpha,\theta,\mathbf{a}^2) - \phi_{\mathbf{a}^2}(\mathbf{P}(\alpha,\theta))}{PSF(\alpha,\theta,\mathbf{a}^2)}\\
&\qquad+ \sum_{\mathbf{a}\neq\mathbf{a}^1,\mathbf{a}^2} \frac{PSF(\alpha,\theta,\mathbf{a}) - \phi_{\mathbf{a}}(\mathbf{P}(\alpha,\theta))}{PSF(\alpha,\theta,\mathbf{a}^2)}\\
&\quad= \frac{PSF(\alpha,\theta,\mathbf{a}^2) - \phi_{\mathbf{a}^2}(\mathbf{P}(\alpha,\theta))}{PSF(\alpha,\theta,\mathbf{a}^2)}\\
&\qquad+ \sum_{\mathbf{a}\neq\mathbf{a}^1,\mathbf{a}^2} \frac{PSF(\alpha,\theta,\mathbf{a})}{PSF(\alpha,\theta,\mathbf{a}^2)}\,\frac{PSF(\alpha,\theta,\mathbf{a}) - \phi_{\mathbf{a}}(\mathbf{P}(\alpha,\theta))}{PSF(\alpha,\theta,\mathbf{a})}.
\end{aligned}$$

Since $\sqrt{\theta}\frac{PSF(\alpha,\theta,\mathbf{a})}{PSF(\alpha,\theta,\mathbf{a}^2)}$ converges to zero as $\theta \to \infty$ for $\mathbf{a} \neq \mathbf{a}^1, \mathbf{a}^2$, it follows from (7.50) that

$$\sqrt{\theta}\frac{\phi_{\mathbf{a}^1}(\mathbf{P}(\alpha,\theta)) - PSF(\alpha,\theta,\mathbf{a}^1)}{PSF(\alpha,\theta,\mathbf{a}^2)} \text{ and } \sqrt{\theta}\frac{PSF(\alpha,\theta,\mathbf{a}^2) - \phi_{\mathbf{a}^2}(\mathbf{P}(\alpha,\theta))}{PSF(\alpha,\theta,\mathbf{a}^2)}$$

have the same limit, $-H_2^{\alpha}$, in distribution as θ tends to infinity. Since H_2^{α} and $-H_2^{\alpha}$ have the same distribution, we get (7.52).

□

7.4 Notes

The study of the asymptotic behavior of the Poisson–Dirichlet distribution for large θ has a long history. It was first mentioned in [182] that the large-θ limit corresponds to a fixed mutation rate per nucleotide site within the locus, with a large number of sites. Watterson and Guess [183] obtained asymptotic results for the means of the most common alleles in the neutral model.

The result in Theorem 7.2 is essentially obtained in Griffiths [91] where a similar fluctuation theorem was obtained for a K-allele model in the limit as K and θ both go to infinity. The proof presented here follows the idea in [91]. Theorem 7.4 first appeared in Handa [100] where a different proof is given.

Theorem 7.6 appears in Chapter VI of [110]. A slightly different form appears as Exercise 18 of Chapter 3 in [62]. The results of both Theorem 7.8 and Theorem 7.9 appear in [114]. The proofs here are novel. Theorem 3.5 in [73] stated an incorrect fluctuation limit of the Dirichlet process.

Griffiths [91] obtained the asymptotic normality of the homozygosity of order two for the K-allele model in the limit as K and θ both go to infinity. The results in Theorem 7.10 and the one-parameter part of Theorem 7.12 were obtained in [118]. Theorem 7.11 first appeared in [100] but the proof here is different. The two-parameter part of Theorem 7.12 seems to be new. The proof here follows the idea in [118] with some modifications in the details.

Chapter 8
Large Deviations for the Poisson–Dirichlet Distribution

As seen in Chapter 1, the parameter θ in the Poisson–Dirichlet distribution is the scaled population mutation rate in the context of population genetics. When the mutation rate is small, there is a tendency for only a few alleles to have high frequencies and dominate the population. On the other hand, large values of θ correspond to the situation where the proportions of different alleles are evenly spread. Noting that θ is proportional to the product of certain effective population size and individual mutation rate, large values of θ also correspond to a population with fixed individual mutation rate and large effective population size. The large deviations are established in this chapter for the Poisson–Dirichlet distribution and its two-parameter counterpart when θ tends to either zero or infinity. Also included are several applications that provide motivation for the large deviation results. Appendix B includes basic terminology and results of large deviation theory.

8.1 Large Mutation Rate

This section is devoted to establishing LDPs (large deviation principles, see Appendix B) for the Poisson–Dirichlet distribution and its two-parameter counterpart when the parameter θ tends to infinity. Since the state spaces involved are compact, rate functions obtained are automatically good rate functions.

8.1.1 The Poisson–Dirichlet Distribution

Recall that the Poisson–Dirichlet distribution with parameter θ, denoted by Π_θ, is the law of $\mathbf{P}(\theta) = (P_1(\theta), P_2(\theta), \ldots)$. It is a probability on the space

$$\nabla_\infty = \left\{ (p_1, p_2, \ldots) : p_1 \geq p_2 \geq \cdots \geq 0, \sum_{j=1}^{\infty} p_j = 1 \right\},$$

S. Feng, *The Poisson–Dirichlet Distribution and Related Topics*,
Probability and its Applications, DOI 10.1007/978-3-642-11194-5_8,

and has the GEM representation through the descending order statistics of

$$V_1 = U_1, \; V_n = (1-U_1)\cdots(1-U_{n-1})U_n, \; n \geq 2, \tag{8.1}$$

with $\{U_n : n \geq 1\}$ being a sequence of i.i.d. $Beta(1,\theta)$ random variables. For notational convenience, Π_θ will also denote the extension of the Poisson–Dirichlet distribution to the closure ∇ of ∇_∞. The main result of this subsection is the LDP on the space ∇ for $\{\Pi_\theta : \theta > 0\}$ as θ tends to infinity. This will be established through a series of lemmas.

Lemma 8.1. *For any $n \geq 1$, let $Z_n(\theta) = \max\{U_1,\ldots,U_n\}$. Then the family $\{Z_n(\theta) : \theta > 0\}$ satisfies an LDP on $[0,1]$ with speed $1/\theta$ and rate function*

$$I(x) = \begin{cases} \log \frac{1}{1-x}, & x \in [0,1) \\ \infty, & \textit{else}. \end{cases} \tag{8.2}$$

Proof. Let

$$\begin{aligned} \Lambda(\lambda) &= \operatorname*{ess\,sup}_{y\in[0,1]}\{\lambda y + \log(1-y)\} \\ &= \begin{cases} \lambda - 1 - \log\lambda, & \lambda > 1 \\ 0, & \text{else}, \end{cases} \end{aligned} \tag{8.3}$$

where ess sup denotes the essential supremum. Then clearly $\Lambda(\lambda)$ is finite for all λ, and is differentiable. By direct calculation, we have

$$E[e^{\theta\lambda Z_n}] = \int_0^1 \exp\{\theta F_\theta(y)\}\,dy,$$

where

$$\begin{aligned} F_\theta(y) &= \lambda y + \frac{\log n + \log\theta}{\theta} \\ &\quad + \frac{n-1}{\theta}\log[1-(1-y)^\theta] + \frac{\theta-1}{\theta}\log(1-y). \end{aligned} \tag{8.4}$$

Therefore

$$\lim_{\theta\to\infty} \log\{E[e^{\theta\lambda Z_n}]\}^{\frac{1}{\theta}} = \Lambda(\lambda),$$

which, combined with Theorem B.8 (with $\varepsilon = 1/\theta$), implies the lemma. □

Lemma 8.2. *For any $k \geq 1$, let $n_k(\theta)$ denote the integer part of θ^k. Then the family $\{Z_{n_k(\theta)}(\theta) : \theta > 0\}$ satisfies an LDP with speed $1/\theta$ and rate function I defined in (8.2).*

Proof. Choosing $n = n_k(\theta)$ in (8.4), we get

$$F_\theta(y) = \lambda y + \frac{(\log n_k(\theta) + \log\theta)}{\theta}$$
$$+ \frac{n_k(\theta) - 1}{\theta}\log[1 - (1-y)^\theta] + \frac{\theta - 1}{\theta}\log(1-y).$$

For any ε in $(0, 1/2)$, and $\lambda \geq 0$, we have

$$\begin{aligned}\Lambda(\lambda) &= \lim_{\theta\to\infty}\frac{1}{\theta}\log E[e^{\theta\lambda U_1}]\\ &\leq \limsup_{\theta\to\infty}\frac{1}{\theta}\log E[e^{\theta\lambda Z_{n_k(\theta)}}]\\ &\leq \max\{\lambda\varepsilon, \operatorname*{ess\,sup}_{y\geq\varepsilon}[\lambda y + \log(1-y)]\},\end{aligned}$$

where the last inequality follows from the fact that for y in $[\varepsilon, 1]$,

$$\lim_{\theta\to\infty}\theta^l\log[1-(1-y)]^\theta = 0 \text{ for any } l \geq 1.$$

Letting ε go to zero, it follows that

$$\Lambda(\lambda) = \lim_{\theta\to\infty}\frac{1}{\theta}\log E[e^{\theta\lambda Z_{n_k(\theta)}}].$$

For negative λ, we have

$$\limsup_{\theta\to\infty}\frac{1}{\theta}\log E[e^{\theta\lambda Z_{n_k(\theta)}}] \geq \lim_{\delta\to 0}\operatorname*{ess\,sup}_{y\geq\delta}\{\lambda y + \log(1-y)\} = 0 = \Lambda(\lambda).$$

The lemma follows by another application of Theorem B.8. □

Lemma 8.3. *For any $n \geq 1$, let $W_n = (1-U_1)(1-U_2)\cdots(1-U_n)$. Then for any $\delta > 0$,*

$$\limsup_{\theta\to\infty}\frac{1}{\theta}\log\mathbb{P}\{W_{n_2(\theta)} \geq \delta\} = -\infty. \tag{8.5}$$

Proof. By direct calculation,

$$\mathbb{P}\{W_{n_2(\theta)} \geq \delta\} = \mathbb{P}\left\{\theta\sum_{j=1}^{n_2(\theta)}\log(1-U_j) \geq \theta\log\delta\right\}$$
$$\leq e^{\theta\log\frac{1}{\delta}}(E[e^{\theta\log(1-U_1)}])^{n_2(\theta)} = e^{\theta\log\frac{1}{\delta}}\left(\frac{1}{2}\right)^{n_2(\theta)}$$
$$= \exp\left[\theta\log\frac{1}{\delta} - (\theta^2 - 1)\log 2\right].$$

The lemma follows by letting θ go to infinity. □

The next lemma establishes the LDP for $\{P_1(\theta) : \theta > 0\}$.

Lemma 8.4. *The family* $\{P_1(\theta) : \theta > 0\}$ *satisfies an LDP on* $[0,1]$ *with speed* $1/\theta$ *and rate function I, given by* (8.2).

Proof. Use the GEM representation for $P_1(\theta)$ and set $\hat{P}_1(\theta) = \max\{V_1,\ldots,V_{n_2(\theta)}\}$. Then clearly $P_1(\theta) \geq \hat{P}_1(\theta)$. For any $\delta > 0$, it follows from Lemma 8.3 that

$$\limsup_{\theta\to\infty} \frac{1}{\theta} \log \mathbb{P}\{P_1(\theta) - \hat{P}_1(\theta) > \delta\} \leq \limsup_{\theta\to\infty} \frac{1}{\theta} \log \mathbb{P}\{W_{n_2(\theta)} > \delta\} = -\infty.$$

In other words, $P_1(\theta)$ and $\hat{P}_1(\theta)$ are exponentially equivalent, and thus have the same LDPs, provided one of them has an LDP. By definition, we have

$$U_1 = Z_1(\theta) \leq \hat{P}_1(\theta) \leq Z_{n_2(\theta)}.$$

Applying Lemma 8.1, Lemma 8.2, and Corollary B.9, we conclude that the law of $\hat{P}_1(\theta)$ satisfies an LDP on space $[0,1]$ with speed $1/\theta$ and rate function I.

□

For any $n \geq 1$, let

$$\overline{\nabla}_n = \left\{(p_1,\ldots,p_n) : 0 \leq p_n \leq \ldots \leq p_1, \sum_{k=1}^{n} p_k \leq 1\right\}. \tag{8.6}$$

Lemma 8.5. *For fixed* $n \geq 2$, *the family* $\{(P_1(\theta),\ldots,P_n(\theta)) : \theta > 0\}$ *satisfies an LDP on the space* $\overline{\nabla}_n$ *with speed* $1/\theta$ *and rate function*

$$S_n(p_1,\ldots,p_n) = \begin{cases} \log \frac{1}{1-\sum_{k=1}^n p_k}, & (p_1,\ldots,p_n) \in \overline{\nabla}_n,\ \sum_{k=1}^n p_k < 1 \\ \infty, & \textit{else}. \end{cases} \tag{8.7}$$

Proof. Since $\overline{\nabla}_n$ is compact, by Theorem B.6, the family

$$\{(P_1(\theta),\ldots,P_n(\theta)) : \theta > 0\}$$

satisfies a partial LDP. By Theorem 2.6, the density function $g_1^\theta(p)$ of $P_1(\theta)$ and the joint density function $g_n^\theta(p_1,\ldots,p_n)$ of $(P_1(\theta),\ldots,P_n(\theta))$ satisfy the relations

$$g_1^\theta(p)p(1-p)^{1-\theta} = \theta \int_0^{(p/(1-p))\wedge 1} g_1^\theta(x)\,dx,$$

and

$$g_n^\theta(p_1,\ldots,p_n)$$
$$= \begin{cases} \frac{\theta^n(1-\sum_{k=1}^n p_k)^{\theta-1}}{p_1\cdots p_n} \int_0^{(p_n/(1-\sum_{k=1}^n p_k))\wedge 1} g_1^\theta(u)\,du, & (p_1,\ldots,p_n) \in \overline{\nabla}_n^\circ \\ 0, & \text{else}, \end{cases} \tag{8.8}$$

where

$$\overline{\nabla}_n^{\circ} = \left\{ (p_1,\ldots,p_n) \in \nabla_n : 0 < p_n < \cdots < p_1 < 1, \sum_{k=1}^{n} p_k < 1 \right\}.$$

Clearly $\overline{\nabla}_n$ is the closure of $\overline{\nabla}_n^{\circ}$. Now for any $(p_1,\ldots,p_n) \in \overline{\nabla}_n$ and $\delta > 0$, let

$$G((p_1,\ldots,p_n);\delta) = \{(q_1,\ldots,q_n) \in \overline{\nabla}_n : |q_k - p_k| < \delta, k = 1,\ldots,n\},$$
$$F((p_1,\ldots,p_n);\delta) = \{(q_1,\ldots,q_n) \in \overline{\nabla}_n : |q_k - p_k| \le \delta, k = 1,\ldots,n\}.$$

Then the family $\{G((p_1,\ldots,p_n);\delta) : \delta > 0, (p_1,\ldots,p_n) \in \overline{\nabla}_n\}$ is a base for the topology of $\overline{\nabla}_n$. First assume that $(p_1,\ldots,p_n)$ is in $\overline{\nabla}_n^{\circ}$. Then we can choose δ so that $F((p_1,\ldots,p_n);\delta)$ is a subset of $\overline{\nabla}_n^{\circ}$. By (8.8), for any $(q_1,\ldots,q_n)$ in $F((p_1,\ldots,p_n);\delta)$

$$g_n^{\theta}(q_1,\ldots,q_n) \le \frac{\theta^n(1-\sum_{k=1}^n (p_k-\delta))^{\theta-1}}{(p_1-\delta)\cdots(p_n-\delta)}, \tag{8.9}$$

which implies

$$\limsup_{\theta\to\infty} \frac{1}{\theta} \log \mathbb{P}_{n,\theta}\{F((p_1,\ldots,p_n);\delta)\} \le -\log \frac{1}{1-\sum_{k=1}^n (p_k-\delta)}, \tag{8.10}$$

where $\mathbb{P}_{n,\theta}$ denotes the law of $(P_1(\theta),\ldots,P_n(\theta))$. Therefore

$$\limsup_{\delta\to 0} \limsup_{\theta\to\infty} \frac{1}{\theta} \log \mathbb{P}_{n,\theta}\{F((p_1,\ldots,p_n);\delta)\} \le -S_n(p_1,\ldots,p_n). \tag{8.11}$$

Clearly (8.11) also holds for $(p_1,\ldots,p_n) = (0,\ldots,0)$. For other points outside $\overline{\nabla}_n^{\circ}$, there are two possibilities. Either $p_n = 1 - \sum_{i=1}^{n-1} p_i > 0$ or $p_l = 0$ for some $1 < l \le n$. Since the estimate (8.9) holds in the first case, (8.11) holds too. In the second case, let

$$k = \inf\{l : 1 < l < n, p_l = 0\}.$$

Then we have $S_n(p_1,\ldots,p_n) = S_k(p_1,\ldots,p_k)$ and (8.11) follows from the upper bound for

$$\mathbb{P}_{k,\theta}\{F((p_1,\ldots,p_k);\delta)\}$$

and the fact that

$$\mathbb{P}_{n,\theta}\{F((p_1,\ldots,p_n);\delta)\} \le \mathbb{P}_{k,\theta}\{F((p_1,\ldots,p_k);\delta)\}.$$

Next we turn to the lower bound. First note that if $\sum_{k=1}^n p_k = 1$, the lower bound is trivially true since $S_n(p_1,\ldots,p_n) = \infty$. Hence we assume that $\sum_{k=1}^n p_k < 1$.

If $(p_1,\ldots,p_n)$ is in $\overline{\nabla}_n^{\circ}$, then we can choose δ so that

$$0 < \delta < \frac{1-\sum_{k=1}^n p_k}{n}$$

and

$$G((p_1,\ldots,p_n);\delta) \subset \overline{\nabla}_n^\circ.$$

Using (8.8) again, one has that for any $(q_1,\ldots,q_n)$ in $G((p_1,\ldots,p_n);\delta)$

$$g_n^\theta(q_1,\ldots,q_n) \geq \theta^n \left(1-\sum_{k=1}^n (p_k+\delta)\right)^{\theta-1} \int_0^{((p_n-\delta)/(1-\sum_{k=1}^n (p_k-\delta)))\wedge 1} g_1(u)\,du,$$

which implies

$$\begin{aligned}
&\liminf_{\theta\to\infty} \frac{1}{\theta} \log \mathbb{P}_{n,\theta}\{G((p_1,\ldots,p_n);\delta)\}\\
&\geq -\log \frac{1}{1-\sum_{k=1}^n (p_k+\delta)}\\
&- \inf\left\{I(p) : p < \frac{p_n-\delta}{1-\sum_{k=1}^n (p_k-\delta)} \wedge 1\right\}\\
&= -\log \frac{1}{1-\sum_{k=1}^n (p_k+\delta)},
\end{aligned}$$

where in the second line, we used the LDP obtained in Lemma 8.4. Letting δ go to zero, yields

$$\liminf_{\delta\to 0} \liminf_{\theta\to\infty} \frac{1}{\theta} \log \mathbb{P}_{n,\theta}\{G((p_1,\ldots,p_n);\delta)\} \geq -S_n(p_1,\ldots,p_n). \tag{8.12}$$

If $(p_1,\ldots,p_n)$ is not in $\overline{\nabla}_n^\circ$, then there is $1 \leq l \leq n$ such that $p_l = 0$. Choosing $\varepsilon > 0$ and $\delta' > 0$ so that

$$\begin{aligned}
&(p_1(\varepsilon),\ldots,p_n(\varepsilon)) \equiv (p_1+\varepsilon,\ldots,p_n+\varepsilon) \in \overline{\nabla}_n^\circ\\
&G((p_1(\varepsilon),\ldots,p_n(\varepsilon));\delta') \subset G((p_1;\ldots,p_n),\delta) \cap \overline{\nabla}_n^\circ,
\end{aligned}$$

one gets

$$\begin{aligned}
&\liminf_{\theta\to\infty} \frac{1}{\theta} \log \mathbb{P}_{n,\theta}\{G((p_1,\ldots,p_n);\delta)\} \qquad (8.13)\\
&\quad \geq \liminf_{\theta\to\infty} \frac{1}{\theta} \log \mathbb{P}_{n,\theta}\{G((p_1(\varepsilon),\ldots,p_n(\varepsilon));\delta')\}.
\end{aligned}$$

Taking limits in the order of $\delta' \to 0$, $\varepsilon \to 0$ and $\delta \to 0$, and taking into account the continuity of $S_n(p_1,\ldots,p_n)$, it follows that (8.12) holds in this case. The lemma now follows from (8.11), (8.12) and Theorem B.6.

□

Lemma 8.6. *For $k \geq 2$, the family $\{P_k(\theta) : \theta > 0\}$ satisfies an LDP on $[0,1]$ with speed $1/\theta$ and rate function*

$$I^k(p) = \begin{cases} \log \frac{1}{1-kp}, & p \in [0, 1/k) \\ \infty, & else. \end{cases} \tag{8.14}$$

Thus for any $k \geq 1$, the LDP for $\{P_1(\theta) : \theta > 0\}$ is the same as the LDP for the family $\{kP_k(\theta) : \theta > 0\}$.

Proof. For any $k \geq 2$, define the projection map

$$\phi_k : \overline{\nabla}_k \longrightarrow [0,1], \ (p_1, p_2, \ldots, p_k) \mapsto p_k.$$

Clearly ϕ_k is continuous, and Lemma 8.5 combined with the contraction principle implies that the law of $P_k(\theta)$ satisfies an LDP on $[0,1]$ with speed θ and rate function

$$I'(p) = \inf\{S_k(p_1, \ldots, p_k) : p_1 \geq \cdots \geq p_k = p\}.$$

For $p > 1/k$, the infimum is over empty set and is thus infinity. For p in $[0, 1/k]$, the infimum is achieved at the point $p_1 = p_2 = \cdots = p_k = p$. Hence $I'(p) = I^k(p)$ and the result follows.

□

Now we are ready to prove the LDP for $\{\Pi_\theta : \theta > 0\}$.

Theorem 8.1. *The family $\{\Pi_\theta : \theta > 0\}$ satisfies an LDP on the space ∇ with speed $1/\theta$ and rate function*

$$S(\mathbf{p}) = \begin{cases} \log \frac{1}{1-\sum_{k=1}^{\infty} p_k}, & \mathbf{p} = (p_1, p_2, \ldots) \in \nabla, \ \sum_{k=1}^{\infty} p_k < 1 \\ \infty, & else. \end{cases} \tag{8.15}$$

Proof. Due to the compactness of ∇, it suffices, by Theorem B.6, to verify (B.7) for the family $\{\Pi_\theta : \theta > 0\}$. The topology on ∇ can be generated by the following metric:

$$d(\mathbf{p}, \mathbf{q}) = \sum_{k=1}^{\infty} \frac{|p_k - q_k|}{2^k},$$

where $\mathbf{p} = (p_1, p_2, \ldots), \mathbf{q} = (q_1, q_2, \ldots)$. For any fixed $\delta > 0$, let $B(\mathbf{p}, \delta)$ and $\bar{B}(\mathbf{p}, \delta)$ denote the respective open and closed balls centered at $\mathbf{p}$ with radius $\delta > 0$. Set $n_\delta = 1 + [\log_2(1/\delta)]$ where $[x]$ denotes the integer part of x. Set

$$G_{n_\delta}(\mathbf{p}; \delta/2) = \{(q_1, q_2, \ldots) \in \nabla : |q_k - p_k| < \delta/2, k = 1, \ldots, n_\delta\}.$$

Then we have

$$G_{n_\delta}(\mathbf{p}; \delta/2) \subset B(\mathbf{p}, \delta).$$

By Lemma 8.5 and the fact that

$$\Pi_\theta\{G_{n_\delta}(\mathbf{p}; \delta/2)\} = \mathbb{P}_{n_\delta, \theta}\{G((p_1, \ldots, p_{n_\delta}); \delta/2)\},$$

we get that

$$\begin{aligned}\liminf_{\theta\to\infty}\frac{1}{\theta}\log\Pi_\theta\{B(\mathbf{p},\delta)\}&\\ &\geq \liminf_{\theta\to\infty}\frac{1}{\theta}\log\mathbb{P}_{n_\delta,\theta}\{G((p_1,\ldots,p_{n_\delta});\delta/2)\}\\ &\geq -S_{n_\delta}(p_1,\ldots,p_{n_\delta})\geq -S(\mathbf{p}).\end{aligned}\tag{8.16}$$

On the other hand, for any fixed $n\geq 1,\delta_1>0$, let

$$F_n(\mathbf{p};\delta_1)=\{(q_1,q_2,\ldots)\in\nabla:|q_k-p_k|\leq\delta_1,k=1,\ldots,n\}.$$

Then we have

$$\Pi_\theta\{F_n(\mathbf{p};\delta_1)\}=\mathbb{P}_{n,\theta}\{F((p_1,\ldots,p_n);\delta_1)\},$$

and, for δ small enough,

$$\bar{B}(\mathbf{p},\delta)\subset F_n(\mathbf{p};\delta_1),$$

which implies that

$$\begin{aligned}\lim_{\delta\to 0}\limsup_{\theta\to\infty}\frac{1}{\theta}\log\Pi_\theta\{\bar{B}(\mathbf{p};\delta)\}&\\ &\leq \limsup_{\theta\to\infty}\frac{1}{\theta}\log\mathbb{P}_{n,\theta}\{F((p_1,\ldots,p_n);\delta_1)\}\\ &\leq -\inf\{S_n(q_1,\ldots,q_n):(q_1,\ldots,q_n)\in F((p_1,\ldots,p_n),\delta_1)\}.\end{aligned}\tag{8.17}$$

Letting δ_1 go to zero, and then n go to infinity, we get the upper bound

$$\lim_{\delta\to 0}\limsup_{\theta\to\infty}\frac{1}{\theta}\log\Pi_\theta\{\bar{B}(\mathbf{p},\delta)\}\leq -S(\mathbf{p}),\tag{8.18}$$

which, combined with (8.16), implies the result. □

8.1.2 The Two-parameter Poisson–Dirichlet Distribution

Many properties of the Poisson–Dirichlet distribution have generalizations in the two-parameter setting. In this subsection, we obtain the two-parameter generalization of the LDP result for Π_θ. The idea in the proof is very similar to that in the one-parameter case.

Let $0<\alpha<1$ and $\theta>0$. Recall that the GEM representation for the two-parameter Poisson–Dirichlet distribution $PD(\alpha,\theta)$ is given by

$$V_1^{\alpha,\theta}=W_1,V_n^{\alpha,\theta}=(1-W_1)\cdots(1-W_{n-1})W_n,n\geq 2,$$

where the sequence $W_1,W_2,\ldots$ is independent and W_i is a $Beta(1-\alpha,\theta+i\alpha)$ random variable. Let $\mathbf{P}(\alpha,\theta)=(P_1(\alpha,\theta),P_2(\alpha,\theta),\ldots)$ denote the rearrangement of $\{V_n^{\alpha,\theta}:n=1,2,\ldots\}$ in descending order.

Lemma 8.7. *The family* $\{P_1(\alpha,\theta) : \theta > 0\}$ *satisfies an LDP on* $[0,1]$ *with speed* $1/\theta$ *and rate function I given by* (8.2).

Proof. By the GEM representation,

$$\mathbb{E}[e^{\lambda\theta W_1}] \leq \mathbb{E}[e^{\lambda\theta P_1(\alpha,\theta)}] \text{ for } \lambda \geq 0$$
$$\mathbb{E}[e^{\lambda\theta W_1}] \geq \mathbb{E}[e^{\lambda\theta P_1(\alpha,\theta)}] \text{ for } \lambda < 0.$$

On the other hand, by Proposition 3.7,

$$\mathbb{E}[e^{\lambda\theta P_1(\alpha,\theta)}] \leq \mathbb{E}[e^{\lambda\theta P_1(\theta)}] \text{ for } \lambda \geq 0$$
$$\mathbb{E}[e^{\lambda\theta P_1(\alpha,\theta)}] \geq \mathbb{E}[e^{\lambda\theta P_1(\theta)}] \text{ for } \lambda < 0.$$

By Lemma 8.4 and an argument similar to that used in the proof of Lemma 8.1, the LDP for the family $\{W_1 : \theta > 0\}$ is the same as the LDP for the family $\{P_1(\theta) : \theta > 0\}$. The lemma then follows from Theorem B.8.

□

Lemma 8.8. *For each* $n \geq 2$, *the family* $\{(P_1(\alpha,\theta),\ldots,P_n(\alpha,\theta)) : \theta > 0\}$ *satisfies an LDP on* $\overline{\nabla}_n$ *with speed* $1/\theta$ *and rate function* S_n *given by* (8.7).

Proof. Let

$$C_{\alpha,\theta,n} = \frac{\Gamma(\theta+1)\Gamma(\frac{\theta}{\alpha}+n)\alpha^{n-1}}{\Gamma(\theta+n\alpha)\Gamma(\frac{\theta}{\alpha}+1)\Gamma(1-\alpha)^n}. \tag{8.19}$$

It follows from Theorem 2.6 and Theorem 3.6 that for any $n \geq 2$, and $(p_1,\ldots,p_n)$ in $\overline{\nabla}_n$,

$$(P_1(\alpha,\theta),P_2(\alpha,\theta),\ldots,P_n(\alpha,\theta))$$

and

$$(P_1(0,\alpha+\theta),P_2(0,\alpha+\theta),\ldots,P_n(0,\alpha+\theta))$$

have respective joint density functions

$$\begin{aligned}&\varphi_n^{\alpha,\theta}(p_1,\ldots,p_n) \\ &= C_{\alpha,\theta,n}\frac{(1-\sum_{i=1}^n p_i)^{\theta+n\alpha-1}}{\prod_{i=1}^n p_i}\mathbb{P}\left(P_1(\alpha,n\alpha+\theta) \leq \frac{p_n}{1-\sum_{i=1}^n p_i}\right),\end{aligned} \tag{8.20}$$

and

$$\begin{aligned}&g_n^{\alpha+\theta}(p_1,\ldots,p_n) \\ &= (\alpha+\theta)^n\frac{(1-\sum_{i=1}^n p_i)^{\theta+\alpha-1}}{\prod_{i=1}^n p_i}\mathbb{P}\left(P_1(0,\alpha+\theta) \leq \frac{p_n}{1-\sum_{i=1}^n p_i}\right).\end{aligned} \tag{8.21}$$

Noting that

$$\lim_{\theta\to\infty}\frac{\log C_{\alpha,\theta,n}}{\theta}=\lim_{\theta\to\infty}\frac{\log(\alpha+\theta)^n}{\theta}=0,$$

it follows from Lemma 8.4 and Lemma 8.7 that for any $(p_1,\ldots,p_n)$ in $\overline{\nabla}_n$,

$$\begin{aligned}\lim_{\theta\to\infty}\frac{1}{\theta}&\log\varphi_n^{\alpha,\theta}(p_1,\ldots,p_n)\\&=\lim_{\theta\to\infty}\frac{1}{\theta}\log g_n^{\alpha+\theta}(p_1,\ldots,p_n)\\&=-S_n(p_1,\ldots,p_n).\end{aligned}\tag{8.22}$$

The lemma now follows from an argument similar to that used in the proof of Lemma 8.5. □

The following theorem follows easily from Lemma 8.8 and the finite-dimensional approximations used in Theorem 8.1.

Theorem 8.2. *The family* $\{PD(\alpha,\theta):\theta>0\}$ *satisfies an LDP on the space* ∇ *with speed* $1/\theta$ *and rate function S defined in* (8.15).

Remarks:
(a) The effective domain of S is $\nabla\setminus\nabla_\infty$, while $PD(\alpha,\theta)$ is concentrated on ∇_∞.
(b) The LDP for (W_n) holds trivially. Since $\mathbf{P}(\alpha,\theta)$ is the image of the independent sequence (W_n) through the composition of the GEM map and the ordering map, one would hope to apply the contraction principle to get the LDP for $\mathbf{P}(\alpha,\theta)$. Unfortunately the ordering map is not continuous on the effective domain of S, and the contraction principle cannot be applied directly here.
(c) The fact that α does not change the LDP is mainly due to the topology used on space ∇.

8.2 Small Mutation Rate

In this section, we establish the LDP for Π_θ when θ tends to zero, and the LDP for $PD(\alpha,\theta)$ when both θ and α converge to zero. As in the large mutation rate case, these results will be obtained through a series of lemmas and the main techniques in the proof are exponential approximation and the contraction principle.

8.2.1 The Poisson–Dirichlet Distribution

Let $U=U(\theta)$ be a $Beta(1,\theta)$ random variable, and

$$a(\theta)=(-\log\theta)^{-1}.\tag{8.23}$$

All results in this subsection involve limits as θ converges to zero, so that $a(\theta) \to 0^+$.

Lemma 8.9. *The family* $\{U(\theta) : \theta > 0\}$ *satisfies an LDP on* $[0,1]$ *with speed* $a(\theta)$ *and rate function*

$$I_1(p) = \begin{cases} 0, \ p = 1 \\ 1, \ else. \end{cases} \tag{8.24}$$

Proof. For any $b < c$ in $[0,1]$, let $\mathbf{I}$ denote one of the intervals $(b,c), [b,c), (b,c]$, or $[b,c]$. It follows from direct calculation that for $c < 1$

$$\lim_{\theta \to 0} a(\theta) \log \mathbb{P}\{U \in \mathbf{I}\} = -\lim_{\theta \to 0} \frac{\log(1-r^\theta)}{\log \theta} = -1,$$

where $r = \frac{1-c}{1-b}$. If $c = 1$, then $\lim_{\theta \to 0} a(\theta) \log P\{U \in \mathbf{I}\} = 0$. These, combined with compactness of $[0,1]$, implies the result.

□

For any $n \geq 2$, let

$$\hat{P}_n(\theta) = \max\{V_1, \ldots, V_n\},$$

where V_k is defined in (8.1).

Lemma 8.10. *For any* $n \geq 2$, *the family* $\{\hat{P}_n(\theta) : \theta > 0\}$ *satisfies an LDP on* $[0,1]$ *with speed* $a(\theta)$ *and rate function*

$$I_n(p) = \begin{cases} 0, \ p = 1 \\ k, \ p \in [\frac{1}{k+1}, \frac{1}{k}), k = 1, 2, \ldots, n-1 \\ n, \ else. \end{cases} \tag{8.25}$$

Proof. Noting that $\hat{P}_n(\theta)$ is the image of $(U_1, \ldots, U_n)$ through a continuous map, it follows from Lemma 8.9, the independence of $(U_1, \ldots, U_n)$, and the contraction principle, that the family $\{\hat{P}_n(\theta) : \theta > 0\}$ satisfies an LDP on $[0,1]$ with speed $a(\theta)$ and rate function

$$\tilde{I}(p) = \inf \left\{ \sum_{i=1}^n I_1(u_i) : 0 \leq u_i \leq 1, 1 \leq i \leq n; \right.$$
$$\left. \max\{u_1, (1-u_1)u_2, \ldots, (1-u_1)\cdots(1-u_{n-1})u_n\} = p \right\}.$$

One can choose $u_i = 1$ for $i = 1, \ldots, n$ to get $\tilde{I}(1) = 0$. If p is in $[1/2, 1)$, then at least one of the u_i is not one. By choosing $u_1 = p, \ u_i = 1, \ i = 2, \ldots, n$, it follows that $\tilde{I}(p) = 1$ for p in $[1/2, 1)$.

For each $m \geq 2$, we have

$$\begin{aligned} &\max\{u_1, (1-u_1)u_2, \ldots, (1-u_1)\cdots(1-u_m)\} \\ &= \max\{u_1, (1-u_1)\max\{u_2, \ldots, (1-u_2)\cdots(1-u_m)\}\}. \end{aligned} \tag{8.26}$$

Noting that

$$\max\{u_1, 1-u_1\} \geq \frac{1}{2},\ 0 \leq u_1 \leq 1,$$

it follows from (8.26) and an induction on m, that for any $1 \leq i \leq m, 0 \leq u_i \leq 1$,

$$\max\{u_1, (1-u_1)u_2, \ldots, (1-u_1)\cdots(1-u_m)\} \geq \frac{1}{m+1}. \tag{8.27}$$

Thus, for $2 \leq k \leq n-1$, and p in $[\frac{1}{k+1}, \frac{1}{k})$, in order for the equality

$$\max\{u_1, (1-u_1)u_2, \ldots, (1-u_1)\cdots(1-u_{n-1})u_n\} = p$$

to hold, it is necessary that $u_1, u_2, \ldots, u_k$ are all less than one. In other words, $\tilde{I}(p) \geq k$. On the other hand, the function $\max\{u_1, (1-u_1)u_2, \ldots, (1-u_1)\cdots(1-u_k)\}$ is a surjection from $[0,1]^k$ onto $[\frac{1}{k+1}, 1]$, so there exist $u_1 < 1, \ldots, u_k < 1$ such that

$$\max\{u_1, (1-u_1)u_2, \ldots, (1-u_1)\cdots(1-u_k)\} = p.$$

By choosing $u_j = 1$ for $j = k+1, \ldots, n$, it follows that $\tilde{I}(p) = k$.

Finally, for p in $[0, \frac{1}{n})$, in order for

$$\max\{u_1, (1-u_1)u_2, \ldots, (1-u_1)\cdots(1-u_{n-1})u_n\} = p$$

to have solutions, each u_i has to be less than one, and hence $\tilde{I}(p) = n$. Therefore, $\tilde{I}(p) = I_n(p)$ for all p in $[0,1]$.

□

Lemma 8.11. *The family $\{P_1(\theta) : \theta > 0\}$ satisfies an LDP on $[0,1]$ with speed $a(\theta)$ and rate function*

$$J_1(p) = \begin{cases} 0, & p = 1 \\ k, & p \in [\frac{1}{k+1}, \frac{1}{k}), k = 1, 2, \ldots \\ \infty, & p = 0. \end{cases} \tag{8.28}$$

Proof. By the GEM representation of $P_1(\theta)$ and direct calculation it follows that for any $\delta > 0$ and any $n \geq 1$

$$\begin{aligned} \mathbb{P}\{P_1(\theta) - \hat{P}_n(\theta) > \delta\} &\leq \mathbb{P}\{(1-U_1)\cdots(1-U_n) > \delta\} \\ &\leq \delta^{-1}\left(\frac{\theta}{1+\theta}\right)^n, \end{aligned}$$

which implies that

$$\limsup_{\theta \to 0} a(\theta) \log \mathbb{P}\{P_1(\theta) - \hat{P}_n(\theta) > \delta\} \leq -n. \tag{8.29}$$

Hence $\{\hat{P}_n(\theta) : \theta > 0\}$ are exponentially good approximations of $\{P_1(\theta) : \theta > 0\}$. By direct calculation,

$$J_1(p) = \sup_{\delta>0} \liminf_{n\to\infty} \inf_{|q-p|<\delta} I_n(q),$$

and for every closed subset F of $[0,1]$

$$\inf_{q\in F} J_1(q) = \limsup_{n\to\infty} \inf_{q\in F} I_n(q).$$

Clearly $J_1(p)$ is a good rate function. By Theorem B.5 we obtain the lemma.

□

For any $n \geq 1$ and any $\delta > 0$, let $\overline{\nabla}_n, \mathbb{P}_{n,\theta}, G((p_1,\ldots,p_n);\delta)$, and $F((p_1,\ldots,p_n);\delta)$ be defined as in Section 8.1.1. Then we have:

Lemma 8.12. *For any fixed $n \geq 2$, the family $\{\mathbb{P}_{n,\theta} : \theta > 0\}$ satisfies an LDP on space $\overline{\nabla}_n$ with speed $a(\theta)$ and rate function*

$$J_n(p_1,\ldots,p_n) = \begin{cases} 0, & (p_1,p_2,\ldots,p_n) = (1,0\ldots,0) \\ l-1, & 2 \leq l \leq n, \sum_{k=1}^{l} p_k = 1, p_l > 0 \\ n + J_1(\frac{p_n}{1-\sum_{i=1}^{n} p_i} \wedge 1), & \sum_{k=1}^{n} p_k < 1, p_n > 0 \\ \infty, & \textit{else}. \end{cases} \tag{8.30}$$

Proof. For any fixed $n \geq 2$, let $g_1^\theta(p)$ and $g_n^\theta(p_1,\ldots,p_n)$ denote the density function of $P_1(\theta)$ and $(P_1(\theta),\ldots,P_n(\theta))$, respectively.

Since $\overline{\nabla}_n$ is compact, to prove the lemma it suffices to verify that for every $(p_1,\ldots,p_n)$ in $\overline{\nabla}_n$,

$$\begin{aligned} &\lim_{\delta\to 0} \liminf_{\theta\to 0} a(\theta) \log \mathbb{P}_{n,\theta}(F((p_1,\ldots,p_n);\delta)) \\ &\quad = \lim_{\delta\to 0} \limsup_{\theta\to 0} a(\theta) \log \mathbb{P}_{n,\theta}(G((p_1,\ldots,p_n);\delta)) \\ &\quad = -J_n(p_1,\ldots,p_n). \end{aligned} \tag{8.31}$$

For any $(p_1,\ldots,p_n)$ in $\overline{\nabla}_n$, define

$$r = r(p_1,\ldots,p_n) = \max\{i : 1 \leq i \leq n, p_i > 0\}, \tag{8.32}$$

where r is defined to be zero if $p_1 = 0$. We divide the proof into several mutually exclusive cases.

Case I: $r = 1$; i.e., $(p_1,\ldots,p_n) = (1,\ldots,0)$.

For any $\delta > 0$,

$$F((1,\ldots,0);\delta) \subset \{(q_1,\ldots,q_n) \in \overline{\nabla}_n : |q_1 - 1| \leq \delta\},$$

and one can choose $\delta' < \delta$ such that

$$\{(q_1,\ldots,q_n) \in \overline{\nabla}_n : |q_1 - 1| < \delta'\} \subset G((1,\ldots,0);\delta).$$

This combined with Lemma 8.11 implies (8.31) in this case.

Case II: $r = n$, $\sum_{k=1}^{n} p_k < 1$.

Choose $\delta > 0$ so that

$$\delta < \min\left\{ p_n, \frac{1-\sum_{i=1}^{n} p_i}{n} \right\}.$$

It follows from (8.8) that for any

$$(q_1,\ldots,q_n) \in F((p_1,\ldots,p_n),\delta) \cap \overline{\nabla}_n^{\circ},$$

$$g_n^{\theta}(q_1,\ldots,q_n) \le \frac{\theta^n (1-\sum_{k=1}^{n}(p_k+\delta))^{\theta-1}}{(p_1-\delta)\cdots(p_n-\delta)} \int_0^{\frac{p_n+\delta}{1-\sum_{k=1}^{n}(p_k+\delta)}\wedge 1} g_1^{\theta}(u)\,du,$$

which, combined with Lemma 8.11, implies

$$\begin{aligned}
&\limsup_{\delta\to 0}\limsup_{\theta\to 0} a(\theta)\log \mathbb{P}_{n,\theta}\{F((p_1,\ldots,p_n);\delta)\}\\
&\le -n + \lim_{\delta\to 0}\limsup_{\theta\to 0} a(\theta)\log \mathbb{P}\left\{P_1(\theta) \le \frac{p_n+\delta}{1-\sum_{k=1}^{n}(p_k+\delta)}\wedge 1\right\} \qquad (8.33)\\
&\le -\left[n + J_1\left(\frac{p_n}{1-\sum_{i=1}^{n} p_i}\wedge 1\right)\right],
\end{aligned}$$

where the right continuity of $J_1(\cdot)$ is used in the last inequality.
On the other hand, consider the subset

$$\tilde{G}(p_1,\ldots,p_n);\delta) \equiv \prod_{i=1}^{n}\left(p_i + \frac{\delta}{2}, p_i+\delta\right) \cap \overline{\nabla}_n^{\circ}$$

of $G((p_1,\ldots,p_n);\delta)$. Using (8.8) again, it follows that for any point $(q_1,\ldots,q_n)$ in the set $\tilde{G}((p_1,\ldots,p_n);\delta)$

$$\begin{aligned}
&g_n^{\theta}(q_1,\ldots,q_n)\\
&\ge \theta^n \frac{(1-\sum_{k=1}^{n}(p_k+\delta/2))^{\theta-1}}{(p_1+\delta)\cdots(p_n+\delta)} \int_0^{((p_n+\delta/2)/(1-\sum_{k=1}^{n}(p_k+\delta/2)))\wedge 1} g_1^{\theta}(u)\,du,
\end{aligned}$$

which, combined with Lemma 8.11, implies

$$\begin{aligned}
&\liminf_{\theta\to 0} a(\theta)\log \mathbb{P}_{n,\theta}\{G((p_1,\ldots,p_n);\delta)\}\\
&\ge \liminf_{\theta\to 0} a(\theta)\log \mathbb{P}_{n,\theta}\{\tilde{G}((p_1,\ldots,p_n);\delta)\}\\
&\ge -n - J_1\left(\frac{p_n+\delta/2}{1-\sum_{i=1}^{n}(p_i+\delta/2)}\wedge 1\right).
\end{aligned}$$

Therefore

$$\liminf_{\delta\to 0}\liminf_{\theta\to\infty} a(\theta)\log \mathbb{P}_{n,\theta}\{G((p_1,\ldots,p_n);\delta)\} \geq -J_n(p_1,\ldots,p_n). \tag{8.34}$$

Case III: $2 \leq r \leq n-1$, $\sum_{i=1}^{r} p_i < 1$ or $r = 0$.

This case follows from the estimate (8.33) and the fact that $J_1(0) = -\infty$.

Case IV: $r = n$, $\sum_{k=1}^{n} p_k = 1$.

Note that for any $\delta > 0$, $F((p_1,\ldots,p_n);\delta)\cap \overline{\nabla}_n^{\circ}$ is a subset of $\{(q_1,\ldots,q_n)\in \overline{\nabla}_n^{\circ} : |q_i - p_i| \leq \delta, i = 1,\ldots,n-1\}$. By applying **Case II** to $(P_1(\theta),\ldots,P_{n-1}(\theta))$ at the point $(p_1,\ldots,p_{n-1})$, we get

$$\begin{aligned}
&\limsup_{\delta\to 0}\limsup_{\theta\to 0} a(\theta)\log \mathbb{P}_{n,\theta}\{F((p_1,\ldots,p_n);\delta)\}\\
&\quad\leq \lim_{\delta\to 0}\limsup_{\theta\to 0} a(\theta)\log \mathbb{P}_{n-1,\theta}\{F((p_1,\ldots,p_{n-1});\delta)\} \qquad (8.35)\\
&\quad\leq -[n-1+J_1(1)] = -(n-1).
\end{aligned}$$

On the other hand, one can choose small $\delta > 0$ so that $\frac{q_n}{1-\sum_{i=1}^{n} q_i} > 1$ for any $(q_1,\ldots,q_n)$ in $G((p_1,\ldots,p_n);\delta)\cap \overline{\nabla}_n^{\circ}$.

Set

$$\tilde{G} = \{(q_1,\ldots,q_n)\in \overline{\nabla}_n^{\circ} : p_i < q_i < p_i + \delta/(n-1), i = 1,\ldots,n-1; p_n - \delta < q_n < p_n\}.$$

Clearly $\tilde{G}$ is a subset of $G((p_1,\ldots,p_n);\delta)$. It follows from (8.8) that for any $(q_1,\ldots,q_n)$ in $\tilde{G}$,

$$g_n^{\theta}(q_1,\ldots,q_n) \geq \frac{\theta^{n-1}[\theta(1-\sum_{i=1}^{n} q_i)^{\theta-1}]}{(p_1+\delta/(n-1))\cdots(p_{n-1}+\delta/(n-1))p_n}.$$

Let

$$A_n = \left\{(q_1,\ldots,q_{n-1})\in \nabla_{n-1} : p_i < q_i < p_i + \delta/(n-1), i = 1,\ldots,n-1, \sum_{j=1}^{n-1} q_j < 1\right\}.$$

Then

$$\begin{aligned}
&\int_{\tilde{G}} \theta\left(1-\sum_{i=1}^{n} q_i\right)^{\theta-1} dq_1\cdots dq_n\\
&\quad= \int_{A_n} dq_1\cdots dq_{n-1}\int_{p_n-\delta}^{p_n\wedge(1-\sum_{i=1}^{n-1} q_i)} \theta\left(1-\sum_{i=1}^{n} q_i\right)^{\theta-1} dq_n\\
&\quad= \int_{A_n}\left(1+\delta-p_n-\sum_{i=1}^{n-1} q_i\right)^{\theta} dq_1\cdots dq_{n-1},
\end{aligned}$$

which converges to a strictly positive number depending only on δ and $(p_1,\ldots,p_n)$, as θ goes to zero. Hence

$$\begin{aligned}\liminf_{\delta\to 0}\liminf_{\theta\to 0} a(\theta)\log \mathbb{P}_{n,\theta}\{G((p_1,\ldots,p_n);\delta)\}\\ \geq \lim_{\delta\to 0}\liminf_{\theta\to 0} a(\theta)\log \mathbb{P}_{n,\theta}\{\tilde{G}\}\\ \geq -(n-1).\end{aligned} \tag{8.36}$$

Case V: $2\leq r\leq n-1$, $\sum_{i=1}^r p_i = 1$.

First note that for any $\delta>0$, $F((p_1,\ldots,p_n);\delta)$ is a subset of

$$\{(q_1,\ldots,q_n)\in\overline{\nabla}_n : |q_i-p_i|\leq\delta, i=1,\ldots,r\}.$$

On the other hand, for each $\delta>0$ one can choose $\delta_0<\delta$ such that for any $\delta'\leq\delta_0$

$$G((p_1,\ldots,p_n);\delta)\supset\{(q_1,\ldots,q_n)\in\overline{\nabla}_n^{\circ}; |q_i-p_i|<\delta', i=1,\ldots,r\}.$$

Thus (8.31) follows from **Case IV** for $(P_1(\theta),\ldots,P_r(\theta))$. Putting together all the cases, we obtain the lemma.

□

For any $n\geq 1$, set

$$L_n=\left\{(p_1,\ldots,p_n,0,0,\ldots)\in\nabla : \sum_{i=1}^n p_i = 1\right\}$$

and

$$L=\bigcup_{i=1}^{\infty} L_i.$$

Now we are ready to state and prove the main result of this subsection.

Theorem 8.3. *The family $\{\Pi_\theta:\theta>0\}$ satisfies an LDP with speed $a(\theta)$ and rate function*

$$J(\mathbf{p})=\begin{cases}0, & \mathbf{p}\in L_1\\ n-1, & \mathbf{p}\in L_n,\ p_n>0,\ n\geq 2\\ \infty, & \mathbf{p}\notin L.\end{cases} \tag{8.37}$$

Proof. For any $\mathbf{p}=(p_1,p_2,\ldots),\mathbf{q}=(q_1,q_2,\ldots)$ in ∇, let $d(\mathbf{p},\mathbf{q})$ be the metric defined in Theorem 8.1. For any fixed $\delta>0$, let $B(\mathbf{p},\delta)$ and $\bar{B}(\mathbf{p},\delta)$ denote the respective open and closed balls centered at $\mathbf{p}$ with radius $\delta>0$.

We start with the case that $\mathbf{p}$ is not in L. For any $k\geq 1,\delta'>0$, set

$$\bar{B}_{k,\delta'}(\mathbf{p})=\{(q_1,q_2,\ldots)\in\nabla : |q_i-p_i|\leq\delta', i=1,\ldots,k\}.$$

Choose $\delta > 0$ so that $2^k\delta < \delta'$. Then

$$\bar{B}(\mathbf{p},\delta) \subset \bar{B}_{k,\delta'}(\mathbf{p}),$$

and

$$\begin{aligned}
\limsup_{\delta\to 0}\limsup_{\theta\to 0} a(\theta)\log \Pi_\theta\{\bar{B}(\mathbf{p},\delta)\} \\
&\leq \limsup_{\theta\to 0} a(\theta)\log \Pi_\theta\{\bar{B}_{k,\delta'}(\mathbf{p})\} \\
&\leq \limsup_{\theta\to 0} \lambda(\theta)\log \mathbb{P}_{k,\theta}\{F((p_1,\ldots,p_k),\delta')\} \qquad (8.38)\\
&\leq -\inf\{J_k(q_1,\ldots,q_k) : (q_1,\ldots,q_k) \in F((p_1,\ldots,p_k),\delta')\}.
\end{aligned}$$

Letting δ' go to zero, and then k go to infinity, we obtain

$$\begin{aligned}
&\lim_{\delta\to 0}\liminf_{\theta\to 0} a(\theta)\log \Pi_\theta\{B(\mathbf{p},\delta)\} \qquad (8.39)\\
&= \lim_{\delta\to 0}\limsup_{\theta\to 0} a(\theta)\log \Pi_\theta\{\bar{B}(\mathbf{p},\delta)\} = -\infty.
\end{aligned}$$

Next consider the case of $\mathbf{p}$ belonging to L. Without loss of generality, we assume that $\mathbf{p}$ belongs to L_n with $p_n > 0$ for some $n \geq 1$.

For any $\delta > 0$, let

$$\begin{aligned}
\tilde{G}(\mathbf{p};\delta) &= \{\mathbf{q} \in \nabla : |q_k - p_k| < \delta, k = 1,\ldots,n\},\\
\tilde{F}(\mathbf{p};\delta) &= \{\mathbf{q} \in \nabla : |q_k - p_k| \leq \delta, k = 1,\ldots,n\}.
\end{aligned}$$

Clearly, $\bar{B}(\mathbf{p},\delta)$ is a subset of $\tilde{F}(\mathbf{p};2^n\delta)$. Since $\sum_{i=1}^n p_i = 1$, it follows that, for any $\delta > 0$, one can find $\delta' < \delta$ such that

$$B(\mathbf{p},\delta) \supset \tilde{G}(\mathbf{p};\delta').$$

Using results on $(P_1(\theta),\ldots,P_n(\theta))$ in **Case V** of the proof of Lemma 8.12, we get

$$\begin{aligned}
&\lim_{\delta\to 0}\liminf_{\theta\to 0} a(\theta)\log \Pi_\theta(B(\mathbf{p},\delta))\\
&= \lim_{\delta\to 0}\limsup_{\theta\to 0} \lambda(\theta)\log \Pi_\theta(\bar{B}(\mathbf{p},\delta)) \qquad (8.40)\\
&= -(n-1).
\end{aligned}$$

Thus the theorem follows from the compactness of ∇ and Theorem B.6.

□

Remarks:
(a) If we consider the rate function J as an "energy" function, then the energy needed to get $n \geq 2$ different alleles is $n-1$. The values of J form a "ladder of energy". The energy needed to get an infinite number of alleles is infinite and thus it is impossible to have infinitely many alleles under a large deviation.
(b) The effective domain of J is clearly L. This is in sharp contrast to the result in Theorem 8.1, where the effective domain of the corresponding rate function associated with a large mutation rate is

$$\left\{\mathbf{p} \in \nabla : \sum_{i=1}^{\infty} p_i < 1\right\}.$$

The two effective domains are disjoint. One is part of the boundary of ∇ and the other is the interior of ∇, and both have no intersections with the set ∇_∞, where Π_θ is concentrated.
(c) By using techniques from the theory of Dirichlet forms, it was shown in [160] that for the infinitely-many-neutral-alleles model, with probability one, there exist times at which the sample path will hit the boundary of a finite-dimensional subsimplex of ∇ or, equivalently, the single point $(1,0,\ldots)$ iff θ is less than one. The intuition here is that it is possible to have only a finite number of alleles in the population if the mutation rate is small; but in equilibrium, with Π_θ probability one, the number of alleles is always infinite as long as θ is strictly positive. In other words, the critical value of θ between having only a finite number of alleles vs. an infinite number of alleles is zero for Π_θ. In physical terms, this sudden change from one to infinity can be viewed as a phase transition. The result in Theorem 8.3 gives more details about this transition.

8.2.2 Two-parameter Generalization

Consider the parameters α and θ in the ranges $0<\alpha<1, \theta+\alpha>0$. For any $\delta>0$, it follows from the GEM representation for $PD(\alpha,\theta)$ that

$$\mathbb{P}\left(V_1^{\alpha,\theta} > 1-\delta\right) \leq \mathbb{P}(P_1(\alpha,\theta) > 1-\delta).$$

By direct calculation, we have

$$\lim_{\alpha+\theta\to 0} \mathbb{P}\left(V_1^{\alpha,\theta} > 1-\delta\right) = 1.$$

Therefore, $PD(\alpha,\theta)$ converges in the space of probability measures on ∇ to $\delta_{(1,0,\ldots)}$ as $\alpha \vee |\theta|$ converges to zero. In this subsection, we establish the LDP associated with this limit. This is a generalization of Theorem 8.3.

Let

$$a(\alpha,\theta) = (-\log(\alpha \vee |\theta|)^{-1}.$$

It is clear that $a(\alpha,\theta)$ converges to zero if and only if $\alpha \vee |\theta|$ converges to zero.

Lemma 8.13. *For each $i \geq 1$, consider the Beta$(1-\alpha, \theta + i\alpha)$ random variable W_i in the GEM representation for PD(α,θ). As $a(\alpha,\theta)$ converges to zero, the family $\{W_i : \alpha+\theta > 0, 0 < \alpha < 1\}$ satisfies an LDP on $[0,1]$ with speed $a(\alpha,\theta)$ and rate function I_1 defined in (8.24).*

Proof. Fix $i \geq 1$. Write the density function of W_i in the form of

$$\frac{\Gamma(\theta+1+(i-1)\alpha)}{\Gamma(1-\alpha)\Gamma(\theta+i\alpha+1)}(\theta+i\alpha)x^{-\alpha}(1-x)^{\theta+i\alpha-1}.$$

Then it follows from direct calculation that for any p in $[0,1)$,

$$\begin{aligned}
&\lim_{\delta\to 0}\liminf_{a(\alpha,\theta)\to 0} a(\alpha,\theta)\log\mathbb{P}\{|W_i - p| < \delta\}\\
&\quad = \lim_{\delta\to 0}\limsup_{a(\alpha,\theta)\to 0} a(\alpha,\theta)\log\mathbb{P}\{|W_i - p| \leq \delta\}\\
&\quad = \lim_{a(\alpha,\theta)\to 0} a(\alpha,\theta)\log(\theta+i\alpha) = -1.
\end{aligned}$$

For $p = 1$,

$$\begin{aligned}
&\lim_{\delta\to 0}\liminf_{a(\alpha,\theta)\to 0} a(\alpha,\theta)\log\mathbb{P}\{|W_i - 1| < \delta\}\\
&\quad = \lim_{\delta\to 0}\limsup_{a(\alpha,\theta)\to 0} a(\alpha,\theta)\log\mathbb{P}\{|W_i - 1| \leq \delta\}\\
&\quad = \lim_{\delta\to 0}\lim_{a(\alpha,\theta)\to 0} a(\alpha,\theta)\log(1-\delta)^{\theta+i\alpha} = 0.
\end{aligned}$$

The lemma now follows from Theorem B.6. □

Lemma 8.14. *The family $\{P_1(\alpha,\theta) : \alpha+\theta > 0, 0 < \alpha < 1\}$ satisfies an LDP on $[0,1]$ as $a(\alpha,\theta)$ converges to zero with speed $a(\alpha,\theta)$ and rate function $J_1(\cdot)$ defined in (8.28).*

Proof. For any $n \geq 1$, let

$$\hat{P}_n(\alpha,\theta) = \max\{V_1^{\alpha,\theta},\ldots,V_n^{\alpha,\theta}\}.$$

Then by direct calculation

$$\mathbb{P}\{P_1(\alpha,\theta)-\hat{P}_n(\alpha,\theta)>\delta\} \leq \mathbb{P}\{(1-W_1)\cdots(1-W_n)\geq\delta\}$$
$$\leq \delta^{-1}\prod_{i=1}^{n}\frac{\theta+i\alpha}{\theta+i\alpha+1-\alpha},$$

which leads to

$$\limsup_{\alpha\vee|\theta|\to 0} a(\alpha,\theta)\log\mathbb{P}\{P_1(\alpha,\theta)-\hat{P}_n(\alpha,\theta)>\delta\}\leq -n.$$

The remainder of the proof uses arguments similar to those used in the proofs of Lemma 8.10 and Lemma 8.11.

□

Theorem 8.4. *The family* $\{\Pi_{\alpha,\theta}:\alpha+\theta>0, 0<\alpha<1\}$ *satisfies an LDP on* ∇ *as* $\alpha\vee|\theta|$ *converges to zero with speed* $a(\alpha,\theta)$ *and rate function* J *given in (8.37).*

Proof. It suffices to establish the LDP for the finite-dimensional marginal distributions since the infinite-dimensional LDP can be derived from the finite-dimensional LDP through approximation. For any $n\geq 2$, let

$$\varphi_n^{\alpha,\theta}(p_1,\ldots,p_n), \text{ and } g_n^{\alpha+\theta}(p_1,\ldots,p_n)$$

be the respective joint density functions of

$$(P_1(\alpha,\theta),P_2(\alpha,\theta),\ldots,P_n(\alpha,\theta)) \text{ and } (P_1(\alpha+\theta),P_2(\alpha+\theta),\ldots,P_n(\alpha+\theta))$$

given in (8.20) and (8.21).

Since

$$\lim_{a(\alpha,\theta)\to 0} a(\alpha,\theta)\log(\alpha+\theta)=-1,$$

and

$$\lim_{a(\alpha,\theta)\to 0} a(\alpha,\theta)\log C_{\alpha,\theta,n}=-n,$$

it follows that for every $(p_1,\ldots,p_n)$ in $\overline{\nabla}_n^{\circ}$

$$\begin{aligned}\lim_{a(\alpha,\theta)\to 0} & a(\alpha,\theta)\log\varphi_n^{\alpha,\theta}(p_1,\ldots,p_n)\\ &=\lim_{a(\alpha,\theta)\to 0} a(\alpha,\theta)\log g_n^{\alpha+\theta}(p_1,\ldots,p_n)\\ &=-J_n(p_1,\ldots,p_n).\end{aligned}$$

This, combined with Lemma 8.11 and Lemma 8.14, implies that the family

$$(P_1(\alpha,\theta),P_2(\alpha,\theta),\ldots,P_n(\alpha,\theta))$$

satisfies an LDP as $a(\alpha,\theta)$ converges to zero with speed $a(\alpha,\theta)$ and rate function $J_n(p_1,\ldots,p_n)$ defined in (8.30).

□

8.3 Applications

Several applications of Theorem 8.1 and Theorem 8.3 will be considered in this section. For any $n \geq 2$, let

$$\varphi_n(\mathbf{p}) = \sum_{i=1}^{\infty} p_i^n.$$

Then $\varphi_n(\mathbf{P}(\theta))$ is the population homozygosity of order n defined in Section 7.3. Since

$$\varphi_n(\mathbf{P}(\theta)) \leq P_1^{n-1}(\theta),$$

it follows that $\varphi_n(\mathbf{P}(\theta))$ converges to zero as θ tends to infinity. Our next theorem describes the large deviations of $\varphi_n(\mathbf{P}(\theta))$ from zero.

Theorem 8.5. *The family $\{\varphi_n(\mathbf{P}(\theta)) : \theta > 0\}$ satisfies an LDP on $[0,1]$ as θ tends to infinity, with speed $1/\theta$ and rate function $I(x^{1/n})$, where I is given by* (8.2).

Proof. For any $n \geq 2$, the map

$$\varphi_n : \nabla \longrightarrow [0,1], \; \mathbf{p} \mapsto \sum_{i=1}^{\infty} p_i^n$$

is continuous. By Theorem 8.1 and the contraction principle, the family $\{\varphi_n(\mathbf{P}(\theta)) : \theta > 0\}$ satisfies an LDP on $[0,1]$, as θ tends to infinity, with speed $1/\theta$ and rate function

$$\bar{I}(x) = \inf\{S(\mathbf{p}) : \mathbf{p} \in \nabla, \varphi_n(\mathbf{p}) = x\}.$$

Since for any $\mathbf{p}$ in ∇, we have

$$\sum_{i=1}^{\infty} p_i \geq (\varphi_n(\mathbf{p}))^{1/n} = x^{1/n},$$

it follows that $S(\mathbf{p}) \geq I(x^{1/n})$, and thus $\bar{I}(x) \geq I(x^{1/n})$. On the other hand, by choosing

$$\mathbf{p} = (x^{1/n}, 0, \ldots),$$

one obtains $\bar{I}(x) \leq I(x^{1/n})$. Hence $\bar{I}(x) = I(x^{1/n})$, and the result follows.

□

Remarks:
(a) The LDP obtained here describes the deviations of $\varphi_n(\mathbf{P}(\theta))$ from zero. The result in Theorem 7.10 shows that $\frac{\theta^{n-1}}{\Gamma(n)}\varphi_n(\mathbf{P}(\theta))$ converges to one in probability as θ tends to infinity. However, our original motivation was to study the large deviations of $\frac{\theta^{n-1}}{\Gamma(n)}\varphi_n(\mathbf{P}(\theta))$ from one, which is still an open problem.
(b) The Gaussian structure of Theorem 7.10 seems to indicate that the LDP for the family $\{\frac{\theta^{n-1}}{\Gamma(n)}\varphi_n(\mathbf{P}(\theta)) : \theta > 0\}$ will hold with a speed of $1/\theta$; but the following

calculations lead to a different answer. Assume that an LDP holds for the family with speed $a(\theta)$ and a good rate function I. Then for any constant $c > 0$,

$$\begin{aligned}
\mathbb{P}\left\{\frac{\theta^{n-1}}{\Gamma(n)}\varphi_n(\mathbf{P}(\theta)) \geq 1+c\right\} &\geq \mathbb{P}\left\{\frac{\theta^{n-1}}{\Gamma(n)}U_1^n \geq 1+c\right\}\\
&= \mathbb{P}\left\{U_1 \geq \left(\frac{\Gamma(n)(1+c)}{\theta^{n-1}}\right)^{1/n}\right\}\\
&= \left[\left(1-\frac{(\Gamma(n)(1+c))^{1/n}}{\theta^{(n-1)/n}}\right)^{\theta^{(n-1)/n}}\right]^{\theta^{1/n}},
\end{aligned}$$

which implies that

$$\inf_{x \geq 1+c} I(x) = 0 \text{ if } \lim_{\theta\to\infty}\frac{a(\theta)}{\theta^{1/n}} = \infty.$$

Since c is arbitrary, I is zero over a sequence that goes to infinity, which contradicts the fact that $\{x : I(x) \leq M\}$ is compact for every positive M. Hence the LDP speed, if it exists, cannot grow faster than $\theta^{1/n}$.
(c) For r in $[0,1/2)$, the quantity $\theta^r(\frac{\theta^{n-1}}{\Gamma(n)}\varphi_n(\mathbf{P}(\theta)) - 1)$ converges to zero in probability as θ tends to infinity. Large deviations associated with this limit for $r \in (0,1/2)$ are called the *moderate deviation principle* for $\{\frac{\theta^{n-1}}{\Gamma(n)}\varphi_n(\mathbf{P}(\theta)) : \theta > 0\}$. Recent work in [73] indicates that the LDP, corresponding to $r = 0$, may not hold for the family $\{\frac{\theta^{n-1}}{\Gamma(n)}\varphi_n(\mathbf{P}(\theta)) : \theta > 0\}$.

When θ tends to zero, $\varphi_n(\mathbf{P}(\theta))$ converges to one. The LDP associated with this limit is established in the next theorem.

Theorem 8.6. *The family $\{\varphi_n(\mathbf{P}(\theta)) : \theta > 0\}$ satisfies an LDP on $[0,1]$, as θ tends to zero, with speed $a(\theta)$ given in (8.23), and rate function*

$$\hat{J}(p) = \begin{cases} 0, & p = 1\\ k-1, & p \in [\frac{1}{k^{n-1}}, \frac{1}{(k-1)^{n-1}}), k = 2,\ldots\\ \infty, & p = 0. \end{cases} \tag{8.41}$$

Thus in terms of large deviations, $\varphi_n(\mathbf{P}(\theta))$ behaves the same as $P_1^{n-1}(\theta)$.

Proof. Due to Theorem 8.3 and the contraction principle, it suffices to verify that

$$\hat{J}(p) = \inf\{J(\mathbf{q}) : \mathbf{q} \in \nabla, \varphi_n(\mathbf{q}) = p\} = \inf\{S(\mathbf{q}) : \mathbf{q} \in L, \varphi_n(\mathbf{q}) = p\}.$$

For $p = 1$, it follows by choosing $\mathbf{q} = (1,0,\ldots)$ that

$$\inf\{S(\mathbf{q}) : \mathbf{q} \in \nabla, \varphi_n(\mathbf{q}) = p\} = 0.$$

For $p = 0$, there does not exist $\mathbf{q}$ in L such that $\varphi_n(\mathbf{q}) = p$. Hence

$$\inf\{S(\mathbf{q}) : \mathbf{q} \in L, \varphi_n(\mathbf{q}) = p\} = \infty.$$

For any $k \geq 2$, the minimum of $\sum_{i=1}^{k} q_i^n$ over L_k is $k^{-(n-1)}$, which is achieved when all q_i's are equal. Hence for

$$p \in [k^{-(n-1)}, (k-1)^{-(n-1)}),$$

we have

$$\inf\{S(\mathbf{q}) : \mathbf{q} \in \nabla, \varphi_n(\mathbf{q}) = p\} = k - 1 = \hat{J}(p).$$

□

Let $C(\nabla)$ be the set of all continuous functions on ∇, and $\lambda(\theta)$ be a non-negative function of θ. For every f in $C(\nabla)$, define a new probability $\Pi_{\lambda,\theta}^{f}$ on ∇ as

$$\Pi_{\lambda,\theta}^{f}(d\mathbf{p}) = \frac{e^{\lambda(\theta)f(\mathbf{p})}}{\mathbb{E}^{\Pi_\theta}\left[e^{\lambda(\theta)f(\mathbf{p})}\right]}\Pi_\theta(d\mathbf{p}). \tag{8.42}$$

The case of $f(\mathbf{p}) = s\varphi_2(\mathbf{p})$ corresponds to the symmetric selection model. If $s > 0$, then homozygotes have selective advantage over heterozygotes and the model is said to have *underdominant* selection. The case of $s < 0$ is the opposite of $s > 0$ and the model is said to have *overdominant* selection. The Poisson–Dirichlet distribution Π_θ corresponds to $s = 0$. The general distribution considered here is simply a mathematical generalization to these selection models.

Theorem 8.7. *Let $S(\mathbf{p})$ be defined as in (8.15) and assume that*

$$\lim_{\theta\to\infty} \frac{\lambda(\theta)}{\theta} = c \in [0, +\infty).$$

Then the family $\{\Pi_{\lambda,\theta}^{f} : \theta > 0\}$ satisfies an LDP on ∇, as θ tends to infinity, with speed $1/\theta$ and rate function

$$S_{c,f}(\mathbf{p}) = \sup_{\mathbf{q}\in\nabla}\{cf(\mathbf{q}) - S(\mathbf{q})\} - (cf(\mathbf{p}) - S(\mathbf{p})). \tag{8.43}$$

Proof. By Theorem 8.1 and Theorem B.1,

$$\begin{aligned}
\lim_{\theta\to\infty} \frac{1}{\theta} \log \mathbb{E}^{\Pi_\theta}\left[e^{\lambda(\theta)f(\mathbf{p})}\right] &= \lim_{\theta\to\infty} \frac{1}{\theta} \log \mathbb{E}^{\Pi_\theta}\left[e^{\theta\frac{\lambda(\theta)}{\theta}f(\mathbf{p})}\right] \\
&= \sup_{\mathbf{q}}\{cf(\mathbf{q}) - S(\mathbf{q})\}.
\end{aligned}$$

This, combined with the continuity of f, implies that for any $\mathbf{p}$ in ∇

$$
\begin{aligned}
\liminf_{\delta\to 0}\liminf_{\theta\to\infty}\frac{1}{\theta}\log \Pi^f_{\lambda,\theta}\{d(\mathbf{p},\mathbf{q})<\delta\} &\geq -\sup_{\mathbf{q}}\{cf(\mathbf{q})-S(\mathbf{q})\}\\
&+\liminf_{\delta\to 0}\liminf_{\theta\to\infty}\left\{\frac{\lambda(\theta)}{\theta}(f(\mathbf{p})-\delta')+\frac{1}{\theta}\log \Pi_\theta\{d(\mathbf{p},\mathbf{q})<\delta\}\right\}\\
&\geq -S_{c,f}(\mathbf{p}),
\end{aligned}
$$

where δ' converges to zero as δ goes to zero. Similarly we have

$$
\begin{aligned}
\limsup_{\delta\to 0}\limsup_{\theta\to\infty}\frac{1}{\theta}\log \Pi^f_{\lambda,\theta}\{d(\mathbf{p},\mathbf{q})\leq\delta\} &\leq -\sup_{\mathbf{q}}\{cH(\mathbf{q})-S(\mathbf{q})\}\\
&+\limsup_{\delta\to 0}\limsup_{\theta\to\infty}\left\{\frac{\lambda(\theta)}{\theta}(f(\mathbf{p})+\delta')+\frac{1}{\theta}\log \Pi_\theta\{d(\mathbf{p},\mathbf{q})\leq\delta\}\right\}\\
&\leq -S_{c,f}(\mathbf{p}).
\end{aligned}
$$

This, combined with the compactness of ∇ and Theorem B.6, implies the result. □

Theorem 8.8. *Assume that*

$$
\lim_{\theta\to\infty}\frac{\lambda(\theta)}{\theta}=\infty \tag{8.44}
$$

and that f achieves its maximum at a single point $\mathbf{p}_0$. Then the family $\{\Pi^f_{\lambda,\theta}\}$ satisfies an LDP on ∇, as θ tends to infinity, with speed $1/\theta$ and rate function

$$
S_{\infty,f}(\mathbf{p})=\begin{cases}0, & \textit{if } \mathbf{p}=\mathbf{p}_0\\ \infty, & \textit{else}.\end{cases} \tag{8.45}
$$

Proof. Without loss of generality we assume that $\sup_{\mathbf{p}\in\nabla}f(\mathbf{p})=0$; otherwise we can multiply both the numerator and the denominator in the definition of $\Pi^f_{\lambda,\theta}$, by $e^{-\lambda(\theta)f(\mathbf{p}_0)}$.

For any $\mathbf{p}\neq\mathbf{p}_0$, choose δ small enough such that

$$
d_1=\sup_{d(\mathbf{p},\mathbf{q})\leq\delta}f(\mathbf{q})<2d_2=2\inf_{d(\mathbf{p}_0,\mathbf{q})\leq\delta}f(\mathbf{q})<0.
$$

Then by direct calculation

$$
\begin{aligned}
&\limsup_{\theta\to\infty}\frac{1}{\theta}\log \Pi^f_{\lambda,\theta}\{d(\mathbf{p},\mathbf{q})\leq\delta\}\\
&=\limsup_{\theta\to\infty}\frac{1}{\theta}\log\frac{\int_{\{d(\mathbf{p},\mathbf{q})\leq\delta\}}e^{\lambda(\theta)f(\mathbf{q})}\Pi_\theta(d\mathbf{q})}{\mathbb{E}^{\Pi_\theta}[e^{\lambda(\theta)f(\mathbf{q})}]}\\
&\leq\limsup_{\theta\to\infty}\frac{1}{\theta}\log\left[e^{\lambda(\theta)(d_1-d_2)}\frac{\Pi_\theta\{d(\mathbf{p},\mathbf{q})\leq\delta\}}{\Pi_\theta\{d(\mathbf{p}_0,\mathbf{q})\leq\delta\}}\right]\\
&=-\infty,
\end{aligned}
$$

and

$$
\begin{aligned}
&\liminf_{\theta\to\infty}\frac{1}{\theta}\log \Pi^{f}_{\lambda,\theta}\{d(\mathbf{p}_0,\mathbf{q})<\delta\}\\
&=\liminf_{\theta\to\infty}\frac{1}{\theta}\log\frac{\int_{\{d(\mathbf{p}_0,\mathbf{q})<\delta\}} e^{\lambda(\theta)f(\mathbf{q})}\,\Xi_\theta(d\mathbf{q})}{\mathbb{E}^{\Pi_\theta}[e^{\lambda(\theta)f(\mathbf{q})}]}\\
&=\liminf_{\theta\to\infty}\frac{1}{\theta}\log\left[1-\frac{\int_{\{d(\mathbf{p}_0,\mathbf{q})\ge\delta\}} e^{\lambda(\theta)f(\mathbf{q})}\,\Pi_\theta(d\mathbf{q})}{\mathbb{E}^{\Pi_\theta}[e^{\lambda(\theta)f(\mathbf{q})}]}\right]\\
&\ge\liminf_{\theta\to\infty}\frac{1}{\theta}\log\left[1-\frac{\int_{\{d(\mathbf{p}_0,\mathbf{q})\ge\delta\}} e^{\lambda(\theta)f(\mathbf{q})}\,\Pi_\theta(d\mathbf{q})}{\int_{\{d(\mathbf{p}_0,\mathbf{q})\le\delta_1\}} e^{\lambda(\theta)f(\mathbf{q})}\,\Pi_\theta(d\mathbf{q})}\right]\\
&\ge\liminf_{\theta\to\infty}\frac{1}{\theta}\log\left[1-\frac{\exp\{\lambda(\theta)[e_1(\delta)-e_2(\delta_1)]\}}{\Pi_\theta\{d(\mathbf{p}_0,\mathbf{q})\le\delta\}}\right]\\
&=0,\ \text{by choosing small enough } \delta_1,
\end{aligned}
$$

where

$$
e_1(\delta)=\sup_{d(\mathbf{p}_0,\mathbf{q})\ge\delta} f(\mathbf{q}),\ e_2(\delta_1)=\inf_{d(\mathbf{p}_0,\mathbf{q})\le\delta_1} f(\mathbf{q}).
$$

Again the result follows from the compactness of ∇ and Theorem B.6.

□

Consider the overdominant selection model with $f(\mathbf{p})=-\varphi_2(\mathbf{p})$. If $\lambda(\theta)=o(\theta)$, then the selection model has the same LDP as the neutral model. At the critical scale θ, the rate function begins to depend on the selection and a phase transition occurs. The next result shows that a similar critical phenomenon also exists in the underdominant selection model.

Corollary 8.9 *For $f(\mathbf{p})=\phi_2(\mathbf{p})$,*

$$
S_{c,f}(\mathbf{p})=\begin{cases} S(\mathbf{p}), & \text{if } \lim_{\theta\to\infty}\frac{\lambda(\theta)}{\theta}=0\\ -cf(\mathbf{p})+S(\mathbf{p}), & \text{if } \lim_{\theta\to\infty}\frac{\lambda(\theta)}{\theta}=c\le c_0\\ g(c)-cf(\mathbf{p})+S(\mathbf{p}), & \text{if } \lim_{\theta\to\infty}\frac{\lambda(\theta)}{\theta}=c\in(c_0,+\infty), \end{cases} \tag{8.46}
$$

where, for $c\ge c_0$,

$$
g(c)=\log\left(\frac{1-\sqrt{1-2/c}}{2}\right)+c\left(\frac{1+\sqrt{1-2/c}}{2}\right)^2,
$$

and $c_0>2$ solves the equation

$$
g(c_0)=0.
$$

Proof. The key step is to calculate

$$g(c) = \sup_{\mathbf{q}\in\nabla}\left\{c\sum_{i=1}^{\infty} q_i^2 + \log(1-\sum_{i=1}^{\infty} q_i)\right\}.$$

Write $\sum_{i=1}^{\infty} q_i$ as x. Then for any given x, the maximum of $\sum_{i=1}^{\infty} q_i^2$ is x^2. Hence

$$g(c) = \sup_{x\in[0,1]} \{cx^2 + \log(1-x)\}.$$

Let c_0 satisfy

$$\log\left(\frac{1-\sqrt{1-2/c_0}}{2}\right) + c_0\left(\frac{1+\sqrt{1-2/c_0}}{2}\right)^2 = 0.$$

Then it follows by direct calculation that $c_0 > 2$, and

$$g(c) = \begin{cases} 0, & \text{if } c \le c_0\\ \log\left(\frac{1-\sqrt{1-2/c}}{2}\right) + c\left(\frac{1+\sqrt{1-2/c}}{2}\right)^2, & \text{if } c > c_0. \end{cases}$$

□

Applying Theorem 8.3 and Theorem B.1 to $\Pi^f_{\lambda,\theta}$, we obtain the following result when the mutation rate is small.

Theorem 8.10. *Let $J(\mathbf{p})$ be defined as in (8.37) and $a(\theta)$ be given as in (8.23). For a fixed integer $n \ge 1$, constant s, and $f(\mathbf{p}) = s\varphi_n(\mathbf{p})$, the family $\{\Pi^f_{\lambda,\theta} : \theta > 0\}$ satisfies an LDP on ∇, as θ tends to zero, with speed $a(\theta)$ and rate function*

$$J_{\lambda,s,n}(\mathbf{p}) = \begin{cases} J(\mathbf{p}), & \lim_{\theta\to 0}\lambda(\theta)a(\theta) = 0\\ J(\mathbf{p}) + sc(1-\varphi_n(\mathbf{p})), & \lim_{\theta\to 0}\lambda(\theta)a(\theta) = c > 0,\ s > 0\\ J(\mathbf{p}) + |s|c\varphi_n(\mathbf{p}) & \\ -\inf\{\frac{|s|c}{m^{n-1}} + m - 1 : m \ge 1\}, & \lim_{\theta\to 0}\lambda(\theta)a(\theta) = c > 0,\ s < 0. \end{cases}$$

Proof. Theorem 8.3, combined with Varadhan's lemma and the Laplace method, implies that the family $\{\Pi^f_{\lambda,\theta} : \theta > 0\}$ satisfies an LDP on ∇, as θ converges to zero, with speed $a(\theta)$ and rate function

$$\sup_{\mathbf{q}\in\nabla}\{sc\varphi_n(\mathbf{q}) - J(\mathbf{q})\} - (sc\varphi_n(\mathbf{p}) - J(\mathbf{p})).$$

The case $c = 0$ is clear. For $c > 0,\ s > 0$,

$$\sup_{\mathbf{q}\in\nabla}\{sc\varphi_n(\mathbf{q}) - J(\mathbf{q})\}$$

achieves its maximum sc at $\mathbf{q} = (1,0,\ldots)$. For $c > 0$ and $s < 0$,

$$\sup_{\mathbf{q}\in\nabla}\{sc\varphi_n(\mathbf{q})-J(\mathbf{q})\} = -\inf_{\mathbf{q}\in\nabla}\{|s|c\varphi_n(\mathbf{q})+J(\mathbf{q})\} = -\inf_{m\geq 1}\left\{\frac{|s|c}{m^{n-1}}+m-1\right\}.$$

□

It is clear from the theorem that selection has an impact on the rate function only when the selection intensity $\lambda(\theta)$ is proportional to $a(\theta)^{-1}$. Consider the case of $\lambda(\theta) = a(\theta)^{-1}$. Then for $s > 0$, the homozygote has selective advantage, and the small-mutation-rate limit is $(1,0,\ldots)$ and diversity disappears. The energy $J_{\lambda,s,n}(\mathbf{p})$ needed for a large deviation from $(1,0,\ldots)$ is larger than the neutral energy $J(\mathbf{p})$. For $s < 0$, heterozygotes have selection advantage. Since $J_{\lambda,s,n}(\cdot)$ may reach zero at a point that is different from $(1,0,\ldots)$, several alleles can coexist in the population when the selection intensity goes to infinity and θ tends to zero. A concrete case of allele coexistence is included in the next result.

Corollary 8.11 *Let*

$$n=2,\ \lambda(\theta)=2a(\theta)^{-1},$$

and

$$l_k=\frac{k(k+1)}{2},\ \mathbf{p}_k=\Bigg(\underbrace{\frac{1}{k+1},\ldots,\frac{1}{k+1}}_{k+1},\ldots\Bigg)\in\nabla_\infty,\ k=0,1,\ldots.$$

Assume that $s<0$. Then for $-l_{k+1}<s<-l_k$, $k\geq 0$, the equation

$$J_{\lambda,s,2}(\mathbf{p})=0$$

has a unique solution $\mathbf{p}_k$. For $k\geq 1$ and $s=-l_k$, the equation

$$J_{\lambda,s,2}(\mathbf{p})=0$$

has two solutions $\mathbf{p}_{k-1}$ and $\mathbf{p}_k$.

Proof. For any $m\geq 1$, let

$$N_m=\frac{2|s|}{m}+m-1.$$

Then

$$N_m-N_{m+1}=\frac{2|s|-m(m+1)}{m(m+1)},$$

and for any $k\geq 0$ and s in $(-l_{k+1},-l_k)$,

$$N_m-N_{m+1}>0 \ \text{ for } m\leq k,$$
$$N_m-N_{m+1}<0 \ \text{ for } m\geq k+1.$$

Hence

$$\inf_{m\geq 1}\left\{\frac{2|s|}{m}+m-1\right\}=\inf_{\mathbf{q}\in\nabla}\{|s|c\varphi_n(\mathbf{q})+J(\mathbf{q})\}=N_{k+1}$$

and is attained at the unique point $\mathbf{q} = \mathbf{p}_k$. In other words, $\mathbf{p}_k$ is the unique solution to the equation

$$J_{\lambda,s,2}(\mathbf{p}) = 0.$$

For $k \geq 1$ and $s = -l_k$, we have

$$\begin{aligned} N_m - N_{m+1} &> 0 \quad \text{for } m < k, \\ N_k - N_{k+1} &= 0, \\ N_m - N_{m+1} &< 0 \quad \text{for } m \geq k+1, \end{aligned}$$

which shows that

$$\inf_{\mathbf{q} \in \nabla} \{|s| c \varphi_n(\mathbf{q}) + J(\mathbf{q})\} = N_k = N_{k+1}$$

and is attained at $\mathbf{p}_{k-1}$ and $\mathbf{p}_k$.

□

Remark: It is worth noting that $\{l_k : k \geq 0\}$ are the death rates of Kingman's coalescent.

8.4 Notes

The motivations for the study of large deviations for the Poisson–Dirichlet distribution come from the work in Gillespie [87], where simulations were done for several models to study the role of population size in population-genetics models of molecular evolution. One of the models is an infinite-alleles model with selective overdominance or heterozygote advantage. It was observed and conjectured that if the selection intensity and the mutation rate get large at the same speed, the behavior looks like that of a neutral model. A rigorous proof of this conjecture was obtained in [119] through the study of Gaussian fluctuations. The results in Theorem 8.1 and Theorem 8.7 provide an alternate proof of this conjecture.

The material in Section 8.1.1 comes from [23]. The results in Section 8.1.2 are based on the work in [71]. The LDP for a small mutation rate in Section 8.2.1 can be found in [72]. The LDP in Section 8.2.2 is from [74].

The underdominant and overdominant distributions are special cases of Theorem 4.4 in [65]. Corollary 8.9 is from [117], where the infinitely many alleles model with homozygote advantage was studied. In [73], moderate deviation principles were established for both the Poisson–Dirichlet distribution and the homozygosity. These results are generalized to the two-parameter setting in [74].

Chapter 9
Large Deviations for the Dirichlet Processes

The Dirichlet process and its two-parameter counterpart are random, purely atomic probabilities with masses distributed according to the Poisson–Dirichlet distribution and the two-parameter Poisson–Dirichlet distribution, respectively. The order by size of the masses does not matter, and the GEM distributions can then be used in place of the corresponding Poisson–Dirichlet distributions in the definition. When θ increases, these masses spread out evenly among the support of the type's measure. Eventually the support is filled up and the types measure emerges as the deterministic limit. This resembles the behavior of empirical distributions for large samples. The focus of this chapter is the LDPs for the one- and two-parameter Dirichlet processes when θ tends to infinity. Relations of these results to Sanov's theorem will be discussed. For simplicity, the space is chosen to be $E = [0,1]$ in this chapter even though the results hold on any compact metric space. The diffuse probability ν_0 on E will be fixed throughout the chapter.

9.1 One-parameter Case

Recall that $M_1(E)$ denotes the space of probability measures on E equipped with the weak topology. Recall that the Dirichlet process with parameter θ is a random measure defined as

$$\Xi_{\theta,\nu_0} = \sum_{i=1}^{\infty} P_i(\theta)\delta_{\xi_i},$$

where $\mathbf{P}(\theta) = (P_1(\theta), P_2(\theta), \ldots)$ has the Poisson–Dirichlet distribution with parameter θ, and independent of $\mathbf{P}(\theta)$, $\xi_1, \xi_2, \ldots$ are i.i.d. with common distribution ν_0. The law of Ξ_{θ,ν_0} is denoted by Π_{θ,ν_0}.

Theorem 9.1. *As θ tends to infinity, Ξ_{θ,ν_0} converges in probability to ν_0, in the space $M_1(E)$.*

Proof. Let $C(E)$ be the space of continuous functions on E equipped with the topology of uniform convergence. The compactness of E guarantees existence of a count-

S. Feng, *The Poisson–Dirichlet Distribution and Related Topics*, Probability and its Applications, DOI 10.1007/978-3-642-11194-5_9,

able dense subset $\{f_i : i \geq 1\}$ of $C(E)$. For any μ, ν in $M_1(E)$, define

$$d_w(\mu,\nu) = \sum_{i=1}^{\infty} \frac{|\langle \mu - \nu, f_i\rangle| \wedge 1}{2^i}.$$

Then d_w is a metric generating the weak topology on $M_1(E)$. For any f in $C(E)$,

$$\mathbb{E}[\langle \Xi_{\theta,\nu_0}, f\rangle] = \langle \nu_0, f\rangle,$$

and, by the Ewens sampling formula,

$$\begin{aligned}\mathbb{E}[(\langle \Xi_{\theta,\nu_0}, f\rangle)^2] &= \langle \nu_0, f^2\rangle \mathbb{E}[\varphi_2(\mathbf{P}(\theta))] + \langle \nu_0, f\rangle^2 \mathbb{E}\left[\sum_{i\neq j} P_i(\theta)P_j(\theta)\right]\\ &= \left\langle \frac{1}{\theta+1}\nu_0, f^2\right\rangle + \frac{\theta}{\theta+1}\langle \nu_0, f\rangle^2.\end{aligned}$$

Thus for any $\delta > 0$, $r \geq 1$ satisfying $2^{1-r} < \delta$,

$$\begin{aligned}\mathbb{P}\{d_w(\Xi_{\theta,\nu_0}, \nu_0) > \delta\} &\leq \sum_{i=1}^{r} \mathbb{P}\left\{|\langle \Xi_{\theta,\nu_0} - \nu_0, f_i\rangle| \geq \frac{2^{i-1}\delta}{r}\right\}\\ &\leq \sum_{i=1}^{r} \left(\frac{r}{2^{i-1}\delta}\right)^2 \mathbb{E}[(\langle \Xi_{\theta,\nu_0} - \nu_0, f_i\rangle)^2]\\ &= \frac{1}{\theta+1}\sum_{i=1}^{r} \left(\frac{r}{2^{i-1}\delta}\right)^2 [\langle \nu_0, f_i^2\rangle - \langle \nu_0, f_i\rangle^2],\end{aligned}$$

which leads to the result.

□

Next, the LDP for the family $\{\Xi_{\theta,\nu_0} : \theta > 0\}$ is established through a series of lemmas.

Lemma 9.1. *For any $n \geq 1$, let $B_1,\ldots,B_n$ be a measurable partition of the set E satisfying*

$$p_i = \nu_0(B_i) > 0,\ i = 1,\ldots,n.$$

Then the family $\{(\Xi_{\theta,\nu_0}(B_1),\ldots,\Xi_{\theta,\nu_0}(B_n)) : \theta > 0\}$ satisfies an LDP on the space Δ_n defined in (5.1), as θ tends to infinity, with speed $1/\theta$ and rate function

$$H(\mathbf{p}|\mathbf{q}) = \sum_{i=1}^{n} p_i \log \frac{p_i}{q_i}. \tag{9.1}$$

Proof. First note that by Theorem 2.24, the distribution of

$$(\Xi_{\theta,\nu_0}(B_1),\ldots,\Xi_{\theta,\nu_0}(B_n))$$

is *Dirichlet*$(\theta p_1,\ldots,\theta p_n)$.

For any Borel measurable subset C of Δ_n, by Stirling's formula

$$
\begin{aligned}
&\log \mathbb{P}\{(\Xi_{\theta,\nu_0}(B_1),\ldots,\Xi_{\theta,\nu_0}(B_n)) \in C\} \\
&= \log\left\{ \frac{\sqrt{2\pi}(\theta)^{\theta-\frac{1}{2}} e^{\frac{\alpha}{12\theta}}}{(\sqrt{2\pi})^n (\theta p_1)^{\theta p_1 - \frac{1}{2}} e^{\frac{\alpha_1}{12\theta p_1}} \cdots (\theta p_n)^{\theta p_n - \frac{1}{2}} e^{\frac{\alpha_n}{12\theta p_n}}} \right. \\
&\quad \left. \times \int_C q_1^{\theta p_1 - 1} \cdots q_n^{\theta p_n - 1} dq_1 \cdots dq_{n-1} \right\} \\
&= \frac{n-1}{2} \log \frac{1}{2\pi} + \frac{1}{2} \log \frac{(\theta p_1)\cdots(\theta p_n)}{\theta} + \frac{1}{12\theta}\left[\alpha - \frac{\alpha_1}{p_1} - \cdots - \frac{\alpha_n}{p_n}\right] \\
&\quad - \theta \sum_{i=1}^n p_i \log p_i + \log \int_C q_1^{\theta p_1 - 1} \cdots q_n^{\theta p_n - 1} dq_1 \cdots dq_{n-1},
\end{aligned}
\tag{9.2}
$$

where $0 < \alpha, \alpha_1, \ldots, \alpha_n < 1$ are some constants.

For any $\varepsilon > 0$, let $C_\varepsilon = \{\mathbf{q} \in C : \min_{1 \le i \le n} q_i \ge \varepsilon\}$. For any measurable function f on Δ_n, $||f||_{L^\theta}$ denotes the L^θ norm of f with respect to the Lebesgue measure m on Δ_n. Choosing θ large enough that $\min_{1 \le i \le n}\{\theta p_i\} > 1$, then

$$
\begin{aligned}
||\chi_C e^{\sum_{i=1}^n p_i \log q_i}||_{L^\theta} &= \left(\int_C q_1^{\theta p_1} \cdots q_n^{\theta p_n} dq_1 \cdots dq_{n-1}\right)^{1/\theta} \qquad (9.3)\\
&\le \left(\int_C q_1^{\theta p_1 - 1} \cdots q_n^{\theta p_n - 1} dq_1 \cdots dq_{n-1}\right)^{1/\theta} = ||\chi_C e^{\sum_{i=1}^n (p_i - 1/\theta)\log q_i}||_{L^\theta} \\
&\le ||\chi_{C_\varepsilon} e^{\sum_{i=1}^n p_i \log q_i}||_{L^\theta} \varepsilon^{-n/\theta} + m(C \setminus C_\varepsilon) \\
&\le ||\chi_C e^{\sum_{i=1}^n p_i \log q_i}||_{L^\theta} \varepsilon^{-n/\theta} + m(C \setminus C_\varepsilon).
\end{aligned}
$$

Letting $\theta \to \infty$, then $\varepsilon \to 0$, we obtain

$$
\lim_{\theta\to\infty} \left(\int_C q_1^{\theta p_1 - 1} \cdots q_n^{\theta p_n - 1} dq_1 \cdots dq_{n-1}\right)^{\gamma} = \operatorname{ess\,sup}\{\chi_C e^{\sum_{i=1}^n p_i \log q_i} : \mathbf{q} \in \Delta_n\}.
\tag{9.4}
$$

For any subset B of Δ_n, we have

$$
\operatorname{ess\,sup}\{\chi_B e^{\sum_{i=1}^n \theta p_i \log q_i} : \mathbf{q} \in \Delta_n\} \le e^{-\inf_{\mathbf{q}\in B} \sum_{i=1}^n p_i \log \frac{1}{q_i}}.
\tag{9.5}
$$

and for any open subset G of Δ_n,

$$
\begin{aligned}
\operatorname{ess\,sup}\{\chi_G e^{\sum_{i=1}^n \theta p_i \log q_i} : \mathbf{q} \in \Delta_n\} &= \operatorname{ess\,sup}\{e^{\sum_{i=1}^n \theta p_i \log q_i} : \mathbf{q} \in G\} \qquad (9.6)\\
&= e^{-\inf_{\mathbf{q}\in G} \sum_{i=1}^n \theta p_i \log \frac{1}{q_i}}.
\end{aligned}
$$

Putting together (9.2), (9.4), and (9.5) yields that for any closed subset B of Δ_n

$$\limsup_{\theta\to\infty}\frac{1}{\theta}\log\mathbb{P}\{(\Xi_{\theta,\nu_0}(B_1),\ldots,\Xi_{\theta,\nu_0}(B_n))\in B\}\le -\inf_{x\in B}H(\mathbf{p}|\mathbf{q}), \tag{9.7}$$

while (9.6), combined with (9.2) and (9.4), implies that for any open subset G of Δ_n

$$\liminf_{\theta\to\infty}\frac{1}{\theta}\log\mathbb{P}\{(\Xi_{\theta,\nu_0}(B_1),\ldots,\Xi_{\theta,\nu_0}(B_n))\in G\}\ge -\inf_{x\in G}H(\mathbf{p}|\mathbf{q}). \tag{9.8}$$

Finally, by continuity, the level set $\{\mathbf{q}\in\Delta_n : H(\mathbf{p}|\mathbf{q})\le c\}$ is compact for any $c\ge 0$.

□

Remark: The rate function $H(\mathbf{p}\,|\,\mathbf{q})$ is the relative entropy of $\mathbf{p}$ with respect to $\mathbf{q}$, defined in (B.10). If $p_i=0$ for some $1\le i\le n$, then $\Xi_{\theta,\nu_0}(A_i)\equiv 0$. By treating $0\log\frac{0}{0}$ as zero, the result of Theorem 9.1 can be generalized to cover these degenerate cases.

Let $\mathfrak{P}$ be the collection of all finite measurable partitions of E by Borel measurable sets. For any $\mu\in M_1(E), \iota=\{B_1,\ldots,B_r\}\in\mathfrak{P}$, define

$$\pi_\iota(\mu)=(\mu(B_1),\ldots,\mu(B_r)).$$

For any $\mu,\nu\in M_1(E)$, the relative entropy $H(\mu|\nu)$ of μ with respect to ν can be written as

$$\begin{aligned}H(\mu|\nu) &= \sup_{g\in C(E)}\left\{\int_0^1 g\,d\mu-\log\int_0^1 e^g d\nu\right\} \\ &= \sup_{g\in B(E)}\left\{\int_0^1 g\,d\nu-\log\int_0^1 e^g d\nu\right\},\end{aligned} \tag{9.9}$$

where $B(E)$ is the set of bounded measurable functions on E.

Lemma 9.2. *For any* $\mu,\nu\in M_1(E)$,

$$H(\mu|\nu)=\sup_{\iota\in\mathfrak{P}}H(\pi_\iota(\mu)|\pi_\iota(\nu)). \tag{9.10}$$

Proof. If μ is not absolutely continuous with respect to ν, then both sides of (9.10) are infinite. Now we assume that $\phi(x)=\frac{d\mu}{d\nu}(x)$ exists. Then for any $\iota=(B_1,\ldots,B_r)\in\mathfrak{P}$, consider the function $g\in B([0,1])$ defined as

$$g(z)=\sum_{i:\mu(B_i)>0}\log\frac{\mu(B_i)}{\nu(B_i)}\chi_{B_i}(z).$$

Clearly

$$H(\pi_\iota(\mu)|\pi_\iota(\nu))=\int_0^1 g(z)d\mu-\log\int_0^1 e^{g(z)}d\nu.$$

By (9.9), $H(\pi_\iota(\mu)|\pi_\iota(\nu)) \leq H(\mu|\nu)$ which implies that

$$\sup_{\iota \in \mathfrak{P}} H(\pi_\iota(\mu)|\pi_\iota(\nu)) \leq H(\mu|\nu).$$

On the other hand, for any $n \geq 1$, let

$$\phi_n(x) = \sum_{k=1}^{n2^n} (k-1)/2^n \chi_{B_k}(x) + n\chi_{A_n}(x),$$

where $B_k = \{z : (k-1)/2^n \leq \phi(z) < k/2^n\}$, $A_n = \{z : \phi(z) \geq n\}$. Then ϕ_n converges point-wise to ϕ. Let $\ell = \{B_1, \ldots, B_{n2^n}, A_n\}$. Then

$$\sup_{\iota \in \mathfrak{P}} H(\pi_\iota(\mu)|\pi_\iota(\nu)) \geq H(\pi_\ell(\mu)|\pi_\ell(\nu)) \geq \int_0^1 \phi_n \log \phi_n d\nu.$$

Letting n go to infinity, (9.10) follows from (B.10) and the monotone convergence theorem.

□

Let $E_\nu = \{x \in E : \nu(\{x\}) = 0\}$. Then Lemma 9.2 can be modified to get:

Lemma 9.3. *For any* $\mu, \nu \in M_1(E)$,

$$H(\mu|\nu) = \sup_{x_1 < x_2 < \cdots < x_k \in E_\nu, k \geq 1} H(\mu^{x_1,\ldots,x_k}|\nu^{x_1,\ldots,x_k}), \tag{9.11}$$

where

$$\mu^{x_1,\ldots,x_k} = (\mu([0,x_1)), \mu([x_1,x_2)), \ldots, \mu(x_k,1]),$$
$$\nu^{x_1,\ldots,x_k} = (\nu([0,x_1)), \nu([x_1,x_2)), \ldots, \nu(x_k,1]).$$

Proof. By Lemma 9.2,

$$H(\mu|\nu) \geq \sup_{x_1 < x_2 < \cdots < x_k \in E_\nu, k \geq 1} H(\mu^{x_1,\ldots,x_k}|\nu^{x_1,\ldots,x_k}). \tag{9.12}$$

On the other hand, the representation (9.9) guarantees that for any $\varepsilon > 0$ there is a continuous function g on E such that

$$H(\mu|\nu) \leq \int g d\mu - \log\left(\int e^g d\nu\right) + \varepsilon.$$

Now choose $x_1^n < x_2^n \cdots < x_{k_n}^n$ in E_ν such that

$$\lim_{n\to\infty} \max_{i=0,\ldots,k_n-1} [|x_{i+1}^n - x_i^n| + \max_{x,y \in [x_i^n, x_{i+1}^n]} |g(x) - g(y)|] = 0.$$

This choice is possible because E_ν is a dense subset of E. Let $x_0^n = 0$. Then

$$
\begin{aligned}
H(\mu|\nu) &\leq \sum_{i=0}^{k_n} g(x_i^n)\mu([x_i^n, x_{i+1}^n)) + g(x_{k_n}^n)\mu([x_{k_n}^n, 1]) \\
&\quad - \log\left[\sum_{i=0}^{k_n-1} e^{g(x_i^n)}\nu([x_i^n, x_{i+1}^n)) + e^{g(x_{k_n}^n)}\nu([x_{k_n}^n, 1])\right] + \varepsilon + \delta_n(g) \\
&\leq \sup_{\alpha_i, i=0,\ldots,k_n} \left\{\sum_{i=0}^{k_n} \alpha_i \mu([x_i^n, x_{i+1}^n)) + \alpha_{k_n}\mu([x_{k_n}^n, 1])\right. \\
&\quad \left. - \log\left[\sum_{i=0}^{k_n-1} e^{\alpha_i}\nu([x_i^n, x_{i+1}^n)) + e^{\alpha_{k_n}}\nu([x_{k_n}^n, 1])\right]\right\} + \varepsilon + \delta_n(g) \\
&= H(\mu^{x_1^n,\ldots,x_{k_n}^n}|\nu^{x_1^n,\ldots,x_{k_n}^n}) + \varepsilon + \delta_n(g),
\end{aligned}
$$

where $\delta_n(g)$ converges to zero as n goes to infinity. As n goes to infinity and ε goes to zero, we get that

$$H(\mu|\nu) \leq \sup_{x_1 < x_2 < \cdots < x_k \in E_\nu, k \geq 1} H(\mu^{x_1,\ldots,x_k}|\nu^{x_1,\ldots,x_k}),$$

which, combined with (9.12), implies the result.

□

Theorem 9.2. *The family* $\{\Xi_{\theta,\nu_0} : \theta > 0\}$ *satisfies an LDP on* $M_1(E)$*, as* θ *tends to infinity, with speed* $1/\theta$ *and rate function*

$$\hat{H}(\nu) = \begin{cases} H(\nu_0|\nu), & \nu \in M_{1,\nu_0}(E) \\ \infty, & \textit{else}, \end{cases} \tag{9.13}$$

where

$$M_{1,\nu_0}(E) = \{\nu \in M_1(E) : \textit{the support of } \nu \textit{ is contained in the support of } \nu_0\}.$$

Proof. For any $\delta > 0$, and ν in $M_1(E)$, let

$$B(\nu,\delta) = \{\mu \in M_1(E) : d_w(\mu,\nu) < \delta\},\ \bar{B}(\nu,\delta) = \{\mu \in M_1(E) : d_w(\mu,\nu) \leq \delta\},$$

with d_w as in the proof of Theorem 9.1. Since the weak topology on $M_1(E)$ is generated by the family

$$\{\mu \in M_1(E) : f \in C(E),\ x \in \mathbb{R},\ \varepsilon > 0,\ |\langle \mu, f\rangle - x| < \varepsilon\},$$

one can find $f_1,\ldots,f_m$ in $C(E)$ and $\varepsilon > 0$ such that

$$\{\mu \in M_1(E) : |\langle \mu, f_j\rangle - \langle \nu, f_j\rangle| < \varepsilon : j = 1,\ldots,m\} \subset B(\nu,\delta).$$

Let

$$M = \sup\{|f_j(x)| : x \in E, j = 1,\ldots,m\},$$

and choose $x_1,\ldots,x_k \in E_\nu$ such that

$$\sup\{|f_j(x)-f_j(y)| : x,y\in[x_i,x_{i+1}], i=0,1,\ldots,k, x_{k+1}=1; j=1,\ldots,m\} < \varepsilon/4.$$

Choose $0<\delta_1<\frac{\varepsilon}{2(k+1)M}$, and define

$$\begin{aligned}O_{x_1,\ldots,x_k}(\nu,\delta_1) = \{&\mu\in M_1(E) : |\mu([x_k,1])-\nu([x_k,1])|<\delta_1,\\ &|\mu([x_i,x_{i+1}))-\nu([x_i,x_{i+1}))|<\delta_1, i=0,\ldots,k-1\},\end{aligned}$$

it follows that for any μ in $O_{x_1,\ldots,x_k}(\nu,\delta_1)$ and any f_j, one has

$$\begin{aligned}|\langle\mu,f_j\rangle-\langle\nu,f_j\rangle| &= \Bigg|\int_{[x_k,1]} f_j(x)(\mu(dx)-\nu(dx))\\ &\quad+\sum_{i=0}^{k-1}\int_{[x_i,x_{i+1})} f_j(x)(\mu(dx)-\nu(dx))\Bigg|\\ &< \frac{\varepsilon}{2}+\sum_{i=0}^{k}|f_j(x_i)|\delta_1<\varepsilon,\end{aligned}$$

which implies that

$$O_{x_1,\ldots,x_k}(\nu,\delta_1)\subset\{\mu\in M_1(E) : |\langle\mu,f_j\rangle-\langle\nu,f_j\rangle|<\varepsilon : j=1,\ldots,m\}\subset B(\nu,\delta).$$

Introduce the map

$$F : M_1(E)\to\Delta_{k+1}, \mu\mapsto(\mu([0,t_1)),\ldots,\mu([t_k,1])).$$

Then $\Pi_{\theta,\nu_0}\circ F^{-1}$ is a Dirichlet distribution with parameters

$$\frac{\theta}{\gamma}(\nu_0([0,x_1)),\ldots,\nu_0([x_k,1])).$$

For any ν in $M_{1,\nu_0}(E)$, it follows from Theorem 9.1 that

$$\begin{aligned}-\hat{H}(\nu) &\le -H(\nu_0^{x_1,\ldots,x_k}|\nu^{x_1,\ldots,x_k})\\ &\le \liminf_{\theta\to\infty}\frac{1}{\theta}\log\Pi_{\theta,\nu_0}\{O_{x_1,\ldots,x_k}(\nu,\delta_1)\}\\ &\le \liminf_{\theta\to\infty}\frac{1}{\theta}\log\Pi_{\theta,\nu_0}\{B(\nu,\delta)\}.\end{aligned} \tag{9.14}$$

Letting δ go to zero, we obtain

$$-\hat{H}(\nu)\le\lim_{\delta\to 0}\liminf_{\theta\to\infty}\frac{1}{\theta}\log\Pi_{\theta,\nu_0}\{B(\nu,\delta)\}. \tag{9.15}$$

If ν does not belong to the set $M_{1,\nu_0}(E)$, then (9.15) holds trivially.

On the other hand, for any $x_1,\dots,x_k$ in E_ν, the boundary points of sets

$$[0,x_1),[x_2,x_3),\dots,[x_k,1]$$

consist of $x_1,\dots,x_k$ only and they all have ν-measure zero. Hence the map F is continuous at ν. This implies that for any $\delta_2>0$, there exists $\delta>0$ such that

$$\bar{B}(\nu,\delta)\subset O_{x_1,\dots,x_k}(\nu,\delta_2).$$

Set

$$\begin{aligned}\bar{O}_{x_1,\dots,x_k}(\nu,\delta_2)=\{\mu\in M_1(E):&|\mu([x_k,1])-\nu([x_k,1])|\le\delta_1,\\ &|\mu([x_i,x_{i+1}))-\nu([x_i,x_{i+1}))|\le\delta_1, i=0,\dots,k-1\}.\end{aligned}$$

Then we have

$$\begin{aligned}&\lim_{\delta\to 0}\limsup_{\theta\to\infty}\frac{1}{\theta}\log\Pi_{\theta,\nu_0}\{\bar{B}(\nu,\delta)\}\\ &\quad\le\lim_{\delta\to 0}\limsup_{\theta\to 0\infty}\frac{1}{\theta}\log\Pi_{\theta,\nu_0}\{\bar{O}_{t_1,\dots,t_k}(\nu,\delta_2)\}.\end{aligned}\tag{9.16}$$

By letting δ_2 go to zero and applying Theorem 9.1 again, we obtain

$$\lim_{\delta\to 0}\limsup_{\theta\to 0\infty}\frac{1}{\theta}\log\Pi_{\theta,\nu_0}\{\bar{B}(\nu,\delta)\}\le -H^{\theta}_{\nu_0^{x_1,\dots,x_k}}(\nu^{x_1,\dots,x_k}).\tag{9.17}$$

Finally, it follows by taking the supremum over the set E_ν and applying Lemma 9.3 that

$$\lim_{\delta\to 0}\limsup_{\theta\to\infty}\frac{1}{\theta}\log\Pi_{\theta,\nu_0}\{\bar{B}(\nu,\delta)\}\le -\hat{H}(\nu),\tag{9.18}$$

which, combined with (9.15), implies the theorem.

□

Remark: If the empirical distribution converges to ν_0 in Sanov's theorem, then the rate function for the corresponding LDP is given by $H(\cdot|\nu_0)$. Consider Ξ_{θ,ν_0} as the empirical distribution weighted according to the Poisson–Dirichlet distribution. The limit is still ν_0 when θ tends to infinity but the rate function for the corresponding LDP is given $H(\nu_0|\cdot)$, which is a dual to the rate function in Sanov's theorem.

Corollary 9.3 *Let F be a continuous function on $M_1(E)$, and set*

$$\Pi^F_{\theta,\nu_0}(d\nu)=C\exp[\theta F(\nu)]\Pi_{\theta,\nu_0}(d\nu),$$

where $C>0$ is the normalizing constant $(\int_{M_1(E)}\exp[\theta F(\nu)]\Pi_{\theta,\nu_0}(d\nu))^{-1}$. Then the family $\{\Pi^F_{\theta,\nu_0}\}$ satisfies an LDP on the space $M_1(E)$, as θ tends to infinity, with speed $1/\theta$ and rate function

$$\hat{H}_F(\nu) = \sup_{\mu \in M_1(E)} \{F(\mu) - \hat{H}(\mu)\} - (F(\nu) - \hat{H}(\nu)).$$

Proof. By Theorem B.1,

$$\begin{aligned}
&\lim_{\theta \to \infty} \frac{1}{\theta} \log C \\
&= - \lim_{\theta \to \infty} \frac{1}{\theta} \log \int e^{\theta F(\mu)} \Pi_{\theta,\nu_0}(d\mu) \\
&= - \sup_{\mu} \{F(\mu) - \hat{H}(\mu)\}.
\end{aligned} \tag{9.19}$$

For any $\nu \in M_1(E)$, it follows from direct calculation that

$$\begin{aligned}
&\lim_{\delta \to 0} \liminf_{\theta \to \infty} \frac{1}{\theta} \log \int_{B(\nu,\delta)} e^{\theta F(\mu)} \Pi_{\theta,\nu_0}(d\mu) \\
&= F(\nu) + \lim_{\delta \to 0} \liminf_{\theta \to \infty} \frac{1}{\theta} \log \Pi_{\theta,\nu_0}\{B(\nu,\delta)\} \\
&= F(\nu) + \lim_{\delta \to 0} \limsup_{\theta \to \infty} \frac{1}{\theta} \log \Pi_{\theta,\nu_0}\{\bar{B}(\nu,\delta)\} \\
&= \lim_{\delta \to 0} \limsup_{\theta \to \infty} \frac{1}{\theta} \log \int_{\bar{B}(\nu,\delta)} e^{\theta F(\mu)} \Pi_{\theta,\nu_0}(d\mu) \\
&= F(\nu) - \hat{H}(\nu),
\end{aligned}$$

which combined with (9.19) implies

$$\begin{aligned}
&\lim_{\delta \to 0} \liminf_{\theta \to \infty} \frac{1}{\theta} \log \Pi^F_{\theta,\nu_0}\{d_w(\mu,\nu) < \delta\} \\
&= \lim_{\delta \to 0} \limsup_{\theta \to \infty} \frac{1}{\theta} \log \Pi^F_{\theta,\nu_0}\{d_w(\mu,\nu) \le \delta\} = -\hat{H}_F(\nu).
\end{aligned}$$

Since $M_1(E)$ is compact, using Theorem B.6 again, we get the result. □

9.2 Two-parameter Case

The LDPs in this section are under the limit of θ tending to infinity. The parameters α and θ will thus be in the range of $0 < \alpha < 1, \theta > 0$. For the given diffuse probability ν_0 in $M_1(E)$, let $\Xi_{\theta,\alpha,\nu_0}$ be the two-parameter Dirichlet process defined in (3.2) with the law of $\Xi_{\theta,\alpha,\nu_0}$ denoted by $\Pi_{\alpha,\theta,\nu_0}$. Without loss of generality, the support of the ν_0 is assumed to be the whole space E. The next result is the two-parameter analog to Theorem 9.1.

Theorem 9.4. *As θ tends to infinity, $\Xi_{\theta,\alpha,\nu_0}$ converges in probability to ν_0 in space $M_1(E)$.*

Proof. For any f in $C(E)$,

$$\mathbb{E}[\langle \Xi_{\theta,\alpha,\nu_0}, f\rangle] = \langle \nu_0, f\rangle,$$

and, by the Pitman sampling formula,

$$\begin{aligned}\mathbb{E}[(\langle \Xi_{\theta,\alpha,\nu_0}, f\rangle)^2] &= \langle \nu_0, f^2\rangle \mathbb{E}[\varphi_2(\mathbf{P}(\alpha,\theta))] + \langle \nu_0, f\rangle^2 \mathbb{E}\left[\sum_{i\neq j} P_i(\alpha,\theta)P_j(\alpha,\theta)\right] \\ &= \left\langle \frac{1-\alpha}{\theta+1}\nu_0, f^2\right\rangle + \frac{\theta+\alpha}{\theta+1}\langle \nu_0, f\rangle^2.\end{aligned}$$

The remaining steps of the proof follow those in Theorem 9.1.

□

Let $\{\sigma_t : t \geq 0\}, \{\gamma_t : t \geq 0,\}$, and $\sigma_{\alpha,\theta}$ be defined as in Proposition 3.7. For notational convenience, the constant C in the Lévy measure of subordinator $\{\sigma_t : t \geq 0\}$ is chosen to be one in the sequel. The following lemma is the two-parameter generalization of Theorem 2.24.

Lemma 9.4. *Let*

$$\gamma(\alpha,\theta) = \frac{\alpha\gamma(\frac{1}{\alpha})}{\Gamma(1-\alpha)}. \tag{9.20}$$

For any $n \geq 1$, and any $0 < t_1 < \cdots < t_n = 1$, let

$$A_1 = [0,t_1],\ A_i = (t_{i-1},t_i],\ i = 2,\ldots,n$$

be a partition of E. For simplicity, the partition $A_1,\ldots,A_n$ will be associated with $t_1,\ldots,t_n$. Set $a_i = \nu_0(A_i), i = 1,\ldots,n$. Introduce the process

$$Y_{\alpha,\theta}(t) = \sigma(\gamma(\alpha,\theta)t), t \geq 0.$$

Then $(\Xi_{\theta,\alpha,\nu_0}(A_1),\ldots,\Xi_{\theta,\alpha,\nu_0}(A_n))$ has the same distribution as

$$\left(\frac{Y_{\alpha,\theta}(a_1)}{\sigma_{\alpha,\theta}},\ldots,\frac{Y_{\alpha,\theta}(\sum_{j=1}^{n} a_j) - Y_{\alpha,\theta}(\sum_{j=1}^{n-1} a_j)}{\sigma_{\alpha,\theta}}\right).$$

Proof. This follows from the subordinator representation for $PD(\alpha,\theta)$ given in Proposition 3.7.

□

Let

$$Z_{\alpha,\theta}(t) = \frac{Y_{\alpha,\theta}(t)}{\theta},$$

$$\mathbf{Z}_{\alpha,\theta}(t_1,\ldots,t_n) = \left(Z_{\alpha,\theta}(a_1),\ldots,Z_{\alpha,\theta}\left(\sum_{j=1}^{n} a_j\right) - Z_{\alpha,\theta}\left(\sum_{j=1}^{n-1} a_j\right)\right).$$

By direct calculation, one has

$$\begin{aligned}\varphi(\lambda) &= \log \mathbb{E}[e^{\lambda \sigma_1}] \\ &= \int_0^\infty (e^{\lambda x} - 1) x^{-(\alpha+1)} e^{-x} dx \\ &= \begin{cases} \frac{\Gamma(1-\alpha)}{\alpha}[1-(1-\lambda)^\alpha], \ \lambda \leq 1 \\ \infty, \qquad\qquad\qquad\qquad \text{else} \end{cases}\end{aligned} \tag{9.21}$$

and

$$\begin{aligned}L(\lambda) &= \lim_{\theta \to \infty} \frac{1}{\theta} \log \mathbb{E}[e^{\lambda \gamma_1}] \\ &= \begin{cases} \log(\frac{1}{1-\lambda}), \ \lambda < 1 \\ \infty, \qquad\qquad \text{else.} \end{cases}\end{aligned} \tag{9.22}$$

For any real numbers $\lambda_1, \ldots, \lambda_n$,

$$\begin{aligned}&\frac{1}{\theta} \log \mathbb{E}[\exp\{\theta \langle (\lambda_1, \ldots, \lambda_n), \mathbf{Z}_{\alpha,\theta}(t_1, \ldots, t_n) \rangle \}] \\ &= \frac{1}{\theta} \log \mathbb{E}\left[\prod_{i=1}^n (\mathbb{E}_{\gamma(1/\alpha)}[\exp\{\lambda_i \sigma_1\}]^{\frac{\alpha a_i}{\Gamma(1-\alpha)}\gamma(1/\alpha)})\right] \\ &= \frac{1}{\theta} \log \mathbb{E}\left[\exp\left\{\left(\sum_{i=1}^n \frac{\alpha \nu_0(A_i)}{\Gamma(1-\alpha)} \varphi(\lambda_i)\right) \gamma\left(\frac{1}{\alpha}\right)\right\}\right] \\ &\to \Lambda(\lambda_1, \ldots, \lambda_n) = \frac{1}{\alpha} L\left(\frac{\alpha}{\Gamma(1-\alpha)} \sum_{i=1}^n \nu_0(A_i) \varphi(\lambda_i)\right), \ \theta \to \infty.\end{aligned} \tag{9.23}$$

For $(y_1, \ldots, y_n)$ in $\mathbb{R}_+^n$, set

$$\begin{aligned}&J_{t_1,..,t_n}(y_1, \ldots, y_n) \\ &= \sup_{(\lambda_1,\ldots,\lambda_n) \in \mathbb{R}^n} \left\{\sum_{i=1}^n \lambda_i y_i - \Lambda(\lambda_1, \ldots, \lambda_n)\right\} \\ &= \sup_{\lambda_1,\ldots,\lambda_n \in (-\infty,1]^n} \left\{\sum_{i=1}^n \lambda_i y_i + \frac{1}{\alpha} \log\left[\sum_{i=1}^n \nu_0(A_i)(1-\lambda_i)^\alpha\right]\right\}.\end{aligned} \tag{9.24}$$

Theorem 9.5. *The family* $\{\mathbf{Z}_{\alpha,\theta}(t_1, \ldots, t_n) : \theta > 0\}$ *satisfies an LDP on the space* $\mathbb{R}_+^n$*, as* θ *tends to infinity, with speed* $1/\theta$ *and good rate function (9.24).*

Proof. First note that both of the functions φ and L are essentially smooth. Let

$$\mathscr{D}_\Lambda = \{(\lambda_1, \ldots, \lambda_n) : \Lambda(\lambda_1, \ldots, \lambda_n) < \infty\}, \ \mathscr{D}_\Lambda^\circ = \text{interior of } \mathscr{D}_\Lambda.$$

It follows from (9.22) and (9.23) that

$$\mathscr{D}_\Lambda = \left\{(\lambda_1,\ldots,\lambda_n) : \sum_{i=1}^n \nu_0(A_i)\frac{\alpha}{\Gamma(1-\alpha)}\varphi(\lambda_i) < 1\right\}.$$

The fact that ν_0 has support E implies that $\nu_0(A_i) > 0$ for $i = 1,\ldots,n$, and

$$\begin{aligned}\mathscr{D}_\Lambda &= \left\{(\lambda_1,\ldots,\lambda_n) : \sum_{i=1}^n \nu_0(A_i)[1-(1-\lambda_i)^\alpha] < 1\right\}\\ &= \{(\lambda_1,\ldots,\lambda_n) : \lambda_i \le 1, i = 1,\ldots,n\}\setminus\{(1,\ldots,1)\},\\ \mathscr{D}^\circ_\Lambda &= \{(\lambda_1,\ldots,\lambda_n) : \lambda_i < 1, i = 1,\ldots,n\}.\end{aligned}$$

Clearly the function Λ is differentiable on $\mathscr{D}^\circ_\Lambda$ and

$$\begin{aligned}&grad(\Lambda)(\lambda_1,\ldots,\lambda_n)\\ &= \frac{1}{\Gamma(1-\alpha)}L'\left(\frac{\alpha}{\Gamma(1-\alpha)}\sum_{i=1}^n \nu_0(A_i)\varphi(\lambda_i)\right)(\nu_0(A_1)\varphi'(\lambda_1),\ldots,\nu_0(A_n)\varphi'(\lambda_n)).\end{aligned}$$

If a sequence $(\lambda_1^m,\ldots,\lambda_n^m)$ in $\mathscr{D}^\circ_\Lambda$ converges to a boundary point of $\mathscr{D}^\circ_\Lambda$ as m converges to infinity, then at least one coordinate of the sequence approaches one. Since the interior of $\{\lambda : \varphi(\lambda) < \infty\}$ is $(-\infty, 1)$ and φ is essentially smooth, it follows that Λ is essentially smooth. The theorem then follows from Theorem B.8. □

For $(y_1,\ldots,y_n)$ in $\mathbb{R}^n_+$ and $(x_1,\ldots,x_n)$ in E^n, define

$$F(y_1,..,y_n) = \begin{cases}\frac{1}{\sum_{k=1}^n y_k}(y_1,\ldots,y_n), & \sum_{k=1}^n y_k > 0\\ (0,\ldots,0), & (y_1,\ldots,y_n) = (0,\ldots,0)\end{cases}$$

and

$$I_{t_1,\ldots,t_n}(x_1,\ldots,x_n) = \inf\{J_{t_1,\ldots,t_n}(y_1,\ldots,y_n) : F(y_1,\ldots,y_n) = (x_1,\ldots,x_n)\}. \quad (9.25)$$

Clearly

$$I_{t_1,\ldots,t_n}(x_1,\ldots,x_n) = +\infty, \text{ if } \sum_{k=1}^n x_k \ne 1.$$

For $(x_1,\ldots,x_n)$ in $\mathbb{R}^n_+$ satisfying $\sum_{k=1}^n x_k = 1$, we have

$$\begin{aligned}&I_{t_1,\ldots,t_n}(x_1,\ldots,x_n)\\ &= \inf\left\{J_{t_1,\ldots,t_n}(ax_1,\ldots,ax_n) : a = \sum_{k=1}^n y_k > 0\right\}\\ &= \inf\left\{\sup_{(\lambda_1,\ldots,\lambda_n)\in(-\infty,1]^n}\left\{a\sum_{i=1}^n \lambda_i x_i + \frac{1}{\alpha}\log\left[\sum_{i=1}^n \nu_0(A_i)(1-\lambda_i)^\alpha\right]\right\} : a > 0\right\}.\end{aligned}$$

Further calculations yield

$$
\begin{aligned}
&I_{t_1,\ldots,t_n}(x_1,\ldots,x_n)\\
&= \inf\Big\{\sup_{(\lambda_1,\ldots,\lambda_n)\in(-\infty,1]^n}\Big\{a-\log a \\
&-\sum_{i=1}^n a(1-\lambda_i)x_i + \frac{1}{\alpha}\log\Big[\sum_{i=1}^n \nu_0(A_i)[a(1-\lambda_i)]^\alpha\Big]\Big\} : a>0\Big\},\\
&= \inf\{a-\log a : a>0\} + \sup_{(\tau_1,\ldots,\tau_n)\in R_+^n}\left\{\frac{1}{\alpha}\log\left[\sum_{i=1}^n \nu(A_i)\tau_i^\alpha\right] - \sum_{i=1}^n \tau_i x_i\right\}\\
&= \sup_{(\tau_1,\ldots,\tau_n)\in R_+^n}\left\{\frac{1}{\alpha}\log\left[\sum_{i=1}^n \nu_0(A_i)\tau_i^\alpha\right] + 1 - \sum_{i=1}^n \tau_i x_i\right\}.
\end{aligned}
\tag{9.26}
$$

Theorem 9.6. *The family* $(\Xi_{\theta,\alpha,\nu_0}(A_1),\ldots,\Xi_{\theta,\alpha,\nu_0}(A_n))$ *satisfies an LDP on the space* E^n*, as* θ *tends to infinity, with speed* $1/\theta$ *and rate function*

$$
I_{t_1,\ldots,t_n}(x_1,\ldots,x_n) = \begin{cases} \sup_{(\tau_1,\ldots,\tau_n)\in R_+^n}\{\frac{1}{\alpha}\log[\sum_{i=1}^n \nu_0(A_i)\tau_i^\alpha] \\ \qquad +1-\sum_{i=1}^n \tau_i x_i\}, & \sum_{k=1}^n x_k = 1 \\ \infty, & \textit{else}. \end{cases} \tag{9.27}
$$

Proof. Since $J_{t_1,\ldots,t_n}(0,\ldots,0)=\infty$, the function F is continuous on the effective domain of $J_{t_1,\ldots,t_n}$. The theorem then follows from Lemma 9.4 and the contraction principle (Theorem B.2).

□

Remark: Since the effective domain of $I_{t_1,\ldots,t_n}(x_1,\ldots,x_n)$ is contained in Δ_n, by Theorem B.3, the result in Theorem 9.6 holds with E^n being replaced by Δ_n.

For each μ in $M_1(E)$, define

$$
I^\alpha(\mu) = \sup_{f\geq 0, f\in C(E)}\left\{\frac{1}{\alpha}\log\left(\int_0^1 (f(x))^\alpha \nu_0(dx)\right) + 1 - \int_0^1 f(x)\mu(dx)\right\}, \tag{9.28}
$$

Lemma 9.5. *For any* μ *in* $M_1(E)$*,*

$$
I^\alpha(\mu) = \sup_{0<t_1<t_2<\cdots<t_n=1;n\geq 1}\{I_{t_1,\ldots,t_n}(\mu(A_1),\ldots,\mu(A_n))\}. \tag{9.29}
$$

Proof. It follows from Tietze's extension theorem and Luzin's theorem, that the set $C(E)$ can be replaced with $B(E)$ in the definition of I^α. This implies that

$$
I^\alpha(\mu) \geq \sup\{I_{t_1,\ldots,t_n}(\mu(A_1),\ldots,\mu(A_n)) : 0<t_1<t_2<\cdots<t_n=1; n=1,2,\ldots\}.
$$

On the other hand, for each non-negative f in $C(E)$, let

$$
t_i = \frac{i}{n},\ \tau_i = f(t_i),\ i=1,\ldots,n.
$$

Then

$$\frac{1}{\alpha}\log\left(\int_0^1 (f(x))^\alpha \nu_0(dx)\right) - \int_0^1 f(x)\mu(dx)$$
$$= \lim_{n\to\infty}\left\{\frac{1}{\alpha}\log\left[\sum_{i=1}^n \nu_0(A_i)\tau_i^\alpha\right] - \sum_{i=1}^n \tau_i\mu(A_i)\right\}$$
$$\leq \sup\{I_{t_1,\ldots,t_n}(\mu(A_1),\ldots,\mu(A_n)) : 0 < t_1 < t_2 < \cdots < t_n = 1; n = 1,2,\ldots\},$$

which implies

$$I^\alpha(\mu) \leq \sup\{I_{t_1,\ldots,t_n}(\mu(A_1),\ldots,\mu(A_n)) : 0 < t_1 < t_2 < \cdots < t_n = 1; n = 1,2,\ldots\}.$$

□

Remark: It follows from the proof of Lemma 9.5 that the supremum in (9.29) can be taken over all partitions with $t_1,\ldots,t_{n-1}$ being continuity points of μ. By monotonically approximating non-negative $f(x)$ with strictly positive functions from above, it follows from the monotone convergence theorem that the supremum in (9.28) can be taken over strictly positive bounded functions; i.e.,

$$I^\alpha(\mu) = \sup_{f>0, f\in C(E)}\left\{\frac{1}{\alpha}\log\left(\int_0^1 (f(x))^\alpha \nu_0(dx)\right) + 1 - \int_0^1 f(x)\mu(dx)\right\}, \quad (9.30)$$

We are now ready to prove the main result of this section.

Theorem 9.7. *The family $\{\Xi_{\theta,\alpha,\nu_0} : \theta > 0\}$ satisfies an LDP, as θ tends to infinity, with speed $1/\theta$ and rate function I^α.*

Proof. Let $\{f_j(x) : j = 1,2,\ldots\}$ be the countable dense (in the supremum norm) subset of $C(E)$ used in the definition of the metric d_w on $M_1(E)$. Since $M_1(E)$ is compact, it suffices to show that

$$\begin{aligned}
&\lim_{\delta\to 0}\liminf_{\theta\to\infty}\frac{1}{\theta}\log\mathbb{P}\{d_w(\mu,\nu) < \delta)\} \\
&= \lim_{\delta\to 0}\limsup_{\theta\to\infty}\frac{1}{\theta}\log P\{d_w(\mu,\nu) \leq \delta)\} \qquad (9.31)\\
&= -I^\alpha(\mu).
\end{aligned}$$

Choose a partition $t_1,\ldots,t_n$ and m large enough so that

$$\{\nu \in M_1(E) : |\langle\nu, f_j\rangle - \langle\mu, f_j\rangle| < \delta/2 : j = 1,\ldots,m\} \subset \{d_w(\nu,\mu) < \delta\}.$$

and

$$\sup\{|f_j(x) - f_j(y)| : x,y \in A_i, i = 1,\ldots,n; j = 1,\ldots,m\} < \delta/8.$$

For $\delta_1 < \frac{\delta}{4n}$, define

$$V_{t_1,\ldots,t_n}(\mu,\delta_1) = \{(y_1,\ldots,y_n) \in \Delta_n : |y_i - \mu(A_i)| < \delta_1, i = 1,\ldots,n\}.$$

Introduce the map

$$\Psi_{t_1,\ldots,t_n} : M_1(E) \to \Delta_n, \nu \mapsto (\nu(A_1),\ldots,\nu(A_n)).$$

Then clearly

$$\Psi^{-1}_{t_1,\ldots,t_n}(V_{t_1,\ldots,t_n}(\mu,\delta_1)) \subset \{d_w(\nu,\mu) < \delta\}.$$

Next consider partitions $t_1,\ldots,t_n$ such that $t_1,\ldots,t_{n-1}$ are continuity points of μ. Since $\Psi_{t_1,\ldots,t_n}$ is continuous at μ, for any $\delta_2 > 0$, one can choose $\delta > 0$ small enough such that

$$\{d_w(\nu,\mu) \leq \delta\} \subset \Psi^{-1}_{t_1,\ldots,t_n}\{V_{t_1,\ldots,t_k}(\mu,\delta_2)\}.$$

Following an argument similar to that used in the proof of Theorem 9.2, and taking into account the remark after Lemma 9.5, we obtain (9.31).

□

9.3 Comparison of Rate Functions

The large mutation LDPs for Π_θ and $PD(\alpha,\theta)$ have the same rate function that is independent of the parameter α; but rate function, I^α, for the LDP of $\Xi_{\theta,\alpha,\nu_0}$ does depend on α. Since the value of a rate function, with a unique zero point, describes the difficulty for the underlying random element to deviate from the zero point, it is natural to compare the rate functions of $\Xi_{\theta,\alpha,\nu_0}$ and Ξ_{θ,ν_0}. The main question is whether the LDP for the two-parameter Dirichlet process is consistent with the LDP for the one-parameter Dirichlet process in terms of the convergence of corresponding rate functions when α converges to zero.

For any μ in $M_1(E)$, let

$$I^0(\mu) = \sup_{f>0, f\in B(E)} \left\{ \int_0^1 \log f(x)\nu_0(dx) + 1 - \int_0^1 f(x)\mu(dx) \right\}. \tag{9.32}$$

The next result shows that I^0 is the rate function for the LDP of Ξ_{θ,ν_0} when θ tends to infinity.

Lemma 9.6.

$$I^0(\mu) = \hat{H}(\mu) = H(\nu_0|\mu). \tag{9.33}$$

Proof. If ν_0 is not absolutely continuous with respect to μ, then $H(\nu_0|\mu) = +\infty$. Let A be a set such that $\mu(A) = 0, \nu_0(A) > 0$ and define

$$f_m(x) = \begin{cases} m, & x \in A \\ 1, & \text{else.} \end{cases}$$

Then

$$I^0(\mu) \geq \nu_0(A)\ \log m \to \infty \text{ as } m \to \infty.$$

Next we assume $\nu_0 \ll \mu$ and denote $\frac{d\nu_0}{d\mu}(x)$ by $\phi(x)$. By definition,

$$H(\nu_0|\mu) = \int_0^1 \phi(x) \log(\phi(x))\, \mu(dx).$$

For any $M > 0$, let $\phi_M(x) = \phi(x) \wedge M$. Let

$$E_1 = \{x \in E : \phi(x) \geq e^{-1}\}, E_2 = E \setminus E_1.$$

Since

$$\lim_{M \to \infty} \log \int_0^1 \phi_M(x)\mu(dx) = 0,$$

and the function $x \log x$ is bounded below, the monotone convergence on E_1 and the dominated convergence theorem on E_2 imply that

$$\begin{aligned} &I^0(\mu) \\ &\geq \lim_{M \to \infty} \left[\int_0^1 \phi_M(x) \log \phi_M(x)\, \nu_0(dx) - \log \int_0^1 \phi_M(x)\, \mu(dx) \right] \\ &= H(\nu_0|\mu). \end{aligned} \tag{9.34}$$

On the other hand, it follows, by letting $f(x) = e^{g(x)}$ in (9.9), that

$$H(\nu_0|\mu) = \sup_{f>0, f \in B(E)} \left\{ \int_0^1 \log f(x)\, \nu_0(dx) - \log \int_0^1 f(x)\, \mu(dx) \right\}.$$

Since

$$\int_0^1 f(x)\, \mu(dx) - 1 \geq \log \int_0^1 f(x)\, \mu(dx),$$

we get that

$$H(\nu_0|\mu) \geq I^0(\mu) \tag{9.35}$$

which combined with (9.34) implies (9.33). □

The following result reveals the monotone structure among the rate functions of the two-parameter Dirichlet process.

Theorem 9.8. *For any μ in $M_1(E)$, $0 \leq \alpha_1 < \alpha_2 < 1$,*

$$I^{\alpha_2}(\mu) \geq I^{\alpha_1}(\mu), \tag{9.36}$$

and for any α in $(0,1)$, there exists μ in $M_1(E)$ satisfying $\nu_0 \ll \mu$ such that

$$I^{\alpha}(\mu) > I^0(\mu). \tag{9.37}$$

Proof. By Hölder's inequality, for any $0 < \alpha_1 < \alpha_2 < 1$,

$$\frac{1}{\alpha_1}\log\left(\int_0^1 (f(x))^{\alpha_1}\,\nu_0(dx)\right) \le \frac{1}{\alpha_2}\log\left(\int_0^1 (f(x))^{\alpha_2}\,\nu_0(dx)\right), \tag{9.38}$$

and the inequality becomes strict if $f(x)$ is not constant almost surely under ν_0. Hence $I^{\alpha}(\mu)$ is non-decreasing in α over $(0,1)$. It follows from the concavity of $\log x$ that

$$\frac{1}{\alpha}\log\left(\int_0^1 (f(x))^{\alpha}\nu_0(dx)\right) \ge \int_0^1 \log f(x)\nu_0(dx),$$

which implies that $I^{\alpha}(\mu) \ge I^0(\mu)$ for $\alpha > 0$.

Next choose μ in $M_1(E)$ such that $\nu_0 \ll \mu$ and $\frac{d\nu_0}{d\mu}(x)$ is not a constant with ν_0 probability one; then $I^{\alpha}(\mu) > I^{\alpha/2}(\mu) \ge I^0(\mu)$ for $\alpha > 0$.

□

The following form of minimax theorem is the key in establishing consistency of the rate functions.

Definition 9.1. Let M and N be two spaces. A function $f(x,y)$ defined on $M \times N$ is convexlike in x if for every x_1, x_2 in M and $0 \le \lambda \le 1$, there exists x in M such that

$$f(x,y) \le \lambda f(x_1,y) + (1-\lambda) f(x_2,y), \text{ for all } y \in N.$$

The function $f(x,y)$ is said to be concavelike in y if for every y_1, y_2 in N and $0 \le \lambda \le 1$, there exists y in N such that

$$f(x,y) \ge \lambda f(x,y_1) + (1-\lambda) f(x,y_2), \text{ for all } x \in M.$$

The function $f(x,y)$ is convex–concavelike if it is convexlike in x and concavelike in y.

Theorem 9.9. (Sion's minimax theorem [164]) *Let M and N be any two topological spaces and $f(x,y)$ be a function on $M \times N$. If M is compact, $f(x,y)$ is convex-concavelike, and, for each y in N, $f(x,y)$ is lower semi-continuous in x, then*

$$\sup_{y\in N}\inf_{x\in M} f(x,y) = \inf_{x\in M}\sup_{y\in N} f(x,y).$$

Now we turn to the problem of consistency.

Theorem 9.10. *For any μ in $M_1(E)$,*

$$\lim_{\alpha\to 0} I^{\alpha}(\mu) = I^0(\mu).$$

Proof. Let $M = [0,1/2], N = \{f \in B(E) : f > 0\}$. For any α in M and f in N, define

$$F(\alpha,f) = \begin{cases} \frac{1}{\alpha}\log(\int_0^1 (f(x))^{\alpha}\,\nu_0(dx)) + 1 - \int_0^1 f(x)\,\mu(dx), & \alpha > 0 \\ \int_0^1 \log f(x)\,\nu_0(dx) + 1 - \int_0^1 f(x)\,\mu(dx), & \alpha = 0. \end{cases}$$

Then clearly

$$I^0(\mu) = \sup_{f\in N} \inf_{\alpha\in M} F(\alpha, f),$$

and, by Theorem 9.8,

$$\lim_{\alpha\to 0} I^\alpha(\mu) = \inf_{\alpha\in M} \sup_{f\in N} F(\alpha, f).$$

For each fixed α, $F(\alpha, f)$ is clearly concave as a function of f from N. For fixed f in N, $F(\alpha, f)$ is continuous in α. Since $F(\alpha, f)$ is monotone in α, it is convexlike. The theorem now follows from Sion's minimax theorem.

□

Remarks:
(a) Both $\Xi_{\theta,\alpha,\nu_0}$ and Ξ_{θ,ν_0} converge to ν_0 for large θ. When θ becomes large, each component of $\mathbf{P}(\theta,\alpha)$ is more likely to be small. The introduction of positive α plays a similar role. Thus the mass in $\Xi_{\theta,\alpha,\nu_0}$ spreads more evenly than the mass in Ξ_{θ,ν_0}. In other words, $\Xi_{\theta,\alpha,\nu_0}$ is "closer" to ν_0 than Ξ_{θ,ν_0}. This observation is made rigorous through the fact that I^α can be strictly bigger than I^0. The monotonicity of I^α in α shows that α can be used to measure the relative "closeness" to ν_0 among all $\Xi_{\theta,\alpha,\nu_0}$ for large θ.
(b) The process $Y_{\alpha,\theta}(t)$ is a process with exchangeable increments. The method here may be adapted to establish LDPs for other processes with exchangeable increments.

Consider the map Φ defined in (5.76) that maps every element in $M_1(E)$ to the descending sequence of masses of all its atoms. One would hope to get the LDP for Π_θ in Chapter 8 from the LDP for Π_{θ,ν_0} through the contraction principle using the map Φ. Unfortunately, Φ is not continuous on the effective domain of $\hat{H}$ as the following example shows. Let $\mu_n = \frac{1}{2}\nu_0 + \frac{1}{2n}\sum_{k=1}^n \delta_{k/n^2}$. Then μ_n converges weakly to $\frac{1}{2}[\nu_0 + \delta_0]$ while $\Phi(\mu_n) = (\frac{1}{2n}, \ldots, \frac{1}{2n}, 0, 0..)$ converges to $(0, \ldots 0, \ldots)$ rather than $(1/2, 0, \ldots) = \Phi(\delta_0)$.

The rate function in Theorem 8.1, however, does have a connection to the relative entropy, as the next result shows.

Theorem 9.11. *Let $S(\mathbf{P})$ be defined in (8.15). For any $m \geq 1, n > m$, set*

$$\nabla_n^{(p_1,\ldots,p_m)} = \{(q_1,\ldots,q_{n-m}) : (p_1,\ldots,p_m,q_1,\ldots,q_{n-m}) \in \nabla_n\}.$$

Let

$$H_n((p_1,\ldots,p_m,q_1,\ldots,q_{n-m}))$$

denote the relative entropy of $(\frac{1}{n},\ldots,\frac{1}{n})$ with respect to $(p_1,\ldots,p_m,q_1,\ldots,q_{n-m})$. Then

$$S(\mathbf{p}) = \lim_{m\to\infty} \lim_{n\to\infty} \inf\{H_n((p_1,\ldots,p_m,q_1,\ldots,q_{n-m})) : (q_1,\ldots,q_{n-m}) \in \nabla_n^{(p_1,\ldots,p_m)}\}.$$

Proof. The equality holds trivially if $p_1 + \cdots + p_m = 1$. Assume that $\sum_{i=1}^m p_i < 1$. Then

$$H_n((p_1,\ldots,p_m,q_1,\ldots,q_{n-m})) = \sum_{i=1}^{m} \frac{1}{n}\log\frac{1}{np_i} + \frac{n-m}{n}\log\frac{1}{n} + \frac{1}{n}\log\frac{1}{q_1\cdots q_{n-m}}.$$

Since $\sum_{i=1}^{n-m} q_i = 1-\sum_{i=1}^{m} p_i$, and the product $q_1\cdots q_{n-m}$ reaches its maximum when $q_1,\ldots,q_{n-m}$ are all equal, it follows that

$$\begin{aligned}
&\inf\{H_n((p_1,\ldots,p_m,q_1,\ldots,q_{n-m})) : (q_1,\ldots,q_{n-m}) \in \nabla_n^{(p_1,\ldots,p_m)}\}\\
&= \frac{m}{n}\log\frac{1}{n} + \frac{1}{n}\sum_{i=1}^{m}\log\frac{1}{p_i} + \frac{n-m}{n}\log\frac{n-m}{n}\\
&+ \frac{n-m}{n}\log\frac{1}{1-\sum_{i=1}^{m} p_i}.
\end{aligned}$$

Letting $n \to \infty$, followed by $m \to \infty$, we obtain the result.

□

9.4 Notes

The LDP result for the one-parameter Dirichlet process first appeared in [135]. Lemma 9.1 and Lemma 9.2 are from [21] and the expression (9.9) can be found in [44]. The remaining results of Section 9.1 are based on the material in [22]. Theorem 9.2, viewed as the reversed form of Sanov's theorem, also appears in [84] and [85] where the posterior distributions with Dirichlet prior are studied. Ganesh and O'Connell [84] (page 202) gave the following nice explanation for the reverse relation of the rate functions: "in Sanov's theorem we ask how likely the empirical distribution is to be close to ν_0, given that the true distribution is ν; whereas in the Bayesian context we ask how likely it is that the true distribution is close to ν_0, given that the empirical distribution is close to ν."

An outline of the construction in Lemma 9.4 can be found on page 254 in [152]. The remaining material in Section 9.2 is from Feng [71].

Both Lemma 9.6 and Theorem 9.8 originated from [71]. The proof of Theorem 9.10 was communicated to me by F.Q. Gao. Theorem 9.11 can be found in [70].

The LDP results presented here are associated with the equilibrium distributions. There are also results on path level LDPs for the FV process. In [21], [68], [22], and [79], the LDPs are studied for the FV process with parent-independent mutation. The LDP for FV process without mutation can be found in [188]. Other closely related results can be found in [159] and [141].

Appendix A
Poisson Process and Poisson Random Measure

Reference [130] is the main source for the material on Poisson process and Poisson random measure.

A.1 Definitions

Definition A.1. Let $(\Omega, \mathscr{F}, \mathbb{P})$ be a probability space, and S be a locally compact, separable metric space with Borel σ-algebra $\mathscr{B}$. The set $\mathscr{S}$ denotes the collection of all countable subsets of S. A *Poisson process* with state space S, defined on $(\Omega, \mathscr{F}, \mathbb{P})$, is a map $\digamma$ from $(\Omega, \mathscr{F}, \mathbb{P})$ to $\mathscr{S}$ satisfying:

(a) for each B in $\mathscr{B}$,

$$N(B) = \#\{\digamma \cap B\}$$

is a Poisson random variable with parameter

$$\mu(B) = \mathbb{E}[N(B)];$$

(b) for disjoint sets $B_1, \ldots, B_n$ in $\mathscr{B}$, $N(B_1), \ldots, N(B_n)$ are independent.

If $B_1, B_2, \ldots$ are disjoint, then, by definition, we have

$$N(\cup_{i=1}^{\infty} B_i) = \sum_{i=1}^{\infty} N(B_i),$$

and

$$\mu(\cup_{i=1}^{\infty} B_i) = \sum_{i=1}^{\infty} \mu(B_i).$$

Hence, N is a random measure, and μ is a measure, both on $(S, \mathscr{B})$. The random measure N can also be written in the following form:

S. Feng, *The Poisson–Dirichlet Distribution and Related Topics*,
Probability and its Applications, DOI 10.1007/978-3-642-11194-5,

$$N = \sum_{\varsigma \in F} \delta_{\varsigma}.$$

Definition A.2. The measure μ is called the *mean measure* of the Poisson process F, and N is called a *Poisson random measure* associated with the Poisson process F. The measure μ is also called the mean measure of N.

Remark: Let $S = [0,\infty)$ and μ be the Lebesgue measure on S. Then the Poisson random measure associated with the Poisson process F with state space S and mean measure μ is just the one-dimensional time-homogeneous Poisson process that is defined as a pure-birth Markov chain with birth rate one. The random set F is composed of all jump times of the process.

Definition A.3. Assume that $S = \mathbb{R}^d$ for some $d \geq 1$. Then the mean measure is also called the *intensity measure*. If there exists a positive constant c such that for any measurable set B,

$$\mu(B) = c|B|, \ |B| = \text{Lebesgue measure of } \ B,$$

then the Poisson process F is said to be *homogeneous* with *intensity* c.

A.2 Properties

Theorem A.1. *Let μ be the mean measure of a Poisson process F with state space S. Then μ is diffuse; i.e., for every x in S,*

$$\mu(\{x\}) = 0.$$

Proof. For any fixed x in S, set $a = \mu(\{x\})$. Then, by definition,

$$\mathbb{P}\{N(\{x\}) = 2\} = \frac{a^2}{2} e^{-a} = 0,$$

which leads to the result.

□

The next theorem describes the close relation between a Poisson process and the multinomial distribution.

Theorem A.2. *Let F be a Poisson process with state space S and mean measure μ. Assume that the total mass $\mu(S)$ is finite. Then, for any $n \geq 1, 1 \leq m \leq n$, and any set partition $B_1,\ldots,B_m$ of S, the conditional distribution of the random vector $(N(B_1),\ldots,N(B_m))$ given $N(S) = n$ is a multinomial distribution with parameters n and*

$$\left(\frac{\mu(B_1)}{\mu(S)},\ldots,\frac{\mu(B_m)}{\mu(S)}\right).$$

Proof. For any partitions $n_1,\dots,n_m$ of n,

$$\begin{aligned}
&\mathbb{P}\{N(B_1)=n_1,\dots,N(B_m)=n_m \mid N(S)=n\}\\
&=\frac{\mathbb{P}\{N(B_1)=n_1,\dots,N(B_m)=n_m\}}{\mathbb{P}\{N(S)=n\}}\\
&=\frac{\prod_{i=1}^m \frac{\mu(B_i)_i^n e^{-\mu(B_i)}}{n_i!}}{\frac{\mu(S)^n e^{-\mu(S)}}{n!}}\\
&=\binom{n}{n_1,\dots,n_m}\prod_{i=1}^m\left(\frac{\mu(B_i)}{\mu(S)}\right)^{n_i}.
\end{aligned}$$

□

Theorem A.3. (Restriction and union)

(1) *Let $\digamma$ be a Poisson process with state space S. Then, for every B in $\mathscr{B}$, $\digamma\cap B$ is a Poisson process with state space S and mean measure*

$$\mu_B(\cdot)=\mu(\cdot\cap B).$$

Equivalently, $\digamma\cap B$ can also be viewed as a Poisson process with state space B with mean measure given by the restriction of μ on B.

(2) *Let $\digamma_1$ and $\digamma_2$ be two independent Poisson processes with state space S, and respective mean measures μ_1 and μ_2. Then $\digamma_1\cup\digamma_2$ is a Poisson process with state space S and mean measure*

$$\mu=\mu_1+\mu_2.$$

Proof. Direct verification of the definition of Poisson process.

□

The remaining theorems in this section are stated without proof. The details can be found in [130].

Theorem A.4. (Mapping) *Let $\digamma$ be a Poisson process with state space S and σ-finite mean measure $\mu(\cdot)$. Consider a measurable map h from S to another locally compact, separable metric space S'. If the measure*

$$\mu'(\cdot)=\mu(f^{-1}(\cdot))$$

is diffuse, then $f(\digamma)=\{f(\varsigma):\varsigma\in\digamma\}$ is a Poisson process with state space S' and mean measure μ'.

Theorem A.5. (Marking) *Let $\digamma$ be a Poisson process with state space S and mean measure μ. The mark of each point ς in $\digamma$, denoted by m_ς, is a random variable, taking values in a locally compact, separable metric space S', with distribution $q(z,\cdot)$. Assume that:*

(1) *for every measurable set B' in S', $q(\cdot,B')$ is a measurable function on $(S,\mathscr{B})$;*
(2) *given $\digamma$, the random variables $\{m_\varsigma:\varsigma\in\digamma\}$ are independent;*

then $\tilde{F} = \{(\varsigma, m_\varsigma) : \varsigma \in F\}$ *is a Poisson process on with state space* $S \times S'$ *and mean measure*

$$\tilde{\mu}(dx, dm) = \mu(dx)q(x, dm).$$

The Poisson process $\tilde{F}$ *is aptly called a marked Poisson process.*

Theorem A.6. (Campbell) *Let* F *be a Poisson process on space* $(S, \mathscr{B})$ *with mean measure* μ*. Then for any non-negative measurable function* f*,*

$$\mathbb{E}\left[\exp\left\{-\sum_{\varsigma \in F} f(\varsigma)\right\}\right] = \exp\left\{\int_S (e^{-f(x)} - 1)\mu(dx)\right\}.$$

If f *is a real-valued measurable function on* $(S, \mathscr{B})$ *satisfying*

$$\int_S \min(|f(\mathbf{x})|, 1)\mu(d\mathbf{x}) < \infty,$$

then for any complex number λ *such that the integral*

$$\int_S (e^{\lambda f(x)} - 1)\mu(dx)$$

converges, we have

$$\mathbb{E}\left[\exp\left\{\lambda \sum_{\varsigma \in F} f(\varsigma)\right\}\right] = \exp\left\{\int_S (e^{\lambda f(x)} - 1)\mu(dx)\right\}.$$

Moreover, if

$$\int_S |f(x)|\mu(dx) < \infty, \tag{A.1}$$

then

$$\mathbb{E}\left[\sum_{\varsigma \in F} f(\varsigma)\right] = \int_S f(x)\mu(dx),$$
$$Var\left[\sum_{\varsigma \in F} f(\varsigma)\right] = \int_S f^2(x)\mu(dx).$$

In general, for any $n \geq 1$*, and any real-valued measurable functions* $f_1, \ldots, f_n$ *satisfying (A.1), we have*

$$\mathbb{E}\left[\sum_{distinct\ \varsigma_1, \ldots, \varsigma_n \in F} f_1(\varsigma_1) \cdots f_n(\varsigma_n)\right] = \prod_{i=1}^{n} \mathbb{E}\left[\sum_{\varsigma_i \in F} f_i(\varsigma_i)\right]. \tag{A.2}$$

Appendix B
Basics of Large Deviations

In probability theory, the law of large numbers describes the limiting average or mean behavior of a random population. The fluctuations around the average are characterized by a fluctuation theorem such as the central limit theorem. The theory of large deviations is concerned with the rare event of deviations from the average. Here we give a brief account of the basic definitions and results of large deviations. Everything will be stated in a form that will be sufficient for our needs. All proofs will be omitted. Classical references on large deviations include [30], [50], [168], and [175]. More recent developments can be found in [46], [28], and [69]. The formulations here follow mainly Dembo and Zeitouni [28]. Theorem B.6 is from [157].

Let E be a complete, separable metric space with metric ρ. Generic elements of E are denoted by x, y, etc.

Definition B.1. A function I on E is called a *rate function* if it takes values in $[0, +\infty]$ and is lower semicontinuous. For each c in $[0, +\infty)$, the set

$$\{x \in E : I(x) \leq c\}$$

is called a level set. The *effective domain* of I is defined as

$$\{x \in E : I(x) < \infty\}.$$

If all level sets are compact, the rate function is said to be *good*.

Rate functions will be denoted by other symbols as the need arises. Let $\{X_\varepsilon : \varepsilon > 0\}$ be a family of E-valued random variables with distributions $\{\mathbb{P}_\varepsilon : \varepsilon > 0\}$, defined on the Borel σ-algebra $\mathscr{B}$ of E.

Definition B.2. The family $\{X_\varepsilon : \varepsilon > 0\}$ or the family $\{\mathbb{P}_\varepsilon : \varepsilon > 0\}$ is said to satisfy a large deviation principle (LDP) on E as ε converges to zero, with speed ε and a good rate function I if

S. Feng, *The Poisson–Dirichlet Distribution and Related Topics*,
Probability and its Applications, DOI 10.1007/978-3-642-11194-5,

$$\text{for any closed set } F,\ \limsup_{\varepsilon\to 0}\ \varepsilon \log \mathbb{P}_\varepsilon\{F\} \le -\inf_{x\in F} I(x), \tag{B.1}$$

$$\text{for any open set } G,\ \liminf_{\varepsilon\to 0}\ \varepsilon \log \mathbb{P}_\varepsilon\{G\} \ge -\inf_{x\in G} I(x). \tag{B.2}$$

Estimates (B.1) and (B.2) are called the upper bound and lower bound, respectively. Let $a(\varepsilon)$ be a function of ε satisfying

$$a(\varepsilon) > 0,\ \lim_{\varepsilon\to 0} a(\varepsilon) = 0.$$

If the multiplication factor ε in front of the logarithm is replaced by $a(\varepsilon)$, then the LDP has speed $a(\varepsilon)$.

It is clear that the upper and lower bounds are equivalent to the following statement: for all $B \in \mathscr{B}$,

$$-\inf_{x\in B^\circ} I(x) \le \liminf_{\varepsilon\to 0} \varepsilon \log \mathbb{P}_\varepsilon\{B\}$$
$$\limsup_{\varepsilon\to 0} \varepsilon \log \mathbb{P}_\varepsilon\{B\} \le -\inf_{x\in \overline{B}} I(x),$$

where B° and $\overline{B}$ denote the interior and closure of B respectively. An event $B \in \mathscr{B}$ satisfying

$$\inf_{x\in B^\circ} I(x) = \inf_{x\in \overline{B}} I(x)$$

is called a I-continuity set. Thus for a I-continuity set B, we have that

$$\lim_{\varepsilon\to 0} \varepsilon \log \mathbb{P}_\varepsilon\{B\} = -\inf_{x\in B} I(x).$$

If the values for ε are only $\{1/n : n \ge 1\}$, we will write P_n instead of $P_{1/n}$.

If the upper bound (B.1) holds only for compact sets, then we say the family $\{\mathbb{P}_\varepsilon : \varepsilon > 0\}$ satisfies the *weak LDP*. To establish an LDP from the weak LDP, one needs to check the following condition which is known as *exponential tightness*: For any $M > 0$, there is a compact set K such that on the complement K^c of K we have

$$\limsup_{\varepsilon\to 0}\ \varepsilon \log \mathbb{P}_\varepsilon\{K^c\} \le -M. \tag{B.3}$$

Definition B.3. The family $\{\mathbb{P}_\varepsilon : \varepsilon > 0\}$ is said to be *exponentially tight* if (B.3) holds.

An interesting consequence of an LDP is the following theorem.

Theorem B.1. (Varadhan's lemma) *Assume that the family $\{\mathbb{P}_\varepsilon : \varepsilon > 0\}$ satisfies an LDP with speed ε and good rate function I. Let f and the family $\{f_\varepsilon : \varepsilon \ge 1\}$ be bounded continuous functions on E satisfying*

$$\lim_{\varepsilon\to 0} \sup_{x\in E} \rho(f_\varepsilon(x), f(x)) = 0.$$

Then

$$\lim_{\varepsilon \to 0} \varepsilon \log E^{\mathbb{P}_\varepsilon}[e^{\frac{f_\varepsilon(x)}{\varepsilon}}] = \sup_{x \in E}\{f(x) - I(x)\}.$$

Remark: Without knowing the existence of an LDP, one can guess the form of the rate function by calculating the left-hand side of the above equation.

The next result shows that an LDP can be transformed by a continuous function from one space to another.

Theorem B.2. (Contraction principle) *Let E, F be complete, separable spaces, and h be a measurable function from E to F. If the family of probability measures $\{\mathbb{P}_\varepsilon : \varepsilon > 0\}$ on E satisfies an LDP with speed ε and good rate function I and the function h is continuous at every point in the effective domain of I, then the family of probability measures $\{P_\varepsilon \circ h^{-1} : \varepsilon > 0\}$ on F satisfies an LDP with speed ε and good rate function, I', where*

$$I'(y) = \inf\{I(x) : x \in E, y = h(x)\}.$$

Theorem B.3. *Let $\{Y_\varepsilon : \varepsilon > 0\}$ be a family of random variables satisfying an LDP on space E with speed ε and rate function I. If E_0 is a closed subset of E, and*

$$\mathbb{P}\{Y_\varepsilon \in E_0\} = 1, \; \{x \in E : I(x) < \infty\} \subset E_0$$

then the LDP for $\{Y_\varepsilon : \varepsilon > 0\}$ holds on E_0.

The next concept describes the situation when two families of random variables are indistinguishable exponentially.

Definition B.4. Let

$$\{X_\varepsilon : \varepsilon > 0\}, \; \{Y_\varepsilon : \varepsilon > 0\}$$

be two families of E-valued random variables on a probability space $(\Omega, \mathscr{F}, \mathbb{P})$. If for any $\delta > 0$ the family $\{\mathscr{P}_\varepsilon : \varepsilon > 0\}$ of joint distributions of $(X_\varepsilon, Y_\varepsilon)$ satisfies

$$\limsup_{\varepsilon \to 0} \varepsilon \log \mathscr{P}_\varepsilon\{\{(x,y) : \rho(x,y) > \delta\}\} = -\infty,$$

then we say that $\{X_\varepsilon : \varepsilon > 0\}$ and $\{Y_\varepsilon : \varepsilon > 0\}$ are *exponentially equivalent* with speed ε.

The following theorem shows that the LDPs for exponentially equivalent families of random variables are the same.

Theorem B.4. *Let $\{X_\varepsilon : \varepsilon > 0\}$ and $\{Y_\varepsilon : \varepsilon > 0\}$ be two exponentially equivalent families of E−valued random variables. If an LDP holds for $\{X_\varepsilon : \varepsilon > 0\}$, then the same LDP holds for $\{Y_\varepsilon : \varepsilon > 0\}$ and vice versa.*

To generalize the notion of exponential equivalence, we introduce the concept of exponential approximation next.

Definition B.5. Consider a family of random variables $\{X_\varepsilon : \varepsilon > 0\}$ and a sequence of families of random variables $\{Y^n_\varepsilon : \varepsilon > 0\}, n = 1, 2, \ldots$, all defined on the same probability space. Denote the joint distribution of $(X_\varepsilon, Y^n_\varepsilon)$ by $\mathscr{P}^n_\varepsilon$. Assume that for any $\delta > 0$,

$$\lim_{n\to\infty} \limsup_{\varepsilon\to 0} \varepsilon \log \mathscr{P}^n_\varepsilon\{\{(x,y) : \rho(x,y) > \delta\}\} = -\infty, \tag{B.4}$$

then the sequence $\{Y^n_\varepsilon : \varepsilon > 0\}$ is called an *exponentially good approximation* of $\{X_\varepsilon : \varepsilon > 0\}$.

Theorem B.5. *Let the sequence of families $\{Y^n_\varepsilon : \varepsilon > 0\}, n = 1, 2, \ldots$, be an exponentially good approximation to the family $\{X_\varepsilon : \varepsilon > 0\}$. Assume that for each $n \geq 1$, the family $\{Y^n_\varepsilon : \varepsilon > 0\}$ satisfies an LDP with speed ε and good rate function I_n. Set*

$$I(x) = \sup_{\delta>0} \liminf_{n\to\infty} \inf_{\{y:\rho(y,x)<\delta\}} I_n(y). \tag{B.5}$$

If I is a good rate function, and for any closed set F,

$$\inf_{x\in F} I(x) \leq \limsup_{n\to\infty} \inf_{y\in F} I_n(y), \tag{B.6}$$

then the family $\{X_\varepsilon : \varepsilon > 0\}$ satisfies an LDP with speed ε and good rate function I.

The basic theory of convergence of sequences of probability measures on a metric space have an analog in the theory of large deviations. Prohorov's theorem, relating compactness to tightness, has the following parallel that links exponential tightness to a partial LDP, defined below.

Definition B.6. A family of probability measures $\{\mathbb{P}_\varepsilon : \varepsilon > 0\}$ is said to satisfy the *partial LDP* if for every sequence ε_n converging to zero there is a subsequence ε'_n such that the family $\{\mathbb{P}_{\varepsilon'_n} : \varepsilon'_n > 0\}$ satisfies an LDP with speed ε'_n and a good rate function I'.

Remark: The partial LDP becomes an LDP if the rate functions associated with different subsequences are the same.

Theorem B.6. (Pukhalskii)
(1) *The partial LDP is equivalent to exponential tightness. Thus the partial LDP always holds on a compact space E.*
(2) *Assume that $\{\mathbb{P}_\varepsilon : \varepsilon > 0\}$ satisfies the partial LDP with speed ε, and for every x in E*

$$\begin{aligned}\lim_{\delta\to 0} \limsup_{\varepsilon\to 0} \varepsilon \log \mathbb{P}_\varepsilon\{\rho(y,x) \leq \delta\} \\ = \lim_{\delta\to 0} \liminf_{\varepsilon\to 0} \varepsilon \log \mathbb{P}_\varepsilon\{\rho(y,x) < \delta\} = -I(x).\end{aligned} \tag{B.7}$$

Then $\{\mathbb{P}_\varepsilon : \varepsilon > 0\}$ satisfies an LDP with speed ε and good rate function I.

For an $\mathbb{R}^d$-valued random variable Y, we define the *logarithmic moment generating function* of Y or its law μ as

$$\Lambda(\lambda) = \log E[e^{\langle \lambda, Y \rangle}] \text{ for all } \lambda \in \mathbb{R}^d \tag{B.8}$$

where $\langle , \rangle$ denotes the usual inner product in $\mathbb{R}^d$. $\Lambda(\cdot)$ is also called the *cumulant generating function* of Y. The Fenchel–Legendre transformation of $\Lambda(\lambda)$ is defined as

$$\Lambda^*(x) := \sup_{\lambda \in \mathbb{R}^d} \{\langle \lambda, x \rangle - \Lambda(\lambda)\}. \tag{B.9}$$

Theorem B.7. (Cramér) *Let $\{X_n : n \geq 1\}$ be a sequence of i.i.d. random variables in $\mathbb{R}^d$. Denote the law of $\frac{1}{n}\sum_{k=1}^n X_k$ by $\mathbb{P}_n$. Assume that*

$$\Lambda(\lambda) = \log E[e^{\langle \lambda, X_1 \rangle}] < \infty \textit{ for all } \lambda \in \mathbb{R}^d.$$

Then the family $\{\mathbb{P}_n : n \geq 1\}$ satisfies an LDP with speed $1/n$ and good rate function $I(x) = \Lambda^*(x)$.

The i.i.d. assumption plays a crucial role in Cramér's theorem. For general situations one has the following Gärtner–Ellis theorem.

Theorem B.8. (Gärtner–Ellis) *Let $\{Y_\varepsilon : \varepsilon > 0\}$ be a family of random vectors in $\mathbb{R}^d$. Denote the law of Y_ε by $\mathbb{P}_\varepsilon$. Define*

$$\Lambda_\varepsilon(\lambda) = \log E[e^{\langle \lambda, Y_\varepsilon \rangle}].$$

Assume that the limit

$$\Lambda(\lambda) = \lim_{n \to \infty} \varepsilon \Lambda_\varepsilon(\lambda/\varepsilon),$$

exists, and is lower semicontinuous. Set

$$\mathscr{D} = \{\lambda \in R^d : \Lambda(\lambda) < \infty\}.$$

If $\mathscr{D}$ has an nonempty interior $\mathscr{D}^\circ$ on which Λ is differentiable, and the norm of the gradient of $\Lambda(\lambda_n)$ converges to infinity, whenever λ_n in $\mathscr{D}^\circ$ converges to a boundary point of $\mathscr{D}^\circ$ (Λ satisfying these conditions is said to be essentially smooth), *then the family $\{\mathbb{P}_\varepsilon : \varepsilon > 0\}$ satisfies an LDP with speed ε and good rate function $I = \Lambda^*$.*

The next result can be derived from the Gärtner–Ellis theorem.

Corollary B.9 *Assume that*

$$\{X_\varepsilon : \varepsilon > 0\}, \{Y_\varepsilon : \varepsilon > 0\}, \{Z_\varepsilon : \varepsilon > 0\}$$

are three families of real-valued random variables, all defined on the same probability space with respective laws

$$\{\mathbb{P}^1_\varepsilon : \varepsilon > 0\}, \{\mathbb{P}^2_\varepsilon : \varepsilon > 0\}, \{\mathbb{P}^3_\varepsilon : \varepsilon > 0\}.$$

If both $\{\mathbb{P}_\varepsilon^1 : \varepsilon > 0\}$ *and* $\{\mathbb{P}_\varepsilon^3 : \varepsilon > 0\}$ *satisfy the assumptions in Theorem B.8 with the same* $\Lambda(\cdot)$*, and with probability one*

$$X \leq Y \leq Z,$$

then $\{\mathbb{P}_\varepsilon^2 : \varepsilon > 0\}$ *satisfies an LDP with speed* ε *and a good rate function given by*

$$I(x) = \sup_{\lambda \in \mathbb{R}} \{\lambda x - \Lambda(\lambda)\}.$$

Infinite-dimensional generalizations of Cramér's theorem are also available. Here we only mention one particular case: Sanov's theorem.

Let $\{X_k : k \geq 1\}$ be a sequence of i.i.d. random variables in $\mathbb{R}^d$ with common distribution μ. For any $n \geq 1$, define

$$\eta_n = \frac{1}{n} \sum_{k=1}^{n} \delta_{X_k}$$

where δ_x is the Dirac measure concentrated at x. The empirical distribution η_n belong to the space $M_1(\mathbb{R}^d)$ of all probability measures on $\mathbb{R}^d$ equipped with the weak topology. A well-known result from statistics says that when n becomes large one will recover the true distribution μ from η_n. Clearly $M_1(\mathbb{R}^d)$ is an infinite dimensional space. Denote the law of η_n, on $M_1(\mathbb{R}^d)$, by $\mathbb{Q}_n$. Then we have:

Theorem B.10. (Sanov) *The family* $\{\mathbb{Q}_n : n \geq 1\}$ *satisfies an LDP with speed* $1/n$ *and good rate function*

$$H(\nu|\mu) = \begin{cases} \int_{R^d} \log \frac{d\nu}{d\mu} d\nu, & \text{if } \nu \ll \mu \\ \infty, & \text{otherwise,} \end{cases} \tag{B.10}$$

where $\nu \ll \mu$ *means that* ν *is absolutely continuous with respect to* μ *and* $H(\nu|\mu)$ *is called the relative entropy of* ν *with respect to* μ.

References

1. M. Abramowitz and I.A. Stegun (1965). *Handbook of Mathematical Functions*. Dover Publ. Inc., New York.
2. D.J. Aldous (1985). *Exchangeability and Related Topics*, Ecole d'Été de Probabilités de Saint Flour, Lecture Notes in Math., Vol. 1117, 1–198, Springer-Verlag.
3. C. Antoniak (1974). Mixtures of Dirichlet processes with applications to Bayesian nonparameteric problems. *Ann. Statist.* **2**, 1152–1174.
4. M. Aoki (2008). Thermodynamic limit of macroeconomic or financial models: one- and two-parameter Poisson–Dirichlet models. *J. Econom. Dynam. Control* **32**, No. 1, 66–84.
5. R. Arratia (1996). Independence of prime factors: total variation and Wasserstein metrics, insertions and deletions, and the Poisson–Dirichlet process. Preprint.
6. R. Arratia, A.D. Barbour and S. Tavaré (1992). Poisson process approximations for the Ewens sampling formula. *Ann. Appl. Probab.* **2**, No. 3, 519–535.
7. R. Arratia, A.D. Barbour and S. Tavaré (1997). Random combinatorial structures and prime factorizations. *Notices of AMS* **44**, No. 8, 903–910.
8. R. Arratia, A.D. Barbour and S. Tavaré (1999). The Poisson–Dirichlet distribution and the scale invariant Poisson process. *Combin. Probab. Comput.* **8**, 407–416.
9. R. Arratia, A.D. Barbour and S. Tavaré (2003). *Logarithmic Combinatorial Structures: A Probabilistic Approach.* EMS Monographs in Mathematics, European Mathematical Society.
10. R. Arratia, A.D. Barbour and S. Tavaré (2006). A tale of three couplings: Poisson–Dirichlet and GEM approximations for random permutations. *Combin. Probab. Comput.* **15**, 31–62.
11. A.D. Barbour, L. Holst and S. Janson (1992). *Poisson Approximation.* Oxford University Press, New York.
12. J. Bertoin (2006). *Random Fragmentation and Coagulation Processes.* Cambridge Studies in Advanced Mathematics, 102. Cambridge University Press, Cambridge.
13. J. Bertoin (2008). Two-parameter Poisson–Dirichlet measures and reversible exchangeable fragmentation-coalescence processes. *Combin. Probab. Comput.* **17**, No. 3, 329–337.
14. P. Billingsley (1968). *Convergence of Probability Measures*, Wiley, New York.
15. P. Billingsley (1972). On the distribution of large prime divisors. *Period. Math. Hungar.* **2**, 283–289.
16. D. Blackwell and J.B. MacQueen (1973). Ferguson distribution via Pólya urn scheme. *Ann. Statist.* **1**, 353–355.
17. M.A. Carlton (1999). Applications of the two-parameter Poisson–Dirichlet distribution. Unpublished Ph.D. thesis, Dept. of Statistics, University of California, Los Angeles.
18. M.F Chen (2005). *Eigenvalues, Inequalities, and Ergodic Theory.* Probability and its Applications (New York). Springer-Verlag, London.
19. S. Coles (2001). *An Introduction to Statistical Modeling of Extreme Values.* Springer-Verlag, London.

S. Feng, *The Poisson–Dirichlet Distribution and Related Topics*,
Probability and its Applications, DOI 10.1007/978-3-642-11194-5,

20. D.A. Dawson (1993). *Measure-valued Markov Processes*, Ecole d'Été de Probabilités de Saint Flour, Lecture Notes in Math., Vol. 1541, 1–260, Springer-Verlag.
21. D.A. Dawson and S. Feng (1998). Large deviations for the Fleming–Viot process with neutral mutation and selection. *Stoch. Proc. Appl.* **77**, 207–232.
22. D.A. Dawson and S. Feng (2001). Large deviations for the Fleming–Viot process with neutral mutation and selection, II. *Stoch. Proc. Appl.* **92**, 131–162.
23. D.A. Dawson and S. Feng (2006). Asymptotic behavior of Poisson–Dirichlet distribution for large mutation rate. *Ann. Appl. Probab.* **16**, No. 2, 562–582.
24. D.A. Dawson and K.J. Hochberg (1979). The carrying dimension of a stochastic measure diffusion. *Ann. Probab.* **7**, 693–703.
25. D.A. Dawson and K.J. Hochberg (1982). Wandering random measures in the Fleming–Viot model. *Ann. Probab.* **10**, 554–580.
26. J.M. DeLaurentis and B.G. Pittel (1985). Random permutations and Brownian motion. *Pacific J. Math.* **119**, 287–301.
27. J. Delmas, J. Dhersin, and A. Siri-Jegousse (2008). Asymptotic results on the length of coalescent trees. *Ann. Appl. Probab.* **18**, No. 3, 997–1025.
28. A. Dembo and O. Zeitouni (1998). *Large Deviations Techniques and Applications*. Second edition. Applications of Mathematics, Vol. 38, Springer-Verlag, New York.
29. B. Derrida (1997). From random walks to spin glasses. *Physica D* **107**, 186–198.
30. J.D. Deuschel and D.W. Stroock (1989). *Large Deviations*. Academic Press, Boston.
31. K. Dickman (1930). On the frequency of numbers containing prime factors of a certain relative magnitude. *Arkiv för Matematik, Astronomi och Fysik* **22**, 1–14.
32. R. Dong, C. Goldschmidt and J. B. Martin (2006). Coagulation–fragmentation duality, Poisson–Dirichlet distributions and random recursive trees. *Ann. Appl. Probab.* **16**, No. 4, 1733–1750.
33. P. Donnelly (1986). Partition structures, Polya urns, the Ewens sampling formula, and the age of alleles. *Theor. Pop. Biol.* **30**, 271–288.
34. P. Donnelly and G. Grimmett (1993). On the asymptotic distribution of large prime factors. *J. London Math. Soc.* **47**, No. 3, 395–404.
35. P. Donnelly and P. Joyce (1989). Continuity and weak convergence of ranked and size-biased permutations on the infinite simplex. *Stoc. Proc. Appl.* **31**, 89–104.
36. P. Donnelly and P. Joyce (1991). Weak convergence of population genealogical processes to the coalescent with ages. *Ann. Prob.* **20**, No. 1, 322–341.
37. P. Donnelly, T.G. Kurtz, and S. Tavaré (1991). On the functional central limit theorem for the Ewens sampling formula. *Ann. Appl. Probab.* **1**, No. 4, 539–545.
38. P. Donnelly and T.G. Kurtz (1996a). A countable representation of the Fleming–Viot measure-valued diffusion. *Ann. Prob.* **24**, No. 2, 698–742.
39. P. Donnelly and T.G. Kurtz (1996b). The asymptotic behavior of an urn model arising in population genetics. *Stoc. Proc. Appl.* **64**, 1–16.
40. P. Donnelly and T.G. Kurtz (1999a). Genealogical processes for the Fleming–Viot models with selection and recombination. *Ann. Appl. Prob.* **9**, No. 4, 1091–1148.
41. P. Donnelly and T.G. Kurtz (1999b). Particle representations for measure-valued population models. *Ann. Prob.* **27**, No. 1, 166–205.
42. P. Donnelly and S. Tavaré (1986). The ages of alleles and a coalescent. *Adv. Appl. Prob.* **12**, 1–19.
43. P. Donnelly and S. Tavaré (1987). The population genealogy of the infinitely-many-neutral alleles model. *J. Math. Biol.* **25**, 381–391.
44. M.D. Donsker and S.R.S. Varadhan (1975). Asymptotic evaluation of certain Markov process expectations for large time, I. *Comm. Pure Appl. Math.* **28**, 1–47.
45. M. Döring and W. Stannat (2009). The logarithmic Sobolev inequality for the Wasserstein diffusion. *Probab. Theory Relat. Fields* **145**, No. 1-2, 189–209.
46. P. Dupuis and R.S. Ellis (1997). *A Weak Convergence Approach to the Theory of Large Deviations*. Wiley, New York.
47. R. Durrett (1996). *Probability: Theory and Examples*. Second edition. Duxbury Press.

48. R. Durrett (2008). *Probability Models for DNA Sequence Evolution*. Second edition. Springer-Verlag, New York.
49. E.B. Dynkin (1994). *An Introduction to Branching Measure-Valued Processes*. Vol. 6, CRM Monographs. Amer Math. Soc., Providence.
50. R. S. Ellis (1985). *Entropy, Large Deviations, and Statistical Mechanics*. Springer, Berlin.
51. S. Engen (1978). *Abundance Models with Emphasis on Biological Communities and Species Diversity*. Chapman-Hall, London.
52. A.M. Etheridge (1991). *An Introduction to Superprocesses*. University Lecture Series, Vol. 20, Amer. Math. Soc., Providence, RI.
53. A.M. Etheridge and P. March (1991). A note on superprocesses. *Probab. Theory Relat. Fields* **89**, 141–147.
54. A. Etheridge, P. Pfaffelhuber, and A. Wakolbinger (2006). An approximate sampling formula under genetic hitchhiking. *Ann. Appl. Probab.* **16**, No. 2, 685–729.
55. S.N. Ethier (1976). A class of degenerate diffusions processes occurring in population genetics. *Comm. Pure. Appl. Math.* **29**, 483–493.
56. S.N. Ethier (1990). The infinitely-many-neutral-alleles diffusion model with ages. *Adv. Appl. Probab.* **22**, 1–24.
57. S.N. Ethier (1992). Eigenstructure of the infinitely-many-neutral-alleles diffusion model. *J. Appl. Probab.* **29**, 487–498.
58. S.N. Ethier and R.C. Griffiths (1987). The infinitely-many-sites model as a measure-valued diffusion. *Ann. Probab.* **15**, No. 2, 515–545.
59. S.N. Ethier and R.C. Griffiths (1993). The transition function of a Fleming–Viot process. *Ann. Probab.* **21**, No. 3, 1571–1590.
60. S.N. Ethier and R.C. Griffiths (1993). The transition function of a measure-valued branching diffusion with immigration. In *Stochastic Processes. A Festschrift in Honour of Gopinath Kallianpur* (S. Cambanis, J. Ghosh, R.L. Karandikar and P.K. Sen, eds.), 71–79. Springer, New York.
61. S.N. Ethier and T.G. Kurtz (1981). The infinitely-many-neutral-alleles diffusion model. *Adv. Appl. Probab.* **13**, 429–452.
62. S.N. Ethier and T.G. Kurtz (1986). *Markov Processes: Characterization and Convergence*, John Wiley, New York.
63. S.N. Ethier and T.G. Kurtz (1992). On the stationary distribution of the neutral diffusion model in population genetics. *Ann. Appl. Probab.* **2**, 24–35.
64. S.N. Ethier and T.G. Kurtz (1993). Fleming–Viot processes in population genetics. *SIAM J. Control and Optimization* **31**, No. 2, 345–386.
65. S.N. Ethier and T.G. Kurtz (1994). Convergence to Fleming–Viot processes in the weak atomic topology. *Stoch. Proc. Appl.* **54**, 1–27.
66. W.J. Ewens (1972). The sampling theory of selectively neutral alleles. *Theor. Pop. Biol.* **3**, 87–112.
67. W.J. Ewens (2004). *Mathematical Population Genetics, Vol. I*, Springer-Verlag, New York.
68. J. Feng (1999). Martingale problems for large deviations of Markov processes. *Stoch. Proc. Appl.* **81**, 165–216.
69. J. Feng and T.G. Kurtz (2006). *Large Deviations for Stochastic Processes*. Amer. Math. Soc., Providence, RI.
70. S. Feng (2007). Large deviations associated with Poisson–Dirichlet distribution and Ewens sampling formula. *Ann. Appl. Probab.* **17**, Nos. 5/6, 1570–1595.
71. S. Feng (2007). Large deviations for Dirichlet processes and Poisson–Dirichlet distribution with two parameters. *Electron. J. Probab.* **12**, 787–807.
72. S. Feng (2009). Poisson–Dirichlet distribution with small mutation rate. *Stoch. Proc. Appl.* **119**, 2082–2094.
73. S. Feng and F.Q. Gao (2008). Moderate deviations for Poisson–Dirichlet distribution. *Ann. Appl. Probab.* **18**, No. 5, 1794–1824.
74. S. Feng and F.Q. Gao (2010). Asymptotic results for the two-parameter Poisson–Dirichlet distribution. *Stoch. Proc. Appl.* **120**, 1159–1177.

75. S. Feng and F.M. Hoppe (1998). Large deviation principles for some random combinatorial structures in population genetics and Brownian motion. *Ann. Appl. Probab.* **8**, No. 4, 975–994.
76. S. Feng and W. Sun (2009). Some diffusion processes associated with two parameter Poisson–Dirichlet distribution and Dirichlet process. *Probab. Theory Relat. Fields*, DOI 10.1007/s00440-009-0238-2.
77. S. Feng, W. Sun, F.Y. Wang, and F. Xu (2009). Functional inequalities for the unlabeled two-parameter infinite-alleles diffusion. Preprint.
78. S. Feng and F.Y. Wang (2007). A class of infinite-dimensional diffusion processes with connection to population genetics. *J. Appl. Prob.* **44**, 938–949.
79. S. Feng and J. Xiong (2002). Large deviation and quasi-potential of a Fleming–Viot process. *Elect. Comm. Probab.* **7**, 13–25.
80. T.S. Ferguson (1973). A Baysian analysis of some nonparametric problems. *Ann. Stat.* **1**, 209–230.
81. R.A. Fisher (1922). On the dominance ratio. *Proc. Roy. Soc. Edin.* **42**, 321–341.
82. W.H. Fleming and M. Viot (1979). Some measure-valued Markov processes in population genetics theory. *Indiana Univ. Math. J.* **28**, 817–843.
83. A. Gamburd (2006). Poisson–Dirichlet distribution for random Belyi surfaces. *Ann. Probab.* **34**, No. 5, 1827–1848.
84. A.J. Ganesh and N. O'Connell (1999). An inverse of Sanov's theorem. *Statist. Probab. Lett.* **42**, No. 2, 201–206.
85. A.J. Ganesh and N. O'Connell (2000). A large-deviation principle for Dirichlet posteriors. *Bernoulli* **6**, No. 6, 1021–1034.
86. J.H. Gillespie (1998). *Population Genetics: A Concise Guide*, The Johns Hopkins University Press, Baltimore and London.
87. J.H. Gillespie (1999). The role of population size in molecular evolution. *Theor. Pop. Biol.* **55**, 145–156.
88. A.V. Gnedin (1998). On convergence and extensions of size-biased permutations. *J. Appl. Prob.* **35**, 642–650.
89. A.V. Gnedin (2004). Three sampling formulas. *Combin. Probab. Comput.* **13**, No. 2, 185–193.
90. V.L. Goncharov (1944). Some facts from combinatorics. *Izvestia Akad. Nauk. SSSR, Ser. Mat.* **8**, 3–48.
91. R.C. Griffiths (1979a). On the distribution of allele frequencies in a diffusion model. *Theor. Pop. Biol.* **15**, 140–158.
92. R.C. Griffiths (1979b). A transition density expansion for a multi-allele diffusion model. *Adv. Appl. Probab.* **11**, 310–325.
93. R.C. Griffiths (1980a). Lines of descent in the diffusion approximation of neutral Wright–Fisher models. *Theor. Pop. Biol.* **17**, 35–50.
94. R.C. Griffiths (1980b). Unpublished note.
95. R.C. Griffiths (1988). On the distribution of points in a Poisson process. *J. Appl. Prob.* **25**, 336–345.
96. L. Gross (1993). Logarithmic Sobolev inequalities and contractivity properties of semigroups. *Dirichlet forms* (Varenna, 1992), 54–88, Lecture Notes in Math., Vol. 1563, Springer-Verlag, Berlin.
97. A. Guionnet and B. Zegarlinski (2003). Lectures on logarithmic Sobolev inequalities. *Séminaire de Probabilités, XXXVI*, 1–134, Lecture Notes in Math., Vol. 1801, Springer-Verlag, Berlin.
98. P.R. Halmos (1944). Random alms. *Ann. Math. Stat.* **15**, 182–189.
99. K. Handa (2005). Sampling formulae for symmetric selection, *Elect. Comm. Probab.* **10**, 223–234.
100. K. Handa (2009). The two-parameter Poisson–Dirichlet point process. *Bernoulli*, **15**, No. 4, 1082–1116.
101. J.C. Hansen (1990). A functional central limit theorem for the Ewens sampling formula. *J. Appl. Probab.* **27**, 28–43.

102. L. Holst (2001). The Poisson–Dirichlet distribution and its relatives revisited. Preprint.
103. F.M. Hoppe (1984). Polya-like urns and the Ewens sampling formula. *J. Math. Biol.* **20**, 91–94.
104. F.M. Hoppe (1987). The sampling theory of neutral alleles and an urn model in population genetics. *J. Math. Biol.* **25**, No. 2, 123–159.
105. R.R. Hudson (1983). Properties of a neutral allele model with intragenic recombination. *Theor. Pop. Biol.* **123**, 183–201.
106. W.N. Hudson and H.G. Tucker (1975). Equivalence of infinitely divisible distributions. *Ann. Probab.* **3**, No. 1, 70–79.
107. T. Huillet (2007). Ewens sampling formulae with and without selection. *J. Comput. Appl. Math.* **206**, No. 2, 755–773.
108. H. Ishwaran and L. F. James (2001). Gibbs sampling methods for stick-breaking priors. *J. Amer. Statist. Assoc.* **96**, No. 453, 161–173.
109. H. Ishwaran and L. F. James (2003). Generalized weighted Chinese restaurant processes for species sampling mixture models. *Statist. Sinica.* **13**, No. 4, 1211–1235.
110. J. Jacod, A. N. Shiryaev (1987). *Limit Theorems for Stochastic Processes*. Springer-Verlag, New York.
111. L.F. James (2003). Bayesian calculus for gamma processes with applications to semiparametric intensity models. *Sankhyā* **65**, No. 1, 179–206.
112. L.F. James (2005a). Bayesian Poisson process partition with an application to Bayesian Lévy moving averages. *Ann. Statist.* **33**, No. 4, 1771–1799.
113. L.F. James (2005b). Functionals of Dirichlet processes, the Cifarelli-Regazzini identity and beta-gamma processes. *Ann. Statist.* **33**, No. 2, 647–660.
114. L.F. James (2008). Large sample asymptotics for the two-parameter Poisson Dirichlet processes. In Bertrand Clarke and Subhashis Ghosal, eds, *Pushing the Limits of Contemporary Statistics: Contributions in Honor of Jayanta K. Ghosh*, 187–199, Institute of Mathematical Statistics, Collections, Vol. 3, Beachwood, Ohio.
115. L.F. James, A. Lijoi, and I. Prünster (2008). Distributions of linear functionals of the two parameter Poisson–Dirichlet random measures. *Ann. Appl. Probab.* **18**, No. 2, 521–551.
116. L.F. James, B. Roynette, and M. Yor (2008). Generalized gamma convolutions, Dirichlet means, Thorin measures, with explicit examples. *Probab. Surv.* **5**, 346–415.
117. P. Joyce and F. Gao (2005). An irrational constant separates selective under dominance from neutrality in the infinite alleles model, Preprint.
118. P. Joyce, S.M. Krone, and T.G. Kurtz (2002). Gaussian limits associated with the Poisson–Dirichlet distribution and the Ewens sampling formula. *Ann. Appl. Probab.* **12**, No. 1, 101–124.
119. P. Joyce, S.M. Krone, and T.G. Kurtz (2003). When can one detect overdominant selection in the infinite-alleles model? *Ann. Appl. Probab.* **13**, No. 1, 181–212.
120. P. Joyce and S. Tavaré (1987). Cycles, permutations and the structure of the Yule process with immigration. *Stoch. Proc. Appl.* **25**, 309–314.
121. S. Karlin and J. McGregor (1967). The number of mutant forms maintained in a population. *Proc. Fifth Berkeley Symp. Math. Statist. and Prob., L. LeCam and J. Neyman, eds*, 415–438.
122. S. Karlin and J. McGregor (1972). Addendum to a paper of W. Ewens. *Theor. Pop. Biol.* **3**, 113–116.
123. D.G. Kendall (1949). Stochastic processes and population growth. *J. Roy. Statist. Soc.* **11**, 230–264.
124. M. Kimura and J.F. Crow (1964). The number of alleles that can be maintained in a finite population. *Genetics* **49**, 725–738.
125. J.C.F. Kingman (1975). Random discrete distributions. *J. Roy. Statist. Soc. B* **37**, 1–22.
126. J.C.F. Kingman (1977). The population structure associated with the Ewens sampling formula. *Theor. Pop. Biol.* **11**, 274–283.
127. J.C.F. Kingman (1980). *Mathematics of Genetics Diversity*. CBMS-NSF Regional Conference Series in Appl. Math. Vol. 34, SIAM, Philadelphia.
128. J.C.F. Kingman (1982). The coalescent. *Stoch. Proc. Appl.* **13**, 235–248.

129. J.C.F. Kingman (1982). On the genealogy of large populations. *J. Appl. Prob.* **19A**, 27–43.
130. J.C.F. Kingman (1993). *Poisson Processes*. Oxford University Press.
131. N. Konno and T. Shiga (1988). Stochastic differential equations for some measure-valued diffusions. *Probab. Theory Relat. Fields* **79**, 201–225.
132. T.G. Kurtz (2000). Particle representations for measure-valued population processes with spatially varying birth rates. *Stochastic Models*, Luis G. Gorostiza and B. Gail Ivanoff, eds., CMS Conference Proceedings, Vol. 26, 299–317.
133. J.F. Le Gall (1999). *Spatial Branching Process, Random Snakes and Partial Differential Equations*. Lectures in Mathematics, ETH Zürich, Birkhäuser.
134. Z. Li, T. Shiga and L. Yao (1999). A reversible problem for Fleming–Viot processes. *Elect. Comm. Probab.* **4**, 71–82.
135. J. Lynch and J. Sethuraman (1987). Large deviations for processes with independent increments. *Ann. Probab.* **15**, 610–627.
136. E. Mayer-Wolf, O. Zeitouni, and M.P.W. Zerner (2002). Asymptotics of certain coagulation-fragmentation process and invariant Poisson–Dirichlet measures. *Electron. J. Probab.* **7**, 1–25.
137. J.W. McCloskey (1965). A model for the distribution of individuals by species in an environment. Ph.D. thesis, Michigan State University.
138. M. Möhle and S. Sagitov (2001). A classification of coalescent processes for haploid exchangeable population models. *Ann. Probab.* **29**, 1547–1562.
139. P.A.P. Moran (1958). Random processes in genetics. *Proc. Camb. Phil. Soc.* **54**, 60–71.
140. C. Neuhauser and S.M. Krone (1997). The genealogy of samples in models with selection. *Genetics* **145**, 519–534.
141. F. Papangelou (2000). The large deviations of a multi-allele Wright–Fisher process mapped on the sphere. *Ann. Appl. Probab.* **10**, No. 4, 1259–1273.
142. C.P. Patil and C. Taillie (1977). Diversity as a concept and its implications for random communities. *Bull. Int. Statist. Inst.* **47**, 497–515.
143. E.A. Perkins (1991). Conditional Dawson-Watanabe superprocesses and Fleming–Viot processes. *Seminar on Stochastic Processes*, Birkhäuser, 142–155.
144. E.A. Perkins (2002). *Dawson-Watanabe Superprocess and Measure-Valued Diffusions*, Ecole d'Été de Probabilités de Saint Flour, Lect. Notes in Math. Vol. 1781, 132–329.
145. M. Perman (1993). Order statistics for jumps of normalised subordinators. *Stoch. Proc. Appl.* **46**, 267–281.
146. M. Perman, J. Pitman and M. Yor (1992). Size-biased sampling of Poisson point processes and excursions. *Probab. Theory Relat. Fields* **92**, 21–39.
147. L.A. Petrov (2009). Two-parameter family of infinite-dimensional diffusions on the Kingman simplex. *Funct. Anal. Appl.* **43**, No. 4, 279–296.
148. J. Pitman (1992). The two-parameter generalization of Ewens' random partition structure. Technical Report 345, Dept. Statistics, University of California, Berkeley.
149. J. Pitman (1995). Exchangeable and partially exchangeable random partitions. *Probab. Theory Relat. Fields* **102**, 145–158.
150. J. Pitman (1996a). Partition structures derived from Brownian motion and stable subordinators. *Bernoulli* **3**, 79–96.
151. J. Pitman (1996b). Random discrete distributions invariant under size-biased permutation. *Adv. Appl. Probab.* **28**, 525–539.
152. J. Pitman (1996c). Some developments of the Blackwell-MacQueen urn scheme. *Statistics, Probability, and Game Theory*, 245–267, IMS Lecture Notes Monogr. Ser. 30, Inst. Math. Statist., Hayward, CA.
153. J. Pitman (1999). Coalescents with multiple collisions. *Ann. Probab.* **27**, 1870–1902.
154. J. Pitman (2002). Poisson–Dirichlet and GEM invariant distributions for split-and-merge transformations of an interval partition. *Combin. Probab. Comput.* **11**, 501–514.
155. J. Pitman (2006). *Combinatorial Stochastic Processes*, Ecole d'Été de Probabilités de Saint Flour, Lecture Notes in Math., Vol. 1875, Springer-Verlag, Berlin.
156. J. Pitman and M. Yor (1997). The two-parameter Poisson–Dirichlet distribution derived from a stable subordinator. *Ann. Probab.* **25**, No. 2, 855–900.

157. A.A. Puhalskii (1991). On functional principle of large deviations. In V. Sazonov and T. Shervashidze, eds, *New Trends in Probability and Statistics*, 198–218. VSP Moks'las, Moskva.
158. S. Sagitov (1999). The general coalescent with asynchronous mergers of ancestral lines *J. Appl. Probab.* **36**, 1116–1125.
159. A. Schied (1997). Geometric aspects of Fleming–Viot and Dawson-Watanabe processes. *Ann. Probab.* **25**, 1160–1179.
160. B. Schmuland (1991). A result on the infinitely many neutral alleles diffusion model. *J. Appl. Probab.* **28**, 253–267.
161. J. Schweinsberg (2000). Coalescents with simultaneous multiple collisions. *Electron. J. Probab.* **5**, 1–50.
162. J.A. Sethuraman (1994). A constructive definition of Dirichlet priors. *Statist. Sinica* **4**, No. 2, 639–650.
163. L.A. Shepp and S.P. Lloyd (1966). Ordered cycle length in a random permutation. *Trans. Amer. Math. Soc.* **121**, 340–351.
164. M. Sion (1958). On general minimax theorem. *Pacific J. Math.* **8**, 171–176.
165. W. Stannat (2000). On the validity of the log-Sobolev inequality for symmetric Fleming–Viot operators. *Ann. Probab.* **28**, No. 2, 667–684.
166. W. Stannat (2002). Long-time behaviour and regularity properties of transition semigroups of Fleming–Viot processes. *Probab. Theory Relat. Fields* **122**, 431–469.
167. W. Stannat (2003). On transition semigroup of (A,Ψ)-superprocess with immigration. *Ann. Probab.* **31**, No. 3, 1377–1412.
168. D.W. Stroock (1984). *An Introduction to the Theory of Large Deviations*. Springer-Verlag, New York.
169. F. Tajima (1983). Evolutionary relationship of DNA sequences in finite populations. *Genetics* **105**, 437–460.
170. S. Tavaré (1984). Line-of-descent and genealogical processes, and their applications in population genetics models. *Theor. Pop. Biol.* **26**, 119–164.
171. S. Tavaré (1987). The birth process with immigration, and the genealogical structure of large populations. *J. Math. Biol.* **25**, 161–168.
172. N.V. Tsilevich (2000). Stationary random partitions of positive integers. *Theor. Probab. Appl.* **44**, 60–74.
173. N.V. Tsilevich and A. Vershik (1999). Quasi-invariance of the gamma process and the multiplicative properties of the Poisson–Dirichlet measures. C. R. Acad. Sci. Paris, t. 329, Série I, 163–168.
174. N.V. Tsilevich, A. Vershik, and M. Yor (2001). An infinite-dimensional analogue of the Lebesgue measure and distinguished properties of the gamma process. *J. Funct. Anal.* **185**, No. 1, 274–296.
175. S.R.S. Varadhan (1984). *Large Deviations and Applications*. SIAM, Philadelphia.
176. A.M. Vershik and A.A. Schmidt (1977). Limit measures arising in the asymptotic theory of symmetric groups, I. *Theory Probab. Appl.* **22**, No. 1, 70–85.
177. A.M. Vershik and A.A. Schmidt (1978). Limit measures arising in the asymptotic theory of symmetric groups, II. *Theory Probab. Appl.* **23**, No. 1, 36–49.
178. J. Wakeley (2006). *An Introduction to Coalescent Theory*. Roberts and Company Publishers.
179. F.Y. Wang (2004). *Functional Inequalities, Markov Semigroups and Spectral Theory*. Science Press, Beijing.
180. G.A. Watterson (1974). The sampling theory of selectively neutral alleles. *Adv. Appl. Prob.* **6**, 463–488.
181. G.A. Watterson (1976). The stationary distribution of the infinitely-many neutral alleles diffusion model. *J. Appl. Prob.* **13**, 639–651.
182. G.A. Watterson (1977). The neutral alleles model, and some alternatives. *Bull. Int. Stat. Inst.* **47**, 171–186.
183. G.A. Watterson and H.A. Guess (1977). Is the most frequent allele the oldest? *Theor. Pop. Biol.* **11**, 141–160.
184. G.A. Watterson (1984). Lines of descent and the coalescent. *Theor. Pop. Biol.* **26**, 119–164.

185. S. Wright (1931). Evolution in Mendelian populations. *Genetics* **16**, 97–159.
186. S. Wright (1949). Adaption and selection. In *Genetics, Paleontology, and Evolution*, ed. G.L. Jepson, E. Mayr, G.G. Simpson, 365–389. Princeton University Press.
187. L. Wu (1995). Moderate deviations of dependent random variables related to CLT. *Ann. Prob.* **23**, No. 1, 420–445.
188. K.N. Xiang and T.S. Zhang (2005). Small time asymptotics for Fleming–Viot processes. *Infin. Dimens. Anal. Quantum Probab. Relat. Top.* **8**, No. 4, 605–630.

Index

S. Feng, *The Poisson–Dirichlet Distribution and Related Topics*,
Probability and its Applications, DOI 10.1007/978-3-642-11194-5,

Breinigsville, PA USA
25 June 2010
240430BV00001B/3/P

9 783642 111938